Victor A. Drits

Electron Diffraction and High-Resolution Electron Microscopy of Mineral Structures

With 126 Figures

Springer-Verlag
Berlin Heidelberg New York
London Paris Tokyo

Professor Dr. VICTOR A. DRITS
Geological Institute of the
Academy of Sciences of the USSR,
7 Pyzhevsky per
109017 Moscow, USSR

Translated from Russian by:

BELLA B. SMOLIAR
Geological Institute of the
Academy of Sciences of the USSR
7 Pyzhevsky per
109017 Moscow, USSR

Title of the original Russian edition:
Strukturnoye issledovaniye mineralov metodami mikrodifraktsii i elektronnoi
mikroskopii vysokogo razresheniya
© by Nauka, Moscow 1981

ISBN-13:978-3-642-71731-4   e-ISBN-13:978-3-642-71729-1
DOI: 10.1007/978-3-642-71729-1

Library of Congress Cataloging-in-Publication Data. Drits, Viktor Anatol'evich. Electron diffraction and high resolution electron microscopy of mineral structures. Translation of: Strukturnoe issledovanie mineralov metodami mikrodifraktsii i elektronnoi mikroskopii vysokogo razresheniia. Bibliography: p. Includes index. 1. Mineralogy, Determinative. 2. Electron microscopy. 3. Optical mineralogy. I. Title. QE369.M5D7513   1987   549'.125   86-33874

© Springer-Verlag Berlin Heidelberg 1987
Softcover reprint of the hardcover 1st edition 1987

Typesetting: K+V Fotosatz GmbH, Beerfelden

2131/3130-543210

Preface

The decision of Springer-Verlag to publish this book in English came as a pleasant surprise. The fact is that I started writing the first version of the book back in 1978. I wished to attract attention to potentialities inherent in selected-area electron diffraction (SAED) which, for various reasons, were not being put to use. By that time, I had at my disposal certain structural data on natural and synthetic minerals obtained using SAED and high-resolution electron microscopy (HREM), and this stimulated my writing this book. There were several aspects concerning these data that I wished to emphasize. First, it was mostly new and understudied minerals that possess the peculiar structural features studied by SAED and HREM. This could interest mineralogists, crystallo-chemists, and crystallographers. Second, the results obtained indicated that, under certain conditions, SAED could be an effective, and sometimes the only possible, method for structure analysis of minerals. This inference was of primary importance, since fine dispersion and poor crystallinity of numerous natural and synthetic minerals makes their structure study by conventional diffraction methods hardly possible. Third, it was demonstrated that in many cases X-ray powder diffraction analysis of dispersed minerals ought to be combined with SAED and local energy dispersion analysis. This was important, since researchers in structural mineralogy quite often ignored, and still ignore even the simplest information which is readily available from geometrical analysis of SAED patterns obtained from microcrystals. Thus it is commonly neglected that SAED, as compared with X-ray powder diffraction, is a simpler technique for determination of unit-cell parameters and revealing superperiodicity, commensurate and incommensurate modulated structures, intergrowth, and other types of structural imperfection. At the same time, there are numerous examples of invalid structural and crystal-chemical conclusions based on X-ray powder diffraction alone.

SAED structure analysis of minerals, however, was shown to be possible only if there were clear understanding of the processes resulting from the interaction between electrons and the crystal. Specifically, the relationship between reflection intensities and structure factors should be established correctly.

Electron diffraction (ED) is closely associated with HREM. The progress in HREM and, in particular, in HREM of minerals, has been especially successful since 1971. The development and undeniable advantages of HREM, however, do not at all mean that the latter can be used to solve all kinds of structural problems and that the role of SAED may be restricted to geometrical analysis of diffraction patterns. In other words, HREM and SAED supplement each other, but HREM cannot replace SAED.

The above considerations have predetermined the composition of the book, which consists of two main parts. The first part deals with theoretical and methodological foundations of ED and HREM, and with their relative merits for mineral structure studies. The results obtained by a number of well-known scientists such as Bethe, Blackman, Cowley, Heidenreich, Hirsch, Vainshtein, and others have been used. I have attempted to present this material in such a way that readers who have no special training in this field but are familiar with the fundamentals of crystallography and crystal chemistry of minerals could grasp the contents. My aim was that the reader should grasp the general outline of the physical foundations of the methods involved, their advantages and limitations, problems in structure analysis and methods for interpreting experimental data, and that he should be able to read modern periodicals on this subject. The second part contains both the original and the reported results of SAED and HREM structure studies of minerals having mostly layer and chain structures. Special attention is given to methodological procedures for the solution of specific structural problems.

In the new version of the book, the main contents have remained unchanged. At the same time, I have included new experimental data on mineral structure studies by SAED and HREM in which I have taken part. This concerns, in particular, the structures and structural peculiarities of manganese minerals. I have also taken account of the most important, to my point of view, theoretical, methodological and experimental results published in the literature in recent years. I was especially pleased to include those works where SAED was used as a method for structure analysis. I have additionally included a chapter on oblique-texture electron diffraction (OTED). Books on this method, *Structural electronography* by Vainshtein (1956) and *Electron diffraction analysis of clay mineral structures* by Zvyagin (1967), were published in English earlier. However, the advantages of OTED for the structure studies of finely dispersed minerals, and especially of phyllosilicates, have been largely underestimated in the West. The chapter in question shows that an increase in precision for the estimation of reflection intensities provides a new level in structure refinements of complex polycrystalline substances. Examples are given of the advantages of OTED over X-ray powder diffraction for the study of extremely poorly crystallized minerals such as smectites.

In the first part of the book, the chapter devoted to the formation of phase contrast in HREM has been extended: direct crystal structure determination methods based on the combination of HREM and SAED data are considered; the concept of a pseudo-weak phase-object, simulation of HREM images and the advantages of high-resolution high-voltage electron microscopy, etc. are discussed.

The section "Structural modulations resulting from the lateral misfit of octahedral and tetrahedral sheets in phyllosilicates" has been included in the second part of the book. Structural features of chain silicates revealed by HREM are here discussed in greater detail.

The writing of this book became possible owing to my successful collaboration with a number of colleagues in the Geological Institute (GIN) of the USSR Academy of Sciences, as well as in other institutes. In this connection, I would like to thank A. L. Dmitrik, GIN, Moscow, who worked together with me for many years, spending hours and hours at the electron microscope to obtain the experimental material required.

Dr. N. I. Organova, IGEM, Moscow, and I have used SAED for the first time to determine a number of mineral structures unknown before. Her contribution is gratefully acknowledged. Particular thanks are due to Academician F. V. Chukhrov and Dr. A. I. Gorshkov, IGEM, Moscow, with whom I have been successfully working for the past decade on the crystal chemistry of manganese minerals, using SAED as the main means for structure analysis.

Prof. Iu. I. Goncharov (Polytechnical Inst., Belgorod) and Dr. I. P. Hadji (VNIISIMS, Moscow) were the first to synthesize new-type triple-chain silicates of various compositions, and Dr. N. D. Zakharov (IKAN, Moscow) obtained the HREM images of these structures that are given in this book. I am deeply grateful to them all. In particular I thank Dr. I. P. Hadji with whom I worked intensively on the problems of interpretation of SAED patterns from various chain silicates.

I particularly wish to express my deep gratitude to Prof. S. Guggenheim, University of Illinois, Chicago, USA, for providing unpublished data and a number of original photographs. These exceptionally elegant results have formed a basis for the section on modulated structures. I am also grateful to Dr. A. Plançon, Orléans University, France and Dr. S. I. Tsipursky, GIN, Moscow for their contribution to OTED studies.

For a number of years I have been a constantly admirer of the excellent works of Prof. P. R. Buseck, Arizona State University, USA and Prof. Dr. R. Veblen, The John Hopkins University,

Maryland, USA, and I am pleased to thank them for providing a list of their papers and numerous preprints. Although the size of the book has not allowed including all their data even in brief, many of their results have been referred to.

Prof. B. B. Zvyagin, IGEM, Moscow, kindly agreed to read the page proofs of Chapters 9, 10, and 11. He found a number of errors and made valuable comments, for which I express my sincere gratitude.

Thanks are due to L. G. Daynyak, A. L. Sokolova, G. V. Karpova, G. V. Sokolova and E. V. Pokrovskaya for kind assistance in the preparation of the manuscript.

VICTOR A. DRITS

Contents

Contents   XI

       Mixed-Chain Minerals**   239

11.1 New Problems in the Structure Study of Chain Silicates ...   239
11.2 Pyroxenes and Amphiboles: Idealized Structures   240
11.3 Fluorocupfferite $Mg_7 [Si_8O_{22}F_2]$, a New Amphibole Variety   245
11.4 Crystal Structures of Triple-Chain Silicates   248
11.5 New Minerals Having Regular Mixed-Chain Structures   252
11.6 Some Methodological Aspects in the Interpretation for
      Point SAED Patterns from Chain Silicates   256
11.7 Direct HREM Observation of the Structural Motif of
      Asbestiform Chain Silicates   269
11.8 Chain-Width Disorder in Chain Silicates   274
11.9 Contrast Distribution in a-Axis HREM Images for Chain-
      Silicate Crystals Having Chain-Width Disorder   275
11.10 Structural Features of Chain Silicates Revealed in c-Axis
      HREM Images   281

      References   285

      Subject Index   295

Introduction

The present book deals with the advantages and limitations of electron diffraction (ED) and high-resolution transmission electron microscopy (HREM) for the structure studies of various minerals.

Although finely dispersed minerals that are the most natural objects for ED are widely abundant, the number of papers on ED structure studies is at present relatively small. One of the reasons for this is that an ED structure study entails certain serious problems some of which are difficult or impossible to avoid. First of all, a correct account of the relationship between the reflection intensity $I(hkl)$ and the corresponding structure amplitude $\Phi(hkl)$ requires knowledge of the nature of interaction between electrons and matter (kinematical, dynamical or intermediate). This is often ambiguous owing to the influence of crystal thickness, degree of structural perfection, complexity of the structure, contents of light and heavy atoms, etc. Calculations of diffraction effects based on the n-beam dynamical theory have shown that for crystals containing relatively heavy atoms the kinematical approximation for the relationship between I and $|\Phi|$ is valid for thicknesses of $20-50$ Å. The phase relationships are violated for even smaller thicknesses. Therefore, crystals having the same structure but differing in thickness (exceeding the "critical" one) should form diffraction patterns differing in the reflection intensity distribution. Furthermore, the intensity distribution is extremely sensible to the crystal orientation with respect to the incident beam. Thus, although we can evaluate only the moduli of structure amplitudes from diffraction data, the calculations mentioned above impose strict limitations on crystal thicknesses suitable for an ED structure study in the kinematical approximation.

Vainshtein (1956) has shown that for an aggregation of numerous disoriented, very small crystallites dynamical diffracted intensities are averaged over crystal thicknesses and orientations. Thus the dynamical effects are substantially reduced and the range of the validity of the kinematical approximation increases. This is the case with, e.g., the oblique-texture electron diffraction (OTED). An OTED pattern is formed with the incident electron beam passing through an inclined preparation consisting of a great number of thin platy crystals lying normal to the texture axis (Zvyagin 1967; Pinsker 1949).

Unfortunately, relatively simple analytical relationships between reflection intensities averaged over crystal thicknesses and orientations and the corresponding structure amplitudes have not been yet found in terms of the n-beam dynamical theory. Therefore, it seems appropriate to take account of satisfactory results of structure refinements (the reliability factor R, the accuracy in atomic coordinates, the dependence of I on the angle of scattering, etc.) which may serve as a criterion for the validity of the kinematical or the dynamical approximation in structure studies utilizing intensities of a few hundreds of reflections. Bearing this in mind, the successful crystal structure studies of fine mosaic films and dispersed minerals that have been carried out for many years in the U.S.S.R. are remarkable. An increase in the accuracy in the intensity measurements up to 10% has ensured structure refinements with $R = 4\%$ and the same precision in atomic coordinates as in single-crystal X-ray structure analysis (Drits 1982; Imamov et al. 1982).

Another method that is used to gain structural information on finely dispersed minerals is the selected-area electron diffraction (SAED). SAED patterns can be obtained from thin crystallites having an area of a few micrometers. However, on the basis of the n-beam dynamical theory it is often believed that SAED is not suitable for structural studies. It is no accident that the workers usually confine their attention to geometrical analysis of SAED patterns.

At the same time, neither X-ray diffraction nor OTED can be used to examine the crystal structures of some finely dispersed and poorly crystallized minerals that are abundant in nature. In this case, SAED is the only source of structural information. A number of workers (Chuhroy and Zvyagin 1966; Drits et al. 1974; Organova et al. 1973b, 1974) have tested the possibilities of SAED for the determination of unknown mineral structures that cannot be solved by other diffraction methods. It is necessary to take account of stronger dynamical effects than in the case of polycrystalline samples, and of all the resultant difficulties in establishing the relationship between I and Φ. Therefore, it could be expected that SAED would at best provide an idealized structure model with a low accuracy in the determination of atomic positions. However, even such studies of dispersed, poorly crystallized minerals are important.

One of the main problems in SAED is the absence of simple and reliable analytical criteria that could be used to determine unambiguously the nature of the interaction between electrons and matter on the basis of point diffraction patterns. For very thin crystals, reproducibility of intensity distribution in SAED patterns obtained from different crystals similarly oriented with respect to the incident beam can serve as such a criterion. Qualitatively, this implies a relatively small contribution of dynamical effects to the formation of the diffraction patterns. If the intensity distribution is sensitive to the crystal orientation and varies for different crystals, the object is apparently unsuitable for structure analysis.

Experimental data given in the present book indicate that under certain conditions SAED can be used for structure determinations.

Owing to the complicated nature of the interaction between electrons and crystals, the transition from $I(hkl)$ to $\Phi(hkl)$ requires taking a comprehensive account of all the factors affecting the diffracted intensity. That is why the kinematical and the dynamical theories are discussed in this book, and the methodological aspects of the use of SAED for structure determinations are treated on this basis. A general outline of the main methods for structure analysis is also given.

The creation of high-resolution electron microscopes equipped with goniometers has ensured the development of HREM, a powerful tool for structure analysis. Iijima (1973) was the first to prove that, under certain conditions, images can be directly interpreted in terms of the object structure. This work promoted a rapid development of HREM, ensuring the study of a fundamentally new world of phenomena that cannot be examined by any other method.

HREM has marked a qualitatively new stage in the study of the real structure of crystalline and quasi-crystalline substances. The most important advantage of HREM for physics of solid state, crystal chemistry and structural mineralogy is that under certain conditions direct relationship between the image and the object structure can be obtained, so that there will be no averaging of structure parameters over the diffracting volume. As the column approximation with the column width equal to the instrumental resolution applies, the contrast variations in the image may correspond "one to one" to the real distribution of electrostatic potential in the crystals under study, including the distribution of point, one-dimensional and two-dimensional defects of different types.

At present HREM is used for direct determinations of crystal structures unknown previously and for revealing the finest deviations of the real structure from the idealized model. Solution of such problems is possible only with HREM imaging under optimum conditions.

In the first part of this book, theoretical foundations of ED and HREM as well as the most important methodological aspects of their application are discussed. In the second part, the results of the application of these methods to various finely dispersed minerals are presented.

CHAPTER 1

Geometrical Features of the Crystal and the Reciprocal Lattices

1.1 Crystal Structure and Crystal Lattice

Crystalline substances have a common feature of internal periodicity. Material particles (atoms, ions, molecules, etc.) in a crystal form a three-dimensional repetitive pattern. Thus the distribution of matter in crystals may be described by a three-dimensional periodic function. Crystal structure is a specific arrangement of material particles in the volume of the crystal. Due to three-dimensional periodicity it is possible to find for each point of the crystal an infinite number of points having the same environment. That is, these points are surrounded by atoms of the same types that are located at the same distances and angles. The most demonstrative way to describe this peculiarity of crystals is the concept of the crystal lattice.

The concept of lattice is a convenient tool to reveal equivalent points in the object, which are called lattice points or nodes. The peculiarity of the lattice is that it coincides with itself if moved parallel to itself along any vector joining a pair of nodes. The operation of the transfer of the lattice parallel to itself from one node to another is called "translation", and so is the vector along which this transfer can be carried out. There is an infinite number of translations in any lattice.

The unit cell formed by vectors a, b, c with a fixed origin contains a specific atomic pattern. If the origin is chosen at another point of the structure, the arrangement of atoms in the unit cell will be changed. However, this "new" atomic pattern still can be used to reproduce the whole structure. While the specific coordinates of atoms within the unit cell depend on the choice of the origin, the lattice is always the same for the given crystal structure. Thus crystal lattice is a geometrical image which may be regarded as a symmetry element that multiplies equivalent points.

An infinite number of specific crystal structures may correspond to the same lattice. Moreover, only 14 types of crystal lattices are possible for all the diversity of real crystal structures, which demonstrates once again the difference between the concepts of structure and lattice.

If the lattice origin is at one of the nodes and the basic vectors a, b, c are chosen, the position of any node is specified by three whole numbers that are called node indices and are written as $[[mnp]]$. This follows from the fact that the radius – vector of a node

$$r_{mnp} = ma + nb + pc \qquad (1.1)$$

is characterized by the components which are equal to numbers m, n, p of the corresponding repeat distances.

If a unit weight is ascribed to each lattice node, the position of all the nodes in the three-dimensional space may be characterized by a lattice function of the type:

$$R(r) = \sum_{m,n,p} \delta(r - r_{mnp}) \,. \tag{1.2}$$

Here

$$\delta(r - r_{mnp}) = \begin{array}{ll} 1; & \text{if} \quad r = r_{mnp} \\ 0; & \text{if} \quad r \neq r_{mnp} \,. \end{array}$$

An infinite number of straight lines and planes may be drawn through the lattice nodes. To describe each set of parallel lines one has to know their orientation with respect to the coordinate system and the repeat distance. These parameters can be specified if in the set of lines a line is chosen passing through the origin, and the indices of the node located on it nearest to origin are found. These indices are written as $[mnp]$.

Each set of parallel equidistant planes is characterized by orientation with respect to the coordinate system and the perpendicular spacing between the nearest planes, which is called interplanar spacing.

To specify the orientation of a set of parallel planes it is sufficient to know that of the plane nearest to the origin. This plane cuts intercepts a/h, b/k, and c/l off the corresponding axes, where h, k, l are integers without a common factor and are written as (hkl).

A plane which intercepts distances of A, B, and C on the given set of axes is represented by the equation

$$X/A + Y/B + Z/C = 1 \,. \tag{1.3}$$

The nth (with respect to the origin) plane (hkl) makes intercepts $A = na/h$, $B = nb/k$ and $C = nc/l$ on the corresponding axes. If the current coordinates of the plane are expressed in terms of fractions of the unit cell parameters,

$$hx + ky + lz = n \,. \tag{1.4}$$

To simplify the description of a set of parallel planes it is sometimes convenient to characterize their orientation by a vector normal to them. Let us find the relationship between the (hkl) plane indices and the (mnp) indices of the corresponding normal. To do this one should join the intersection points of the plane nearest to the origin with the axes by straight lines so that vectors $A = c/l - b/k$; $B = c/l - a/h$ and $C = b/k - a/h$ are obtained (Fig. 1a, b). As vectors A, B and C lie in the (hkl) plane, the normal N should be perpendicular to each of them. As the scalar product of vectors normal to one another is zero

$$(AN) = (c/l - b/k)(ma + nb + pc)$$
$$= m(ac/l - ab/k) + n(cb/l - b^2/k) + p(c^2/l - cb/k) = 0, \tag{1.5}$$

$$(BN) = (c/l - a/h)(ma + nb + pc)$$
$$= m(ac/l - a^2/h) + n(bc/l - ab/h) + p(c^2/l - ac/h) = 0 \,.$$

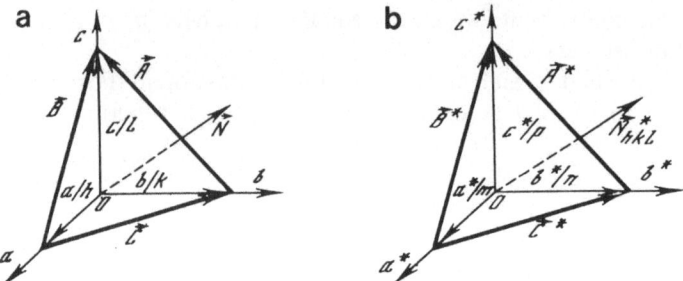

Fig. 1a,b. The plane nearest to the origin and the normal to it; **a** direct lattice; **b** reciprocal lattice

The indices of the direction being determined by ratios $m:n:p$ or $1:n/m:p/m$, Eq. (1.5) is sufficient for the solution of the problem.

The solution of two simultaneous equations is given by

$$\frac{n}{m} = \frac{\begin{vmatrix} (lab-kac)(kc^2-lbc) \\ (la^2-hac)(hc^2-lac) \end{vmatrix}}{\begin{vmatrix} (kbc-lb^2)(kc^2-lbc) \\ (hbc-lab)(hc^2-lac) \end{vmatrix}} ; \quad \frac{p}{m} = \frac{\begin{vmatrix} (kbc-lb^2)(lab-kac) \\ (hbc-lab)(la^2-hac) \end{vmatrix}}{\begin{vmatrix} (kbc-lb^2)(kc^2-lbc) \\ (hbc-lab)(hc^2-lac) \end{vmatrix}} . \quad (1.6)$$

To solve the reverse problem, the (1.5) equations should be transformed as follows:

$$k(mac + nbc + pc^2) = l(mab + nb^2 + pbc),$$
$$h(mac + nbc + pc^2) = l(ma^2 + nab + pac).$$

The hkl indices are then given by

$$h = ma^2 + nab + pac,$$
$$k = mab + nb^2 + pbc, \quad\quad\quad\quad\quad (1.7)$$
$$l = mac + nbc + pc^2 \quad \text{(up to a constant factor)}.$$

The Eqs. (1.6) and (1.7) should be used in interpreting point electron diffraction patterns.

1.2 The Bragg Equation. Reciprocal Lattice. Relationships Between the Indices of Lines and Planes in the Direct and Reciprocal Lattices

The interaction of electrons with the crystal is discussed in detail in Chapter 2. Here we shall carry out an elementary interpretation of wave diffraction, which is necessary to analyze the geometry of diffracted beams. In this specific case the nature of the incident radiation (X-ray, electrons, etc.) is not essential.

Since the atomic pattern in the unit cell may be assigned to each lattice point, instead of considering the interaction of waves with the crystal, one may deal with their interaction with the corresponding lattice, if the scattering ability corresponding to that of the unit cell is imparted to the nodes.

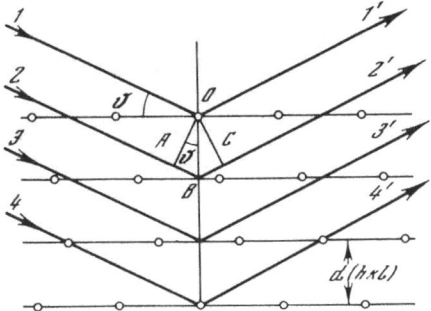

Fig. 2. Wave reflection from a set of parallel planes (hkl)

A crystal lattice can be presented in the form of a set of parallel lattice planes.

Figure 2 shows a two-dimensional section of a crystal lattice so that the traces of a particular set of lattice planes (hkl) with the interplanar spacing d are seen as a system of parallel equally spaced lines. If the incident beam is directed at different angles toward these planes, diffracted rays of noticeable intensity will appear at certain angles similarly to a mirror reflection (Fig. 2). In the rest of directions the scattered waves interfere destructively and their resulting intensity is zero.

Thus each diffracted beam in a diffraction pattern may be regarded as reflected from the corresponding set of lattice planes. In order to determine the value of the angle of incidence for which diffraction takes place, one should take into account that according to the theory of wave optics the rays are scattered in the same phase and amplify one another if the path-length difference is equal to a whole number of wavelengths λ. From Fig. 2 it is seen that the path travelled by the ray 2 is by $AB + BC = 2d\sin\vartheta$ longer then that of the ray 1 ($AB = BC = d\sin\vartheta$), where ϑ is the angle of incidence. Therefore, diffraction will take place if the following equation is satisfied

$$2d(hkl)\sin\vartheta = n\lambda. \tag{1.8}$$

This is the famous Bragg Equation.

A given set of lattice planes may produce diffraction maxima for several definite ϑ angles, as for d, $\lambda = $ const, the path-length difference increases with the increase of ϑ and may be equal not only to λ, but also to 2λ, 3λ, etc. However, it is convenient to assume that the path-length difference is always equal to λ, while the reflection is produced from sets of planes with d-spacings of $d(hkl)/2$, $d(hkl)/3$, etc., as if one, two, three and more parallel and equally spaced imaginary planes were inserted into each interval between the initial planes. Despite the infinite number of lattice plane sets with different values of $d(hkl)$, the total number of reflections is limited, as even for $n = 1$ $d \geq \lambda/2$.

If the nth order reflection from the set of planes (hkl) is regarded as the first order reflection from an imaginary set of planes parallel to the initial one but having a spacing of $d(hkl)/n$, it is evident that these "reflecting" planes should have indices nh nk nl, that are assigned to the diffraction peak involved.

The Bragg Equation has played an extremely important role in the theory and practice of the study of crystalline matter by diffraction methods, as it has ensured revealing the relationship between the geometry of a diffraction pattern and the crystal lattice.

Proceeding from the Bragg idea of the sets of reflecting planes the concept may be introduced of the reciprocal lattice, which permits to analyze relations between crystallographic parameters of the object and its diffraction pattern. We shall first consider lattices with a primitive cell. The reciprocal lattice may be constructed as follows. Each set of planes (hkl) is specified by a vector perpendicular to these planes so that the vector length is inversely proportional to $d(hkl)$.

All the vectors are drawn from the same origin, so that each set of planes is characterized by a pair of normals laid off along opposite directions. The tips of the vectors and the points laid off along each normal at distances $2\,[1/d(hkl)]$, $3\,[1/d(hkl)]$, etc., are the reciprocal lattice points.

The sequence of operations needed to proceed from the direct lattice to the reciprocal one is shown in Fig. 3. As the distances from the origin of the reciprocal lattice to the first, the second, and the nth point are inversely proportional to $d(hkl)$, $d(hkl)/2$, and $d(hkl)/n$, respectively, it is appropriate to relate these nodes to the reflections of the corresponding orders and to assign indices hkl, $2h\,2k\,2l$, $nh\,nk\,nl$ to them. In other words, an imaginary set of the direct lattice planes parallel to the (hkl) plane and having d spacings of $d(hkl)/n$ that produce a diffracted beam of the order $nh\,nk\,nl$ corresponds to a reciprocal lattice node $nh\cdot nk\,nl$. The indices of the reciprocal lattice, in contrast to those of the direct lattice, are written without any brackets.

If there is a set of points with certain indices, the choice of the coordinate system is in fact predetermined, because in order to preserve the indices assigned to the nodes, one ought to choose directions normal to planes (100), (010), and (001) as coordinate axes and to assume the following basic vectors

$$a* = 1/d(100), \quad b* = 1/d(010), \quad c* = 1/d(001).$$

A vector drawn from the original to a reciprocal lattice point with whole-number coordinates is denoted as H_{HKL}. Then

$$H_{hkl} = h\frac{1}{d(001)} + k\frac{1}{d(010)} + l\frac{1}{d(001)} = \frac{1}{d(hkl)},$$

$$H_{HKL} = H\frac{1}{d(001)} + K\frac{1}{d(010)} + L\frac{1}{d(001)} = n\frac{1}{d(hkl)}.$$

H_{hkl} and H_{HKL} are vectors of the same direction, corresponding to the first and the nth nodes relative to the origin, respectively, i.e., $H = nh$, $K = nk$, $L = nl$. Therefore the reciprocal lattice points may be characterized by indices HKL, which under these conditions may take all the possible whole-number values, as hkl for primitive cells are all the possible whole numbers without a common factor. When the relationships between the reciprocal lattice and diffraction patterns are studied (e.g., in the case of indexing), symbols hkl, whose values may have a common factor, are used instead of HKL.

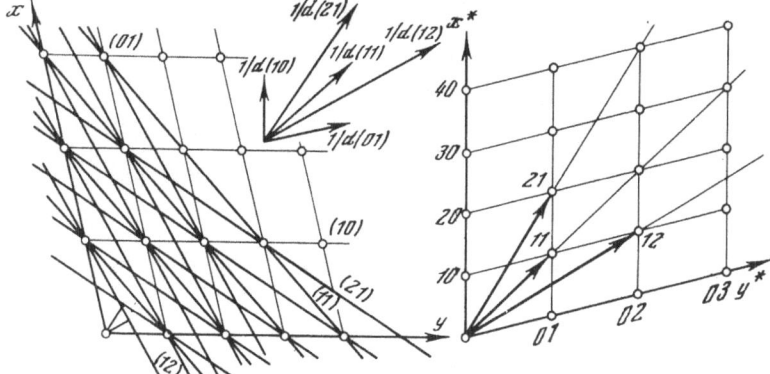

Fig. 3. Sequence of operations required for the construction of the reciprocal lattice

In order to describe the arrangement of points in the reciprocal space one has to choose a coordinate system differing from that of the initial crystal lattice. Let us find the relationship between the basic vectors in the direct and the reciprocal lattices. As $d(100)$ is the distance between the two parallel bc faces of the unit cell equal to the volume of the unit cell V divided by the area of the face $S = |[bc]|$, we have

$$a^* = \frac{1}{d(100)} = \frac{[bc]}{V} = \frac{[bc]}{(a[bc])}. \tag{1.9a}$$

Similarly

$$b^* = \frac{1}{d(010)} = \frac{[ca]}{(b[ca])}, \quad c^* = \frac{1}{d(001)} = \frac{[ab]}{(c[ab])}. \tag{1.9b}$$

It follows from the equations obtained that

$$(aa^*) = (bb^*) = (cc^*) = 1, \tag{1.10}$$

$$(a^*b) = (ab^*) = (ac^*) = (a^*c) = (bc^*) = (b^*c) = 0.$$

Using (1.9), the relationship between the unit cell parameters in the direct lattice and in the reciprocal one can be easily derived (International Tables for X-Ray Crystallography 1962).

In all the cases, the reciprocal lattice belongs to the same symmetry system as the direct one. The concept of the reciprocal lattice might be introduced using first (1.10) and then proving that a reciprocal lattice vector H_{hkl} is perpendicular to the plane (hkl), while its modulus is inverse to $d(hkl)$.

Let us note several important relations between the indices of lattice rows and planes in the direct and the reciprocal lattices, since point electron diffraction patterns may be approximated by planar sections of the reciprocal lattice passing through the origin (see below). As the vector H_{hkl} is normal to the planes (hkl), it is also perpendicular to any vector lying in these planes. For example, if the intersection points of the plane (hkl) with the coordinate axes are joined to produce vectors A, B, and C (Fig. 1a), then $H_{hkl}A = H_{hkl}B = H_{hkl}C = 0$. It is seen in Fig. 1 that

$$H_{hkl}C = (ha* + kb* + lc*)(b/k - a/h) = 0 .$$

Therefore, independently of the system of coordinates, the equation representing the plane (hkl) passing through the origin may be written as

$$H_{hkl}r_{mnp} = hm + kn + lp = 0 \tag{1.11}$$

where $r_{mnp} = ma + nb + pc$ is the radius-vector of a node lying in the plane (hkl).

A similar expression is obtained from (1.4) with $n = 0$.

Equation (1.11) expresses the condition of the parallelism of a lattice row and a lattice plane, i.e., the condition of the fact that the lattice row $[mnp]$ lies in the plane (hkl). Then indices of all planes intersecting along the line $[mnp]$ satisfy Eq. (1.11). The set of planes (hkl) intersecting along the $[mnp]$ direction is a zone, while the $[mnp]$ direction is called the zone axis. Sets of planes (hkl) belonging to the same zone may be replaced by a set of corresponding vectors H_{hkl}. These vectors lie in a plane passing through the origin of the reciprocal lattice. The $[mnp]$ zone axis is perpendicular to this reciprocal lattice plane.

As the indices of all the nodes lying in the reciprocal lattice plane involved satisfy Eq. (1.11), their values can be readily found if the indices $[mnp]$ are known.

On the other hand, if the indices of two reciprocal lattice points lying in the plane passing through the origin ($h_1k_1l_1$ and $h_2k_2l_2$) are known, then the indices of the zone axis $[mnp]$ can be obtained (up to a constant factor).

We have

$$h_1m + k_1n + l_1p = 0 \qquad k_1(n/m) + l_1(p/m) = -h_1 ,$$
$$\text{or}$$
$$h_2m + k_2n + l_2p = 0 \qquad k_2(n/m) + l_2(p/m) = -h_2 .$$

Therefore

$$m:n:p = \begin{vmatrix} k_1l_1 \\ k_2l_2 \end{vmatrix} : \begin{vmatrix} l_1h_1 \\ l_2h_2 \end{vmatrix} : \begin{vmatrix} h_1k_1 \\ h_2k_2 \end{vmatrix} . \tag{1.12}$$

Using (1.11), one may also determine the indices of a plane passing through the origin and containing parallel lines. One must bear in mind, however, that in the case of the reciprocal lattice the indices of the nodes hkl are the current coordinates of the plane (mnp), whereas in the direct lattice the coordinates of the (hkl) plane are the indices of the nodes $[mnp]$. If, for instance, one has to determine the reciprocal lattice plane containing nonparallel vectors $H_{h_1k_1l_1}$ and $H_{h_2k_2l_2}$, the solution will be exactly similar to the one described above. In this case the values mnp obtained with the help of (1.12) will characterize the indices of the reciprocal lattice plane $(mnp)*$ in question. Hence, one may formulate a general rule: A plane in the reciprocal lattice having indices $(mnp)*$ is perpendicular to the direction $[mnp]$ in the crystal lattice, and consequently, all the vectors lying in the plane $(mnp)*$ are normal to the radius-vector r_{mnp}. It is evident that the d-spacing of the set of reciprocal lattice planes $(mnp)*$ is inversely proportional to the distance from the origin to the nearest node $[[mnp]]$.

The expressions obtained may be used to establish a relationship between the indices of planes parallel to each other, one of which belongs to the direct lattice and the other to the reciprocal one. The same problem might be formulated as the determination of the relationship between the indices of the plane (hkl) and those of its normal $[mnp]$ in the direct lattice, or between the indices of the plane $(mnp)^*$ and those of its normal $[hkl]^*$ in the reciprocal lattice. Figure 1b shows a reciprocal lattice plane $(mnp)^*$ which is the nearest to the origin. The plane in question intercepts distances of a^*/m, b^*/n and c^*/p on the corresponding axes. The intersection points of this plane with the axes are joined in such a way that vectors

$$A^* = c^*/p - b^*/n, \quad B^* = c^*/p - a^*/m, \quad C^* = b^*/n - a^*/m$$

are obtained (Fig. 1b). The normal to the plane $(mnp)^*$ is specified by a radius-vector $r = ha + kb + lc$, whereas a set of planes (hkl) is perpendicular to this vector. Hence the (hkl) and the $(mnp)^*$ planes are parallel to each other. Therefore

$$(A^*r^*) = (c^*/p - b^*/n)(ha^* + kb^* + lc^*) = 0 ,$$

$$(B^*r^*) = (c^*/p - a^*/m)(ha^* + kb^* + lc^*) = 0 .$$

These equations may be written as

$$n(ha^*c^* + kb^*c^* + lc^{*2}) = p(ha^*b^* + kb^{*2} + lb^*c^*) ,$$

$$m(ha^*c^* + kb^*c^* + lc^{*2}) = p(ha^{*2} + ka^*b^* + la^*c^*) .$$

Hence, up to a common factor, we have

$$m = ha^{*2} + ka^*b^* + la^*c^* ,$$

$$n = ha^*b^* + kb^{*2} + lb^*c^* , \tag{1.13}$$

$$p = ha^*c^* + kb^*c^* + lc^{*2} .$$

B. B. Zvyagin was the first to derive, in a somewhat different way, equations of the type (1.7) and (1.13), that are widely used for the interpretation of the point electron diffraction patterns (Zvyagin 1968).

1.3 The Ewald Sphere and the Geometrical Interpretation for Diffraction Patterns

There is a simple but very remarkable geometrical interpretation for the Bragg Equation which enables one to express the conditions of diffraction by geometrical representations.

In Fig. 4 a two-dimensional section of a crystal lattice is shown. The initial beam with the wavelength λ falls on the crystal with an incident angle of ϑ, determined by (1.8). Diffracted rays make the same angle with the diffracting planes. The angle between the transmitted and the diffracted beams is 2ϑ. The trace of the reflecting plane bissects this angle.

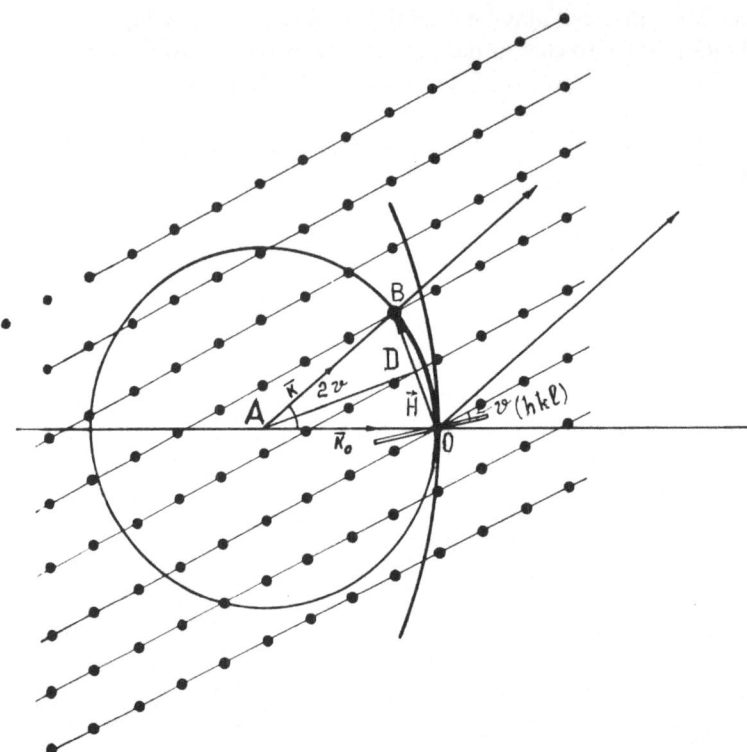

Fig. 4. Geometry of diffraction: reciprocal lattice and reflection sphere applied to X-rays (*1*) and electrons (*2*)

Point O will be considered the origin of the crystal lattice and the directions of the incident beam and the diffracted one will be specified by vectors k_0 and k of the length $1/\lambda$. The segment OA of the length $1/\lambda$ lies along $-k_0$. Another segment of the length $1/\lambda$ is drawn from the point A in the direction of k, while the origin and the tip of the vector k are joined by the vector H. Thus an isosceles triangle OAB is obtained with $<OAB = 2\vartheta$ (the angle between k_0 and k) so that $<OD = DB$, $AD \perp H$ and AD is parallel to the planes (hkl) in question, $(<DAO = <AOP = \vartheta)$. Hence H is perpendicular to the diffracting planes, as it is normal to AD. ADO being a right-angled triangle, $OD = AO \sin \vartheta = \sin \vartheta / \lambda$ and as

$$H = |k - k_0| = 2OD = 2 \sin \vartheta / \lambda \qquad (1.14)$$

we have

$$2 \sin \vartheta / \lambda = n/d(hkl) = 1/d(nh\, nk\, nl) . \qquad (1.15)$$

The combination of (1.14) and (1.15) gives $H = 1/d(nh\, nk\, nl)$. Therefore the vector H is the reciprocal lattice vector so that its tip coincides with the node $nh\, nk\, nl$.

Thus the Bragg Equation is satisfied if the reciprocal lattice vector drawn from the origin into a node may serve as a base of an isosceles triangle with edges of $1/\lambda$ lying along k_0 and k.

A sphere of the radius of $1/\lambda$ is drawn around A (Fig. 4). Any segment joining the origin of the reciprocal lattice with an arbitrary point on the sphere is the base of an isosceles triangle with the third apex in the center of the sphere. However, diffraction will take place only in the case of a reciprocal lattice point lying on the sphere. Generally, this is true only for sufficiently large crystals with no defects.

Thus to find out which lattice planes are in the diffracting position for a given orientation of the crystal lattice with respect to the incident beam, the reciprocal lattice should be constructed and the vector k_0 of the length $1/\lambda$ should be drawn terminating at the origin. A sphere of the radius $1/\lambda$ should be drawn around the other extreme point of this vector. All the nodes of the reciprocal lattice lying on the surface of the sphere will correspond to the direct lattice planes situated in the diffracting position. The sphere thus constructed is called the sphere of reflection or the Ewald sphere (after its originator).

If vectors are drawn from the center of the Ewald sphere to nodes lying on its surface, these directions will specify the directions k of the corresponding diffraction beams. Therefore, Bragg's Law in vector form is:

$$k = k_0 + H_{hkl}.$$

The origin of the reciprocal lattice is always on the surface of the Ewald sphere, whereas the exit point for diffracted beams coincides with the center of this sphere. The relative orientation of the vectors k_0 and k is identical to that of the diffracted and of incident beams displayed in the real diffraction pattern.

When the relative orientation of the crystal and the incident beam is fixed, the wavelength λ and unit cell parameters are essential for diffraction to take place. The greater the unit cell parameters, the smaller the distance between the reciprocal lattice nodes. Therefore the probability for the nodes to cross the Ewald sphere increases. On the other hand, the wavelength determines the curvature of the Ewald sphere. With the decrease of the curvature the number of the reciprocal lattice points that can appear on the surface of the Ewald sphere increases. The X-ray wavelength ranges from 0.7 to 2.5 Å, whereas those of fast electrons are usually less than 0.05 Å. Although the unit cell parameters may vary within wide limits for different crystal structures, they are practically always greater than 3 Å. Thus, in the case of X-rays, the radius of the Ewald sphere is only a little greater than the distance between the nodes, whereas for electron diffraction the Ewald sphere may be often approximated by a plane, which simplifies the geometrical analysis of point electron diffraction pattern.

CHAPTER 2

The Kinematical Theory of Scattering of Electrons by Crystals. Intensity of Diffraction Reflections

2.1 Wave-Like Properties of Electrons

Electrons are negatively charged particles having a wave-like nature. Electrons traveling with a velocity u are characterized by a wavelength λ which is related to the impulse $p = mu$ of electrons as

$$\lambda = h/mu.$$

Due to relativistic effects, the mass of a moving electron m as compared to its rest mass m_0 is described by the equation

$$m = \frac{m_0}{(1 - u^2/c^2)^{1/2}} = \frac{m_0}{(1 - \beta^2)^{1/2}},$$

where $\beta = u/c$ and c is the velocity of light.

The kinetic energy E_{kin} of an electron is determined by the value of the accelerating potential, as

$$E_{kin} = mc^2 - m_0 c^2 = eV. \tag{2.1}$$

Taking the above expressions into account, we have

$$\lambda = \lambda_0 (1 + eV/2 \, m_0 c^2)^{-1/2}, \tag{2.2}$$

where $\lambda_0 = h/(2 m_0 eV)^{1/2} = 12.265 \, V^{-1/2}$ is the nonrelativistic electron wavelength, equal to 0.037 Å with $V = 100$ kV and 0.00867 Å with $V = 1$ MV. It is obvious that the peculiarities in the interaction of an electron beam with a crystal are determined both by the negative charge and by the wave-like nature of electrons. To discuss this in more detail, the phenomena of interference and diffraction should be considered.

2.2 The Kinematical Theory of Scattering of Waves by Crystals

If the scattering object consists of a set of discrete material particles, e.g., atoms or ions, characterized by a definite scattering power, the process of diffraction of waves by a crystal may be treated in the following terms. Each atom affected by the incident beam becomes a source of secondary spherical waves.

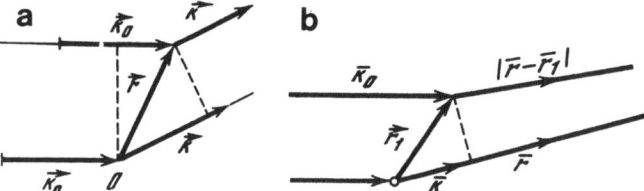

Fig. 5a, b. Path-length difference between the incident wave and the scattered one (Vainshtein 1956); **a** for two scattering centers; **b** for two scattering centers as observed at a long distance

In terms of the kinematical theory it is assumed that
the scattered amplitude is much smaller than that of the incident exciting wave;
the scattering is not accompanied by any change in the wavelength;
multiple scattering effects are absent or negligible.

Secondary waves propagating in all directions with phase relations determined by the geometry of the arrangement of scattering centers interfere with one another. To illustrate the interference of secondary waves consider Fig. 5, showing two scattering atoms. One of these is located at the origin and the position of the other is given by the vector r. The direction of the incident plane wave is specified by vector k_0 and that of the scattered wave by vector k. It can be easily seen that the path traveled by the incident wave to the scattering atom in the position r is by $r \cos \vartheta_0$ longer than that to the atom at the origin. On the contrary, the path of the secondary wave scattered by the atom at the position r is by $r \cos \vartheta$ shorter than that of the wave scattered by that at the origin.

Thus the paths travelled by the beams scattered at the point given by r and at the origin differ by the distance

$$\Delta = r \cos \vartheta_0 - r \cos \vartheta = \lambda (k_0 - k, r) .$$

As the path-length difference determines the phase difference of

$$\gamma = (2 \pi / \lambda) \Delta = 2 \pi (k_0 - k, r) ,$$

the waves scattered at the position r will lag in phase behind those scattered at the origin by this value of γ.

If the phase factor of the wave scattered by the atom at the origin, as observed from the point determined by R, is given by

$$\exp 2 \pi i k (R - u t) ,$$

then for the wave scattered at the position r it is equal to

$$\exp [2 \pi i k (R - u t)] \exp [2 \pi i (k - k_0, r)] . \tag{2.3}$$

When studying diffraction effects we are not interested in temporal characteristics of wave functions because in the case of X-rays and electrons the atomic arrangement in the crystal may be regarded as static.

This is due to the fact that the period of oscillation of electrons is much smaller than that of the thermal vibrations of atoms. For this reason, one may take into account only the second factor in the expression (2.3) that depends on the geometrical arrangement of atoms in the crystal.

If an atom in a position r_j is characterized by a scattering power f_j, the scattered amplitude is given by

$$f_j \exp 2\pi i(k - k_0, r_j) . \tag{2.4}$$

The amplitude of a wave scattered by all atoms in the crystal volume may be written as:

$$A(s) = \sum_j f_j \exp(2\pi i s r_j) , \tag{2.5}$$

where

$$s = k - k_0 .$$

The dimensions of s, as well as those of k and k_0 are inverse to the dimensions of r. While the distribution of scattering centers in the space of the real object is specified by r, the values of the amplitude of scattered waves are described by a function of s in the reciprocal space.

The scattered intensity is equal to $A(s)$ multiplied by its complex conjugate, that is

$$I(s) = \sum_i \sum_j f_i f_j \exp[2\pi i(s, r_j - r_i)]. \tag{2.6}$$

Thus, if scattering powers and the geometrical arrangement are known for a set of atoms involved, the expression (2.6) may be used for an a priori calculation of diffraction effects to be expected. In this case the distribution of scattering centers in the object may be arbitrary.

In crystals, electrons are scattered by an electrostatic potential the values of which may be characterized by a continuous function, $\varphi(r)$. The product $\varphi(r)dV_r$ determines the potential in a volume element dV_r, the position of which is given by the vector r. This value characterizes the scattering power of the volume element dV_r. Naturally, the distribution of $\varphi(r)$ is directly connected with the distribution of scattering centers, as the maxima of $\varphi(r)$ coincide with the centers of atoms. The total amplitude diffracted by an object characterized by a continuous function $\varphi(r)$ is obtained similarly to that in the case of discrete atoms, i.e., by summing the waves diffracted by the whole volume of the crystal:

$$A(s) = \int_V \varphi(r) \exp[2\pi i(k - k_0, r)]dV . \tag{2.7}$$

The expression (2.7) is universal in the sense of its permitting the calculation of scattering amplitude for any s in the case of any type of scattering object (e.g., atom, molecule, crystal) for which $\varphi(r)$ is known. When the distribution of electrostatic potential within the unit cell is defined, its scattering power is given by

$$\Phi(s) = \int_{V_0} \varphi(r) \exp(2\pi i s r)dV_0 . \tag{2.8}$$

Mathematically, expressions (2.7) and (2.8) are Fourier integrals, and therefore, as it will be shown later, the main statements of the theory of electron diffraction may be derived from the theory of Fourier series and integrals.

The object has been hitherto regarded as an arbitrary set of scattering centers or as a continuous distribution of scattering matter. Now the lattice state of a crystal is to be taken into account. If a, b, c are the basic vectors of the unit cell and the origin of the crystal lattice is taken at the origin of a certain fixed unit cell, then the position of the nth unit cell is given by the vector

$$r_n = n_1 a + n_2 b + n_3 c .$$

The atomic pattern is exactly identical for all the unit cells and the position of a jth atom may be defined by a radius-vector r_j which is given by

$$r_j = x_j a + y_j b + z_j c ,$$

where x_j, y_j, z_j are fractional coordinates (i.e., expressed in terms of fractions of unit cell parameters) of the jth atom. Thus, the position of an arbitrary jth atom in the nth unit cell is defined by the sum of vectors r_n and r_j, while, according to (2.4), the scattering amplitude for this atom may be written as

$$f_j \exp[2 \pi i (s, r_n + r_j)] .$$

Carrying out the summation over all the atoms within the unit cell and then over all the unit cells within the crystal volume we obtain

$$A(s) = \sum_{n_1} \sum_{n_2} \sum_{n_3} \sum_j f_j \exp[2 \pi i (s, r_n + r_j)] .$$

As $f_j \exp(2 \pi i s r_j)$ does not depend on n,

$$A(s) = \sum_j f_j \exp 2 \pi i s r_j \sum_{n_1} \sum_{n_2} \sum_{n_3} \exp(2 \pi i s r_n) . \tag{2.9}$$

The first factor in the expression (2.9) signifies the contribution of a unit cell to the scattering of electrons. It is called *the structure amplitude* and is denoted by $\Phi(s)$, that is

$$\Phi(s) = \sum_j f_j \exp[2 \pi i s r_j] . \tag{2.10}$$

Then

$$A(s) = \Phi(s) \sum_{n_1} \sum_{n_2} \sum_{n_3} \exp[2 \pi i s r_n] . \tag{2.11}$$

This is a remarkable and important result showing that the description of diffraction by crystals may be replaced by the description of that from crystal lattices, if the scattering power of the unit cell is assigned to each node. Note that all the nodes of the crystal lattice have the same scattering power in the direction of s. It should be emphasized that this conclusion is of universal significance. Quite often a problem may be simplified if a sufficiently large group of atoms, e.g., a two-dimensional layer, with a finite thickness, is chosen as a scattering center.

The second factor in (2.9) and (2.11) is called *the interference function* and is denoted by

$$D(s) = \sum_n \exp(2\pi i s r_n) \,. \tag{2.12}$$

Thus the total amplitude scattered by a crystal lattice may be regarded as

$$A(s) = \Phi(s)D(s) \,.$$

The vector s may be expressed by continuous fractional coordinates x^*, y^*, z^* in the system defined by vectors a^*, b^*, c^* in the reciprocal space. These vectors are related to vectors a, b, c in the crystal lattice by (1.9) and (1.10), so that

$$s = x^*a^* + y^*b^* + z^*c^* \,. \tag{2.13}$$

It is easy to show that

$$r_n s = x^* n_1 + y^* n_2 + z^* n_3 \,, \tag{2.14}$$

$$r_j s = x^* x_j + y^* y_j + z^* z_j \,. \tag{2.15}$$

Then

$$\Phi(s) = \sum_j f_j \exp[2\pi i(x^* x_j + y^* y_j + z^* z_j)] \,,$$

$$D(s) = \sum_{n_1}\sum_{n_2}\sum_{n_3} \exp[2\pi i(x^* n_1 + y^* n_2 + z^* n_3)] \,.$$

The interference function is more conveniently written as

$$D(x^* y^* z^*) = \sum_{n_1} \exp[2\pi i x^* n_1] \sum_{n_2} \exp[2\pi i y^* n_2] \sum_{n_3} \exp[2\pi i z^* n_3] \,. \tag{2.16}$$

The three factors in (2.16) are geometrical progressions with ratios of $\exp 2\pi i x^*$, $\exp 2\pi i y^*$, and $\exp 2\pi i z^*$, respectively. Therefore,

$$\sum_{n_1=0}^{N_1-1} \exp[2\pi i x^* n_1] = \frac{\exp[2\pi i N_1 x^*]-1}{\exp[2\pi i x^*]-1}$$

$$= \frac{\sin \pi N_1 x^*}{\sin \pi x^*} \exp[-\pi i(N_1-1)x^*] \,.$$

The summation is here over the number of unit cells N_1 along a.

The square of $D(x^* y^* z^*)$ is called *the interference factor* and is given by

$$D^2(x^* y^* z^*) = \frac{\sin^2 \pi N_1 x^*}{\sin^2 \pi x^*} \frac{\sin^2 \pi N_2 y^*}{\sin^2 \pi y^*} \frac{\sin^2 \pi N_3 z^*}{\sin^2 \pi z^*} \,. \tag{2.17}$$

Since $x^* = sa$, $y^* = sb$, $z^* = sc$ [which follows from (2.13) and (1.9)], then

$$D^2(s) = \frac{\sin^2 \pi N_1(sa)}{\sin^2 \pi(sa)} \frac{\sin^2 \pi N_2(sb)}{\sin^2 \pi(sb)} \frac{\sin^2 \pi N_3(sc)}{\sin^2 \pi(sc)} \,. \tag{2.18}$$

Each factor in (2.17) and (2.18) is a periodic function with remarkable properties. The characteristic features of diffraction patterns from crystalline compounds are to a large extent determined by these properties.

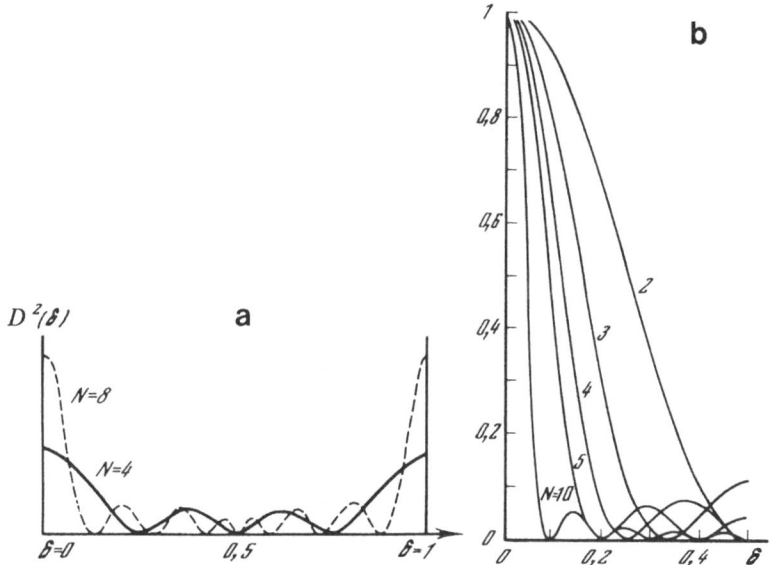

Fig. 6a, b. Interference factor plotted against N: **a** $D^2(x^*)$ curves for $N = 4$ and 8; **b** $D^2(\delta)/N^2$ curves for different N (Güven 1974)

For example, consider the first factor in (2.17):

$$D^2(x^*) = \frac{\sin^2 \pi N_1 x^*}{\sin^2 \pi x^*} . \tag{2.19}$$

When $x^* = 1/N_1, 2/N_1, \ldots (nN_1 + m)/N_1$ (where m and n are whole numbers without a common factor, and $m \neq 0$), $D^2(x^*) = 0$, as the numerator in (2.19) is zero. If $x^* = 0, 1, 2 \ldots h$, both the numerator and the denominator are equal to zero.

Evaluating the indeterminate form we obtain

$$N_1^2 \frac{\cos 2 \pi N_1 x^*}{\cos 2 \pi x^*} .$$

Hence, if $x^* = h$, $D^2(x^*) = D^2(h) = N_1^2$. Thus $D^2(x^*)$ is a periodic function which is equal to N_1^2 for whole-number values of x^*.

Fragments of the two $D^2(x^*)$ curves for $N_1 = 4$ and 8 are shown in Fig. 6. Between the basic maxima of the height of 16 and 64 additional weaker maxima are clearly seen. For $N_1 = 4$ there are two of these between the two neighboring basic maxima, while for $N_1 = 8$ there are six additional maxima. As $D^2(x^*)$ is a periodic function, any basic maximum may be taken as the origin, and thus $D^2(x^*) = D^2(h + \delta) = D^2(\delta)$. Therefore it is convenient to use δ as an argument for D^2 instead of x^*. The value of δ goes from 0 to 1 within the interval between each two successive basic maxima and is expressed in terms of the fractions of the corresponding parameter (a^* in this particular case). $D^2(\delta)$ is given by

$$D^2(\delta) = \frac{\sin^2 \pi N_1 \delta}{\sin^2 \pi \delta} \ . \tag{2.20}$$

The analysis of (2.20) shows that additional maxima appear at $\delta = (2n+1)/2N_1$. The heights of these maxima decrease with the increase of n as $4N_1^2/(2n+1)^2$. The total number of additional maxima is N_1-2. The function $D^2(\delta)$ has centers of symmetry at $\delta = 0$ and $1/2$. Therefore the interval $0 \le \delta \le 1/2$ is sufficient for the construction of $D^2(\delta)$. Güven (1974) proposed to use a normalized function $D^2(\delta)/N^2$, which is equal to unity for $\delta = 0$ regardless of N. Distributions of $D^2(\delta)/N^2$ for different values of N are shown in Fig. 6b. It can be easily seen that the distance from the basic maximum to the first zero point is equal to $1/N_1$ in fractional units and a^*/N_1 in absolute units. It is obvious that it is this value that determines the half-widths of basic maxima. Note that additional maxima are two times narrower than the basic ones.

Thus all the basic maxima of $D^2(x^*)$ arranged with a repeat distance of a^* have the same shape and height regardless of the specific atomic pattern in the unit cell. The half-width of a basic maximum is inversely proportional to the length of the crystal in the direction of $\boldsymbol{a}$, its height is inversely proportional to the squared number of unit cells N^2 in the crystal along this direction, while the area of this maximum is proportional to N. Therefore, for sufficiently large values of N, the intensities of additional maxima may be ignored. Then $D^2(x^*)$ may be considered close to zero for all values of x^* except those equal to whole numbers. It should be born in mind that the interference factor is indifferent to peculiarities of a specific crystal structure if the latter may be described by a crystal lattice.

Note that the extension and the shape of the interference factor maxima are sensitive to certain types of imperfections in crystals (Cowley 1975 b).

It is obvious that the results obtained from the analysis of one of the factors in the three-dimensional interference function $D^2(x^* y^* z^*)$ are equally applicable to the other two. Therefore the function $D^2(x^* y^* z^*)$ for sufficiently large crystals will be nonzero only when the three conditions are satisfied simultaneously: $x^* = h$, $y^* = k$, $z^* = l$, where hkl are whole numbers.

Since diffraction intensity is proportional to $D^2(x^* y^* z^*)$ [see (2.6) and (2.9)], its values differ from zero only in definite directions and for a definite orientation of the crystal lattice relative to the incident beam.

The conditions under which diffraction takes place are given by

$$x^* = s\boldsymbol{a} = (\boldsymbol{k} - \boldsymbol{k}_0)\boldsymbol{a} = \frac{(a\cos\alpha - a\cos\alpha_0)}{\lambda} = h\ ,$$

$$y^* = s\boldsymbol{b} = (\boldsymbol{k} - \boldsymbol{k}_0)\boldsymbol{b} = \frac{(b\cos\beta - b\cos\beta_0)}{\lambda} = k\ , \tag{2.21}$$

$$z^* = s\boldsymbol{c} = (\boldsymbol{k} - \boldsymbol{k}_0)\boldsymbol{c} = \frac{(c\cos\gamma - c\cos\gamma_0)}{\lambda} = l\ ,$$

where $\alpha_0, \beta_0, \gamma_0$ are angles between the incident beam and lattice rows parallel to vectors $\boldsymbol{a}, \boldsymbol{b}, \boldsymbol{c}$, respectively and α, β, γ are those between these vectors and the diffracted beam. In other words, angles α_0, β_0, γ_0 in (2.21) specify the orientation of the crystal lattice with respect to the incident beam, while angles α, β, γ characterize the direction of diffracted beams. Other conditions being fixed, the possible values of α, β, γ are determined by the values of hkl.

Equations in (2.21) correspond to cones formed by diffracted rays scattered by lattice rows parallel to a, b, and c, with solid angles of 2 α, 2 β and 2 γ, respectively. In three-dimensional space a non-zero diffraction intensity appears only in the case where the three cones of diffracted waves scattered by lattice rows intersect along a single direction. This implies simultaneous satisfaction of the three Eqs. (2.21). The corresponding diffracted beam is characterized by the indices hkl. The probability of simultaneous intersection of the three cones along one direction is rather low, as any direction in space is determined, in general, by only two angles, i.e., by intersection of only two cones. Therefore, for fixed values of unit cell parameters, the solution of simultaneous Eqs. (2.21) is possible either for certain α_0, β_0, γ_0 or for a definite choice of the wavelength λ. On the whole, dealing with six angles to find the directions for each diffracted beam is an unreasonably complex problem.

At the same time, the direction of the diffraction vector k may be determined by a single vector equation

$$k - k_0 = ha^* + kb^* + lc^*, \tag{2.22}$$

which is equivalent to the three scalar ones of (2.21). The latter equations may be obtained by multiplying consequently (2.22) by a, b, and c. The right-hand side of (2.22) is equal to a vector terminating in a reciprocal lattice point if hkl are whole numbers that may contain a common factor. The reflecting sphere construction enables one to express the relationship between mutual orientations of the reciprocal lattice vector H_{hkl}, the incident beam (k_0) and the diffracted one (k) in a simple and demonstrative way.

Let us consider another aspect of the interference function, namely, its so-to-say, "internal" structure. The values of $D^2(x^* y^* z^*)$ are equal or close to zero in all points of the reciprocal space except those given by whole number fractional coordinates, i.e., $x^* = h$, $y^* = k$, $z^* = l$. These special points are given by a radius-vector

$$H_{hkl} = ha^* + kb^* + lc^*.$$

Since vectors a, b, c were chosen according to (1.9) and (1.10), the interference function may be regarded as such a reciprocal lattice that diffraction conditions may be satisfied in its nodes.

Diffracted intensity is determined not only by the function $D^2(x^* y^* z^*)$ but also by the structure factor, $|\Phi(x^* y^* z^*)|^2$. It will be shown later that the structure factor is a continuous function which varies rather slowly with x^*, y^*, z^*. Due to the interaction between these two functions, the values of $|\Phi(x^* y^* z^*)|^2$ are practically zero except for a set of points hkl. Thus $|\Phi(x^* y^* z^*)|^2$ contributes to the diffraction pattern by a set of discrete values given by

$$|\Phi(hkl)|^2 = |\sum_j f_j \exp[2\pi i(hx_j + ky_j + lz_j)]|^2. \tag{2.23}$$

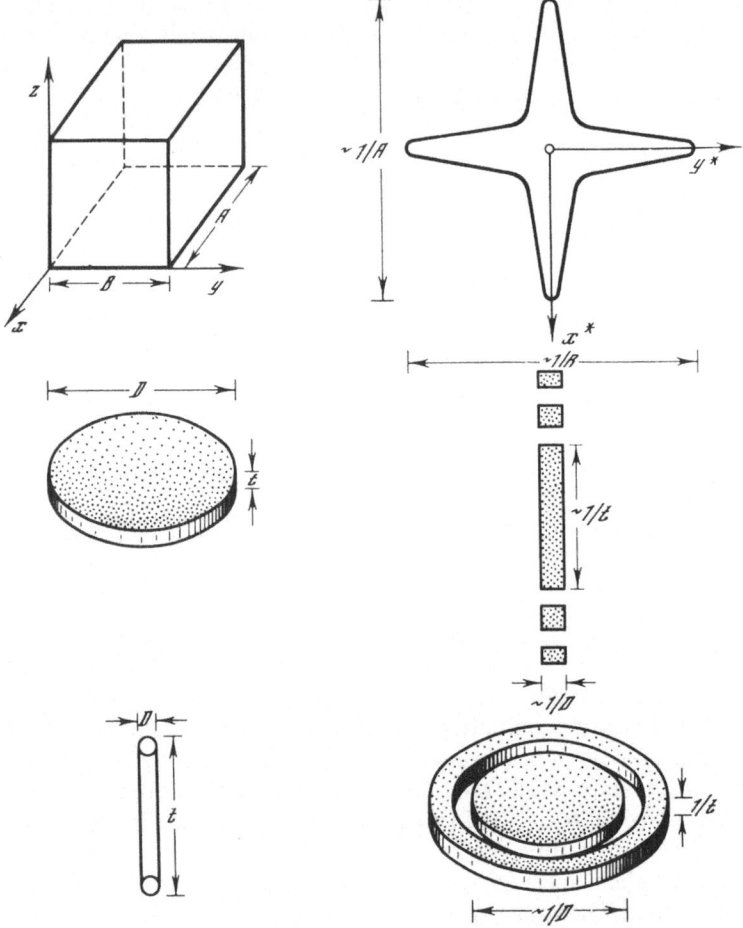

Fig. 7. The relationship between the shape of the crystal and that of the reciprocal lattice point (Hirsh et al. 1977)

In other words, the interference factor turns the diffraction space into a reciprocal lattice, so that each lattice node hkl has a "weight" determined by the corresponding value of $\Phi^2(hkl)$.

The interference factor is often called *the function of the form* or *the form factor*, as extension and shape of the reciprocal lattice nodes are connected directly to the external form of the crystal. Some examples of relations between the shape of a crystal and that of the nodes of the reciprocal lattice are shown in Fig. 7.

Note that all the nodes of the reciprocal lattice are of the same shape which depends only on the size and the shape of the object studied.

Since the reciprocal lattice nodes are of finite dimensions and have a definite shape, one should, when analyzing the intensities of diffraction maxima, take

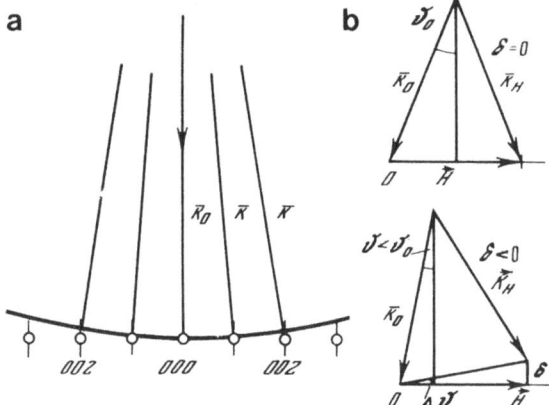

Fig. 8. a The relaxation of diffraction conditions due to extension of reciprocal lattice nodes; **b** The excitation error, resulting from the deviation from the exact Bragg's Law

into account both the maximum intensity and the integrated one, which is concentrated within the volume of a reciprocal lattice node.

For a crystal having the shape of a parallelepiped, the function $D^2(hkl)$ in the centers of the reciprocal lattice nodes is equal to $N_1^2 N_2^2 N_3^2$ and, consequently, the maximum intensity in these points is

$$I_{\max}(hkl) = N^2 \Phi^2(hkl),\tag{2.24}$$

where N is the total number of the unit cells in the crystal. Equation (2.24) is valid for all crystals regardless of their shape.

To produce the value of the integrated intensity, each of the three factors in (2.18) should be integrated between the limits given by the positions of the nearest minima:

$$\int_{-a*/N_1}^{a*/N_1} \frac{\sin^2 \pi N_1 \delta_1}{\sin^2 \pi \delta_1} d\delta_1 \simeq \int_{-\infty}^{\infty} \frac{\sin^2 \pi N_1 \delta_1}{\sin^2 \pi \delta_1} d\delta_1 = a* N_1 = \frac{A_1}{a^2}.\tag{2.25}$$

Then

$$I(hkl) = a* b* c* N_1 N_2 N_3 \Phi^2(hkl) = (V/V_0^2) \Phi^2(hkl)$$

$$= (N/V_0) \Phi^2(hkl),\tag{2.26}$$

where V is the volume of the crystal, and V_0 is the volume of the unit cell.

The finite extension of the reciprocal lattice nodes has serious consequences leading to a relaxation of the strict requirements imposed by the diffraction conditions (2.21). Figure 8 shows a reciprocal lattice row for the case when the scattering object is a thin plate sufficiently extensive in lateral directions. The nodes of the reciprocal lattice in the points with indices hk have a rod-like shape. To satisfy Bragg's Law exactly, the sphere of reflection should cut each reciprocal lattice node precisely at the center. However, owing to the extension of the nodes, diffraction will also occur for a small deviation of the beam from the Bragg angle, ϑ_0.

The Ewald sphere, as shown in Fig. 8, will then intersect the node not at the center, but somewhere in its effective environment where the interference function is non-zero.

In view of the importance of this example for selected-area electron diffraction (SAED), it is worthwhile to consider relative arrangements of the wave vectors k_0 and k with respect to the vector H, when the diffraction conditions satisfy Bragg's Law exactly and not quite exactly. The value of $\delta = \delta_3 c^* = H \operatorname{tg}(\vartheta - \vartheta_0)$ $= H \operatorname{tg} \Delta \vartheta$ (Fig. 8b) is a convenient parameter describing the degree of deviation of the Ewald sphere from the center of a node in the reciprocal lattice.

If the center of a node in the reciprocal lattice lies outside the Ewald Sphere, the angle ϑ between k_0 and the reflecting planes is less than ϑ_0, and $\delta < 0$. If, on the contrary, the center of a node is inside the Ewald Sphere then $\vartheta > \vartheta_0$ and $\delta > 0$.

Thus, a generalized treatment of diffraction from crystal lattices not only has enabled us to formulate conditions defining diffraction directions, but also has led to the concept of the reciprocal lattice in the most natural way.

2.3 Behavior of Electrons in Medium, the Schrödinger Equation, Its Solution in the Kinematical Approximation

Hitherto we have been ignoring the nature of radiation and the specific features in its interaction with matter, since the concepts of wave optics were quite sufficient for the kinematical theory. For geometrical analysis of electron diffraction patterns the results of the kinematical theory are also applicable. However, to evaluate factors affecting intensities of reflections, the physical pattern of the interaction of the radiation with matter should be taken into account. Then electron diffraction may be used for structure studies.

Mechanisms in interaction of charged particles and neutral photons with matter are different, as electrons are scattered by the Coulomb potential, while X-rays are scattered by electron shells in atoms. As a result, the degree of the interaction of electrons with matter differs from that of X-rays. In terms of the kinematical approximation, secondary scattered waves are assumed weak, and their further scattering is ignored. The amplitude for secondary X-ray waves is much weaker than that for electrons which interact with matter much more strongly. Consequently, a weaker agreement with the kinematical theory might be expected for electron diffraction. The problem of the account of secondary scattering of electrons in analytical form is still expecting a complete solution. However, the main peculiarities in electron-diffraction structure analysis may be established even in terms of the kinematical approximation, but with the physical nature of the interaction of electrons with matter taken into account.

The behavior of an electron in the medium is appropriately described by the Schrödinger equation, which is, essentially, a quantum-mechanical analog of the basic wave-optics equation:

$$\nabla^2 \Psi = \frac{\partial^2 \Psi}{\partial x^2} + \frac{\partial^2 \Psi}{\partial y^2} + \frac{\partial^2 \Psi}{\partial z^2} = \frac{1}{u^2} \frac{\partial^2 \Psi}{\partial t^2}, \tag{2.27}$$

where x, y, z are orthogonal coordinates, t is the time, and u is velocity of the propagation of the wave. The function Ψ describes the behavior of the wave if the latter is a solution of the Eq. (2.27).

To find the velocity of electrons, u, expressions $\lambda = h/mu$, and $E = h\nu$ should be applied. The kinetic energy of an electron, $mu^2/2$, is equal to the difference between the full energy and the potential one, i.e., $mu^2/2 = E - U$.

Hence

$$u = \lambda\nu = \frac{E}{[2\,m(E-U)]^{1/2}} \,.$$

Substituting this into (2.27) we obtain

$$\nabla^2\Psi = \frac{2\,m(E-U)}{E^2}\,\frac{\partial^2\Psi}{\partial t^2} \,. \tag{2.28}$$

A plane wave of the type

$$\Psi(r,\,t) = c\exp[2\,\pi i k\,(r-u\,t)] = \Psi(xyz)\exp[-2\,\pi iku\,t]\,, \tag{2.29}$$

may be a solution of (2.28).

$\Psi(xyz)$ is the space component of the wave $\Psi(r,\,t)$. Substituting (2.29) into (2.28) we obtain the Schrödinger Equation

$$\nabla^2\Psi(xyz) + (8\,\pi^2 m/\,h^2)(E-U)\,\Psi(xyz) = 0\,. \tag{2.30}$$

The Eq. (2.30) contains in an explicit form the potential energy of an electron which may be connected directly with the potential of the object. The fact is that an electron in an object with a potential $\varphi(r)$ acquires a potential energy $U(r) = -e\varphi(r)$, and it is due to this that scattering occurs.

$\Psi(r)$, or $\Psi(xyz)$ describes the wave-like properties of an electron depending only on coordinates, and is called *the wave function*. The product of Ψ by the complex conjugate, Ψ^*, multiplied by the volume element dV, determines the probability for finding the electron inside this volume element. The value of $\Psi\Psi^*$ characterizes the electron density, or the number of electrons in a volume unit.

The behavior of an electron in vacuum is the simplest case, as $U(r) = 0$. The solution of (2.30) is then a plane wave given by

$$\Psi_0 = c_0\exp[2\,\pi i k_0 r]\,. \tag{2.31}$$

Substituting (2.31) into (2.30), we have

$$-4\,\pi^2 k_0\,\Psi_0 + \frac{8\,\pi^2 m}{h^2}E\,\Psi_0 = 0 \quad\text{and}\quad k_0 = \frac{1}{\lambda_V} = \frac{\sqrt{2\,mE}}{h}\,.$$

Here $E = eV$, where V is the accelerating voltage. Thus the wavelength in vacuum depends only on V and is equal to $\lambda_V = h/(2\,meV)^{1/2}$.

An electron passing through a boundary of two media, e.g., from vacuum into a crystal having a mean potential φ_0, acquires a potential energy $e\varphi_0$. If the solution is again sought in the form of a plane wave, that is

$$\Psi = c\exp[2\,\pi i k r]\,, \tag{2.32}$$

then after substitution this function into (2.30) we obtain the expression for the wave number in the medium

$$\lambda = 1/k = h/\sqrt{2\,m(E + e\,\varphi_0)}\,. \tag{2.33}$$

In other words, transition of electrons from vacuum into a crystal is accompanied by a change in the wavelength. The difference is, however, very small, because E is usually greater than $e\,\varphi_0$ by a factor of 10^4 or 10^5.

In terms of quantum mechanics, the preservation of charge with the passage of electron waves through an interface requires the following conditions: at the interface the wave functions propagating on both sides of the interface should be continuous and so should be their normal derivatives, $\partial\Psi/\partial z$, that is for $z = 0$

$$\Psi = \Psi_0 \quad \text{and} \quad \partial\Psi/\partial z = \partial\Psi_0/\partial z\,, \tag{2.34}$$

where Ψ_0 and Ψ are the wave functions in vacuum and in the medium, respectively.

The coordinate system may be always chosen in such a way that the wave functions are written as

$$\Psi_0 = c_0 \exp[2\,\pi i(k_{0x}x + k_{0z}z)]\,,$$

$$\Psi = c \exp[2\,\pi i(k_x x + k_z z)]$$

where k_{0x}, k_x are the tangential components of coplanar wave vectors k_0 and k, while k_{0z} and k_z are their normal components. It can be easily seen that satisfaction of the boundary conditions for Ψ and Ψ_0 requires that the value of k should be equal to that of k_0, which contradicts Eq. (2.33). Therefore, one has to assume that a reflected wave should appear resulting from the passage of electrons through the interface, and therefore there should be two waves in vacuum, i.e., $\Psi_0 = c_0 \exp 2\,\pi i(k_{0x}x + k_{0z}z) + c_0' \exp 2\,\pi i(k_{0x}'x + k_{0z}'z)$. Having applied the boundary conditions (2.34) to this case we have

$$c = c_0 + c_0'\,, \quad k_z c = k_{0z}c_0 + k_{0z}'c_0'\,, \quad k_x = k_{0x} = k_{0x}'\,.$$

Figure 9 shows that if the tangential components of the wave vectors are equal, then

$$k_0 \sin\gamma_0 = k_0' \sin\gamma_0' = k \sin\gamma\,.$$

For $k_0 = k_0'$, the angle of incidence γ_0 is equal to the angle of reflection, and the index of refraction μ is given by

$$\mu = \frac{\sin\gamma_0}{\sin\gamma} = \frac{k}{k_0} = \left(\frac{E + e\,\varphi_0}{E}\right)^{1/2} \simeq 1 + \frac{e\,\varphi_0}{2\,E} = 1 + 10^{-5}\,.$$

The reflection factor equal to the ratio of the reflected electron flux density, i_0', to that of the incident electrons, i_0, is given by (Heidenreich 1964)

$$\frac{i_0'}{i_0} = \left|\frac{c_0'}{c_0}\right|^2 = \left(\frac{\mu-1}{\mu+1}\right)^2 \approx 10^{-9}\,.$$

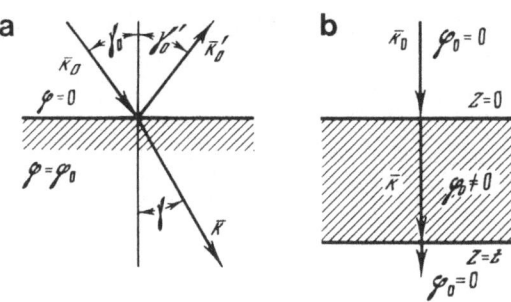

Fig. 9a, b. The values of k and γ for the incident, the reflected, and the transmitted waves; **a** $\gamma \neq 0$; **b** $\gamma = 0$

Thus the amplitude of the reflected wave is in all cases negligible, and therefore the amplitude of the incident wave is practically equal to that of the transmitted wave.

Elastic scattering of electrons in the medium, i.e., that proceeding without a change in the wavelength, may be described, in particular, in terms of the Born method of successive approximations. As $E \gg U$ for fast electrons, the perturbing potential of the crystal is assumed to be so small that the solutions are close to those obtained without it. Therefore, the wave function to be found may be presented as a Born series of wave functions generated by an iterative procedure:

$$\Psi(r) = \Psi_0 + \Psi_1 + \Psi_2 \ldots \tag{2.35}$$

In terms of the kinematical approximation, the solution of the Schrödinger equation may be sought in the form

$$\Psi = \Psi_0 + \Psi_1 , \tag{2.36}$$

where Ψ_0 is the wave function for an electron in vacuum. Substituting (2.36) into (2.30) and taking account of (2.31) we obtain

$$\nabla^2 \Psi_1 + k_0^2 \Psi_1 = (8 \pi^2 me/h^2) \varphi(r) \Psi(r) = K \varphi(r) \Psi(r) ,$$

where $K = 8 \pi^2 me/h^2$.

In the integral form we have (Vainshtein 1956)

$$\Psi(r) = \Psi_0(r) - \frac{K}{4\pi} \int \frac{\exp[2\pi i k |r - r_1|]}{|r - r_1|} \varphi(r_1) \Psi(r_1) dV_{r_1} , \tag{2.37}$$

where r_1 is the vector inside the scattering volume, r is the vector in the point of observation (Fig. 5b).

Under the integral in (2.37) there is the function to be found $\Psi(r_1) = \Psi_0(r_1) + \Psi_1(r_1)$. In terms of the kinematical approximation, $\Psi_1 \ll \Psi_0$ and, taking account of (2.31), we have

$$\Psi(r) = c_0 \exp[2\pi i k_0 r]$$

$$- \frac{K}{4\pi} \int \varphi(r_1) c_0 \exp[2\pi i k_0 r_1] \frac{\exp[2\pi i k |r - r_1|]}{r} dV_{r_1} .$$

As compared to (2.37), here Ψ replaces Ψ_0, and r replaces $|r-r_1|$ in the denominator, as $r \gg r_1$. In addition, as shown in Fig. 5, $k_0 r_1 + k|r-r_1| + kr - kr = (k_0 - k, r_1) + kr$.

The final expression for the wave scattered in the crystal is

$$\Psi_1(r) = \frac{Kc_0}{4\pi} \frac{\exp[2\pi ikr]}{r} \int \varphi(r_1) \exp[2\pi i(k_0-k, r_1)] dV_{r_1} . \tag{2.38}$$

The successive approximations in the series (2.35) may be generated using formulae of the type (Vainshtein 1956)

$$\Psi_n(r) = -\frac{1}{4\pi} \int \frac{\exp[2\pi ik|r-r_1|]}{|r-r_1|} \varphi(r_1) \Psi_{n-1}(r_1) dV_{r_1} .$$

It is obvious that the determination of wave functions by successive approximations needs complicated calculations. The physical meaning of the given approximation is that the perturbing effect of the periodic component of the crystal on the initial incident wave is ignored when the wave function for scattered electrons is calculated according to (2.38). It has been indeed supposed for the derivation of (2.38) that the only source of all secondary waves in the whole scattering volume is the initial plane wave, for which the crystal field potential was assumed to be equal to the mean inner potential φ_0.

Let us consider in more detail the wave function involved. Taking account of (2.38) we have

$$\Psi(r) = c_0 \exp[2\pi ik_0 r] + (c_s'/r) \exp[2\pi ikr] . \tag{2.39}$$

It can be seen that the diffracted wave is a spherical one with an amplitude given by

$$c_s(s) = -\frac{Kc_0}{4\pi r} \int \varphi(r) \exp[2\pi isr] dV_r . \tag{2.40}$$

If the boundary conditions (2.34) are applied, the satisfaction of these will require $c_s = 0$. Otherwise it should be assumed that there is in the crystal at least one more diffracted wave having such parameters as lead to a zero resultant amplitude for diffracted waves on the surface of the crystal. Thus boundary conditions are not satisfied in the kinematical theory, which illustrates once again its approximate nature.

Now we shall show that the spherical wave describing the behavior of an electron in the crystal is characterized by an expression for the amplitude similar to Eq. (2.5). The amplitude analyzed is given by

$$A(s) = K' \int \varphi(r) \exp[2\pi isr] dV_r , \tag{2.41}$$

where $K' = K/4\pi = 2\pi me/h^2$.

The electric potential for the crystal $\varphi(r)$ is the sum of the potentials of electron shells and atomic nuclei. It may be described as a continuous three-dimensional positive function having maxima at the centers of atoms. Therefore the function $\varphi(r)$ may be represented by a superposition of potentials of separate atoms

$$\varphi(r) = \sum_j \varphi_j(r - r_j), \tag{2.42}$$

where φ_j is the atomic potential at the point given by $r - r_j$, and r_j is the radius-vector of the center of the jth atom.

Substituting (2.42) into (2.41) we have

$$A(s) = K' \sum_j \int \varphi_j(r - r_j) \exp[2\pi i(r - r_j, s)] \exp[2\pi i r_j s] dV_r. \tag{2.43}$$

If atomic potentials do not overlap, the scattering power for each atom is given by

$$f_j(s) = K' \int \varphi_j(r - r_j) \exp[2\pi i(r - r_j, s)] dV_r.$$

Then the expression (2.43) may be written

$$A(s) = \sum_j f_j(s) \exp 2\pi i s r_j.$$

This expression for the amplitude of the scattered wave is similar to (2.5), which was derived in terms of the wave optics irrespective of the nature of incident radiation.

Taking account of constant factors that were hitherto ignored [see (2.40)], the amplitude of the spherical wave describing the behavior of an electron scattered by a periodic potential of the crystal may be given by

$$c_s(s) = (c_0/r) \sum f_j(s) \exp 2\pi i s r_j = (c_0/r) \Phi(H) D(H) = c(H). \tag{2.44}$$

Here the summation over the whole set of atoms in the crystal was replaced by summation over the crystal lattice points having scattering powers $\Phi(H)$ [see (2.11)].

The intensity for a scattered spherical wave is given by

$$I(H) = (c_0^2/r^2) \Phi^2(H) D^2(H) = (I_0^2/r^2) \Phi^2(H) D^2(H), \tag{2.45}$$

where I_0 is the incident beam intensity.

2.4 Atomic Scattering Amplitudes, or *f*-Curves

To find amplitudes for electron waves scattered by isolated atoms, the formula (2.41) may be used, so that

$$f_e(s) = K \int_{atom} \varphi(r) \exp[2\pi i s r] dV, \tag{2.46}$$

where $\varphi(r)$ is the atomic potential.

The X-ray scattering amplitude is written similarly

$$f_x(s) = \int \rho(s) \exp[2\pi i s r] dV, \tag{2.47}$$

but here atomic electron density $\rho(r)$ is the scattering matter, and f_x is expressed in terms of electronic units, i.e., with respect to the scattering amplitude of X-rays by an electron.

Thus, provided the atomic potential distribution is known, the values of $f_e(s)$ may be calculated using (2.46) (Vainshtein 1956). The atomic potential is composed by the positive nuclear potential Z/r and the negative potential of the electron cell (Z is the charge on the nucleus).

The charge of the electron shell screens the nuclear potential outside the electron shell and decreases it inside the volume of the atom. The electron charge is spread over the volume of electron shells, while the nuclear charge is localized practically in a point; the resultant potential of a neutral atom is always positive. It is only for negatively charged ions at comparatively long distances from the nucleus that the atomic potential becomes negative. As compared with the Coulomb nuclear potential, the atomic potential falls off more rapidly with r.

The atomic potential distribution, especially in the region of small values of $|s| = 2\sin\vartheta/\lambda$, depends to a considerable extent on the geometry in the distribution of electrons around the nucleus, on the dimensions of the scattering volume, the nature of chemical bonding, and many other factors. As a consequence, the calculation of f_e-curves according to (2.46) is possible only if based on a number of simplifying assumptions and, in particular, on the assumption that the atomic potential is spherically symmetrical (Vainshtein 1956).

Another way to determine atomic scattering factors is based on the relationship between these and X-ray scattering factors f_x. The potential may be expressed in terms of charge density $\rho(r)$ (Hirsh et al. 1965)

$$\varphi(r_i) = \int \frac{\rho(r_j)dV_j}{|r_i - r_j|} . \tag{2.48}$$

Where r_i and r_j are the points in the atom volume, $\rho(r_j) = \rho_+(r_j) - \rho_-(r_j)$, ρ_+ and ρ_- being the values of charge density on the nucleus and in the electron shell, respectively. Substituting (2.48) into (2.46) we obtain

$$f_e = \frac{2\pi me}{h^2} \int \rho(r_j) \exp[2\pi i s r_j] dV_j \int \frac{\exp[2\pi i(r_i - r_j, s)]}{|r_i - r_j|} dV_i . \tag{2.49}$$

Taking (2.47) into account, it is evident that

$$\int [\rho_+(r_j) - \rho_-(r_j)] \exp[2\pi i s r_j] dV_j = e(Z - f_x) .$$

The second integral in (2.49) is equal to $1/4\,\pi s^2$. Finally we obtain

$$f_e(\vartheta) = (me^2/2h^2)(\lambda/\sin\vartheta)^2(Z - f_x) . \tag{2.50}$$

Some general conclusions on the scattering of electrons and X-rays may be drawn from (2.50). In both cases, f decrease continuously with ϑ, and, for a given ϑ, f increases with Z. However, due to the factor $(\lambda/\sin\vartheta)^2$ in (2.50), the atomic scattering factor falls off with ϑ more rapidly than in the case of X-rays. Later this will be shown to improve the convergence of Fourier series, resulting in a considerable decrease in the number of reflections needed, as compared with X-ray structure analysis.

For example, it was shown by Tsipursky and Drits (1977a) that, with 27 structural parameters being refined, atoms may be localized with an e.s.d. of $0.003 - 0.005$ Å, using only about 200 reflections (cf. more than 800 reflections needed for refining similar structure to the same accuracy by X-ray diffraction).

Another peculiarity is that values of f_e are by several orders of magnitude greater than those of f_x for the same atom. In terms of absolute units

$$f_e = 2.38 \cdot 10^{-10} \frac{Z - f_{x_1}}{\left(\dfrac{\sin \vartheta}{\lambda}\right)^2} .$$

For medium scattering angles, $(\sin \vartheta/\lambda = 0.1 - 0.3)$, the ratio $f_e/f_x \approx 10^4 - 10^5$. This explains why dispersed substances are natural objects for electron diffraction, since, roughly speaking, an atom scatters electrons with the same "power" as $10^4 - 10^5$ atoms scatter X-rays. Note also that, for electron diffraction, differences in values of f_e for light and heavy atoms are much less. Whereas the maximum ratio of f_x for the lightest atom and that for the heaviest one is about 100, the corresponding ratio for electrons is three times less. It is evident that relatively small differences in f_e for different atom types facilitate localization of light atoms in the presence of heavy ones.

Vainshtein (1956) and other authors used Eqs. (2.46) and (2.50) for the calculation of f_e-curves.

These data are given in the International Table for X-ray Crystallography (1962), as well as in a number of books (Heidenreich 1964; Hirsh et al. 1965) and manuals.

Atoms in crystals vibrate about certain equilibrium positions. These thermal motions result in the attenuation of the diffraction intensity, since at each given moment atoms are displaced from equilibrium positions, and the phase relations between scattered waves are therefore violated. The degree of attenuation should increase with $\sin \vartheta/\lambda$, or, which is the same, with the decrease of d/n, as the phase disbalance will be greater for smaller d-spacings.

With constant temperature, the amplitude of thermal vibrations depends on the chemical bonding. Therefore different atoms in the same structure have different thermal parameters, so that their contributions to the attenuation of intensity are also different.

Thus, the simplest way to allow for atomic thermal vibrations is to introduce corrections into the distribution of f_e values. Calculations show that f_j^T for the jth atom at the temperature T may be written

$$f_j^T = f_j \exp[- B_j (\sin \vartheta/\lambda)^2] , \tag{2.51}$$

where f_j is the atomic scattering factor for an atom of the type j at rest and B_j is the mean amplitude of its isotropic vibrations at the temperature T.

The real distribution of vibration amplitudes for atoms in crystals is anisotropic. To allow for this, the concept of the thermal ellipsoid is introduced. The axes B_1^j, B_2^j, B_3^j of the ellipsoid characterize the vibration amplitudes for the jth atom, while the angles α_1, α_2, α_3 between these axes and the vector H specify the orientation of the ellipsoid in the space. Thus B_j in (2.51) should be replaced by $B_1^j \alpha_1 + B_2^j \alpha_2 + B_3^j \alpha_3$. The values for isotropic and anisotropic coefficients B are determined in the crystal structure refinement by the least squares method (see below).

2.5 Structure Amplitude and Structure Factor

In terms of the first Born approximation, the wave function describing the scattering of electrons by a unit cell is spherical, as in the case of an isolated atom. The amplitude of this wave is called *the structure amplitude* and is given by

$$\Phi(s) = \int_{\text{unit cell}} \varphi(r) \exp[2\pi i s r] dV.$$

In the reciprocal space specified by vectors a^*, b^*, c^*, vector s is defined in terms of continuous coordinates x^*, y^*, z^* in the general case, and by node indices hkl in the case of the reciprocal lattice.

Since the potential of the unit cell may be regarded as superposition of separate atomic potentials, the expression for the structure amplitude may be written

$$\Phi(hkl) = \sum f_j \exp 2\pi i Hr_j = \sum f_j \exp[2\pi i(hx_j + ky_j + lz_j)], \tag{2.52}$$

where x_j, y_j, z_j are fractional coordinates of the jth atom in the unit cell and f_j is its atomic scattering factor. Summation is taken over all the atoms in the unit cell. Formula (2.52) shows how the structural amplitude is affected by the geometrical arrangement of atoms and by their scattering powers along the diffraction direction specified by indices hkl. Generally, the structure amplitude is a complex quantity, i.e., it may be characterized by the modulus $|\Phi(hkl)|$ and the phase α, as $|\Phi(hkl)| \exp i\alpha_{hkl}$. The physical meaning of this is the following. The amplitude of a spherical wave scattered by the unit cell has an initial phase determined by the inner structure of the unit cell and the scattering direction.

If

$$A(hkl) = \sum f_j \cos[2\pi(hx_j + ky_j + lz_j)], \tag{2.53}$$

$$B(hkl) = \sum f_j \sin[2\pi(hx_j + ky_j + lz_j)],$$

then

$$\Phi(hkl) = A(hkl) + iB(hkl),$$

$$|\Phi(hkl)| = (A^2 + B^2)^{1/2}, \qquad \alpha = \text{arctg}(B/A).$$

Note that the square of the structure amplitude modulus, $|\Phi(hkl)|^2$, is called the *structure factor*. Thus to find $\Phi(hkl)$ one should know the dimensions and the shape of the unit cell, and the coordinates and the scattering powers of atoms contained in it.

If the unit cell is nonprimitive and there are various symmetry elements, it is sufficient to analyze only the symmetrically independent part of the atomic pattern.

In addition, certain transformations of formulae for $\Phi(hkl)$ become possible, so that they are brought into a form convenient for calculations. However, the most important point here is that after these transformations one may easily imagine diffraction features that should be observed from crystals having this or that space symmetry. Analysis of these feature provides the symmetry of the crystal. For example, in centered unit cells, some of the reciprocal lattice points disappear.

The diffraction nature of this effect may be easily explained. For example, in a *C*-centered cell, for each atom at position with coordinates x_j, y_j, z_j there is a corresponding atom at $x_j + 1/2$, $y_j + 1/2$, z_j. Therefore

$$\Phi(hkl) = \sum_{j=1}^{N/2} f_j \exp[2\pi i(hx_j + ky_j + lz_j)]$$

$$+ \sum_{j=1}^{N/2} f_j \exp\left\{2\pi i\left[h\left(x_j + \frac{1}{2}\right) + k\left(y_j + \frac{1}{2}\right) + lz_j\right]\right\}$$

$$= [1 + \exp\pi i(h+k)] \sum_{j=1}^{N/2} f_j \exp 2\pi i(hx_j + ky_j + lz_j)$$

where the summation is over the atoms in a primitive cell only.

Hence reflections having $h + k = 2n + 1$ should be missing in a diffraction pattern from a *C*-centered lattice. Note that systematic absence of reflections with certain indices in a diffraction pattern due to zero values of structure amplitudes is called *extinction*. For *A*-centered cells, reflections with $k + l$ odd will be missing.

In a face-centered cell each atom at xyz is multiplied due to additional nodes. As a result,

$$\Phi(hkl) = [1 + \exp\pi i(h+k) + \exp\pi i(h+l) + \exp\pi i(k+l)] \Phi'(hkl).$$

It is evident that $\Phi(hkl)$ will be nonzero if hkl are either all even or all odd. In a body-centered cell, reflections for $h+k+l$ odd will be missing for similar reasons.

The presence of various symmetry elements in a unit cell permits to transform expressions for structure amplitudes and to reveal symmetry operations relating structure amplitudes or structure factors.

In centrosymmetric structures, for every atom at $x_j y_j z_j$ there is an exactly similar atom at $\bar{x}_j \bar{y}_j \bar{z}_j$.

Hence

$$\Phi(hkl) = \sum_{j=1}^{N/2} f_j \exp 2\pi i(hx_j + ky_j + lz_j) + \sum_{j=1}^{N/2} f_j \exp[-2\pi i(hx_j + ky_j + lz_j)]$$

$$= 2 \sum_{j=1}^{N/2} f_j \cos 2\pi(hx_j + ky_j + lz_j). \tag{2.54}$$

This implies that structure amplitudes for centrosymmetrical crystals are real quantities, since there are only two possible values for their phases, π or 2π, leading to a negative $\Phi(hkl)$ or to a positive one. Thus the phase problem in structure analysis is essentially simplified for centrosymmetrical crystals. In addition, (2.54) implies $\Phi(hkl) = \Phi(\bar{h}\bar{k}\bar{l})$.

If there is a mirror plane *m* normal to the *y* axis, for each at atom x_j, y_j, z_j there is a corresponding atom at x_j, $\bar{y}_j$, z_j. Thus

$$\Phi(hkl) = 2 \sum_{j=1}^{N/2} f_j \cos 2\pi ky_j \exp 2\pi i(hx_j + lz_j).$$

It is obvious that $\Phi(hkl) = \Phi(h\bar{k}l)$. However, since the diffraction pattern is centrosymmetrical, there are four equivalent structure factors $\Phi(hkl)$ $= \Phi(\bar{h}\bar{k}\bar{l}) = \Phi(h\bar{k}l) = \Phi(\bar{h}k\bar{l})$.

On the whole, diffraction data do not permit to distinguish crystals having centers of symmetry from those having mirror planes and/or twofold rotation axes. Screw axes and glide planes lead to zero $|\Phi(hkl)|^2$ with certain hkl. Systematic absence among structure factors with common indices permit to determine the Bravais lattice and to establish the unit cell correctly. Extinctions among reflections hol, okl, hko, hho indicate the presence of glide planes, and information on the presence of screw axes is given by extinctions among ool, oko and hoo reflections. These problems are treated in detail in the International Tables (1962). In SAED, two-dimensional sets of reflections are generally analyzed, the most important diffraction patterns containing reflections hko, hol, okl. These reflections are collected when the initial beam is parallel to c, b, and a, respectively. Therefore reflections of, e.g., the hko type provide information on the projection of the structure along c.

Only two coordinates are used in the projection, and therefore

$$\Phi(hko) = \sum_j f_j \exp 2\,\pi i(hx_j + ky_j)\,. \tag{2.55}$$

A structure projection is characterized by symmetry elements according to which the expression (2.55) is transformed.

Provided the shape and the dimensions of the unit cell as well as the space symmetry and the atomic coordinates are known, structure amplitudes may be calculated using formulae transformed to the most convenient analytical form for each space group (International Tables for X-Ray Crystallography 1962).

To conclude this section, note some peculiarities of the "interaction" between the structure factor and the interference factor in the formation of a diffraction pattern. $\Phi(x^*, y^*, z^*)$ in the diffraction space is continuous, i.e., its values are finite in any point given by the vector $s = x^*a^* + y^*b^* + z^*c^*$. If the cell is centered or contains glide planes or screw axes, $\Phi(x^*y^*z^*)$ is zero in special points in the reciprocal space at positions with whole-number coordinates.

It has been mentioned above that the interference function represents a lattice in the diffraction space. The interaction between the interference function and the structure factor leads to zero intensity for all regions in the diffraction space except those where the basic maxima of $D^2(s)$ are located. On the other hand, the distribution of structure amplitudes in the diffraction space also interacts with the interference function and cancels some of the nodes by its zero values.

2.6 Reflection Intensities in Point Electron Diffraction Patterns in Terms of the Kinematical Approximation

According to the definition of the wave function, the product $\Psi\Psi^*$ is equal to the number of electrons in a unit volume. For a plane wave, $\Psi\Psi^* = C_0 C_0^*$, where C_0 is its amplitude. As it was mentioned, values of the type $\Psi\Psi^*$ or $C_0 C_0^*$ are called the *intensities* of the corresponding waves. In the case of diffraction

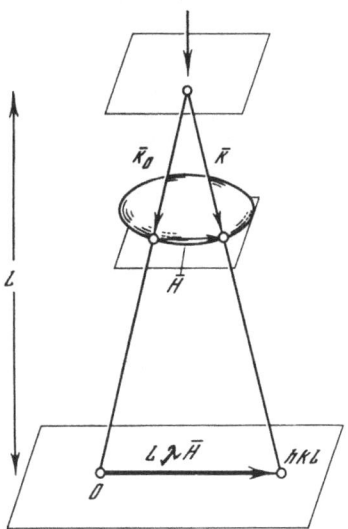

Fig. 10. Arrangement of the object, the screen, the reciprocal lattice vector H, and the diffraction spot on the screen

from crystals it is convenient to characterize each diffracted beam by the total number of electrons passing through the area unit in the time unit. The latter value is the integrated intensity of the given maximum I_I. Let us find the integrated intensity of the diffraction peak in the case of kinematic scattering of electrons by a thin parallel-sided plate. Figure 10 shows the arrangement of the sample, the screen, the reciprocal lattice vector H, and the diffraction spot hkl on the screen. Since the object is a thin plate, the reciprocal lattice nodes may be represented by rods lying along the normal n_0 to the surface of the sample. The intensity distribution within the spot is determined by the distribution of the interference function $D^2(\delta_1, \delta_2)$, i.e., by the lateral dimensions of the plate. If the plate area, illuminated by the incident beam, is large enough, then, according to (2.45) the integrated intensity of the diffraction peak on the screen will be given by

$$I_I = \frac{I_0}{L^2}\, \Phi^2(H) D^2(\delta_3) \int\!\!\int_{-\infty}^{\infty} D^2(\delta_1, \delta_2)\, dx_1\, dx_2 , \tag{2.56}$$

where $r = L$ is the distance between the sample and the screen, $dx_1 = (L\lambda/a)\,d\delta_1$ and $dx_2 = (L\lambda/b)\,d\delta_2$. The integral should be taken over the whole area of the diffraction spot. However, the result will be the same if integration is performed between the limits $-\infty$ and $+\infty$. Taking (2.25) into account, we have

$$I_I(H) = I_0\lambda^2\, \frac{A_1}{a_2}\, \frac{A_2}{b_2}\, \Phi^2(H)\, \frac{\sin^2 \pi\delta_3 N_3}{\sin^2 \pi\delta_3}$$

$$= \lambda^2 I_0 \left(\frac{\Phi(H)}{V_0}\right)^2 \frac{\sin^2 \pi A_3 \delta}{(\pi\delta)^2}\, S , \tag{2.57a}$$

where $\delta = \delta_3/c = \delta_3 c^*$, V_0 is the volume of the unit cell, A_1, A_2, and A_3 are the linear dimensions of the plate, and $S = A_1 A_2$ is the area illuminated by the incident beam.

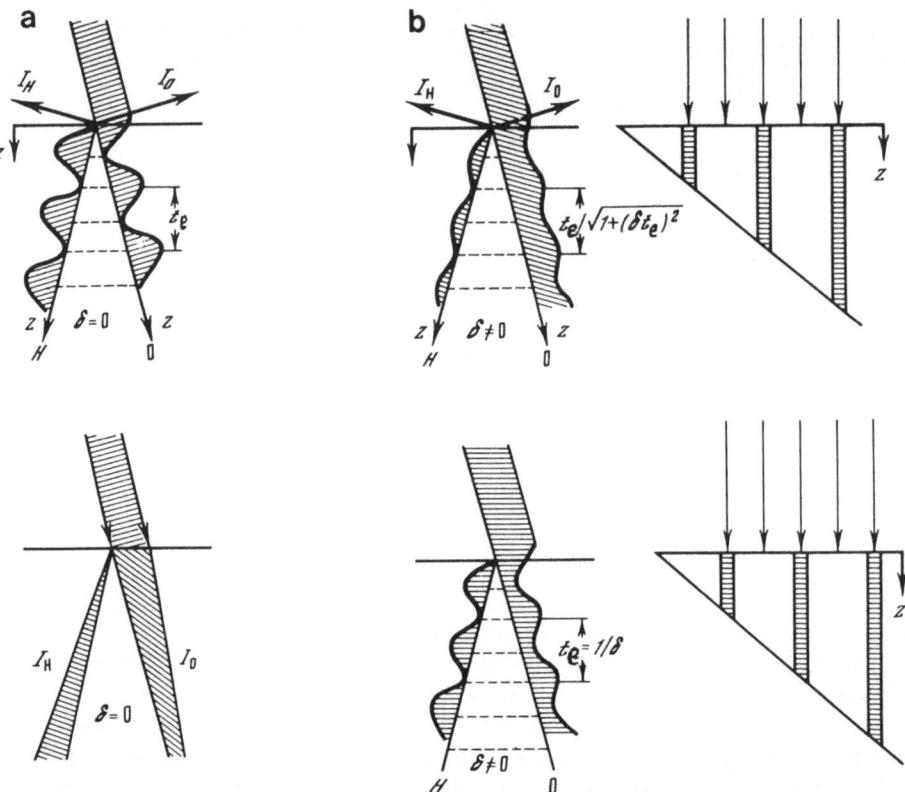

Fig. 11a, b. Dependance of the integrated diffraction intensity on the crystal thickness for the dynamical (above) approximation and the kinematical one (below); **a** $\delta = 0$; **b** $\delta \neq 0$ (Amelinckx 1964)

The relative integrated intensity is given by

$$I_{rel} = \frac{I_I}{I_0 S} = \lambda^2 \frac{\Phi^2(H)}{V_0^2} \frac{\sin^2 \pi \delta A_3}{(\pi \delta)^2} , \qquad (2.57b)$$

where $I_0 S$ is the electron flux having passed through the crystal area S, i.e., the integrated intensity of the incident beam (Vainshtein 1956).

With exact satisfaction of Bragg's Law ($\delta = 0$) and the plate thickness $t = A_3$, we have

$$I_{rel} = \frac{I_I(H)}{I_0 S} = \lambda^2 \frac{\Phi^2(H)}{V_0^2} t^2 . \qquad (2.58)$$

Thus, for the idealized case of a parallel-sided plate, with Bragg's Law satisfied exactly, the relative integrated intensity is proportional to the structure factor and the squared plate thickness. In a real experiment, there may be deviations from these conditions, even in those cases where the kinematical approximation is valid.

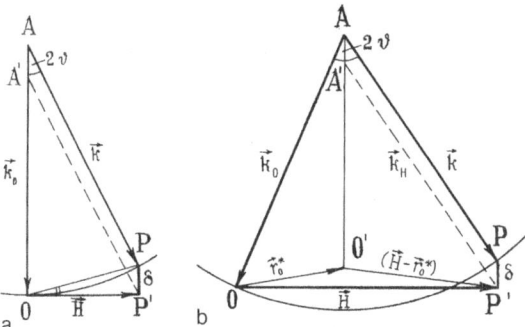

Fig. 12a,b. Effect of the Ewald sphere curvature on the deviation from Bragg's Law with the increase in H. **a** Surface normal, n is parallel to k_0; **b** n is not parallel to k_0

Let us consider the influence of certain factors on I_{rel} (Vainshtein 1956). In a wedge-like crystal, the sample thickness changes continuously. According to (2.57b), the integrated intensity will vary periodically from 0 to $(\lambda^2/\pi^2\delta^2)$ $(\Phi/V_0)^2$ with the period $1/\delta$ (see Fig. 11). If Bragg's Law is satisfied exactly, the increase of I_{rel} should lead, according to (2.58), to a monotonic increase in the integrated intensity according to a parabolic law.

If no limitations are imposed, the value $\lambda[\Phi(H)/V_0]t$ for certain t may turn out greater than unity, leading to $I > I_0S$, which is physically unacceptable. The physical meaning of this is, however, obvious: with the increase of t effects of multiple scattering increase, and the main assumption of the kinematical theory that $I \le I_0S$ is no longer valid. Therefore it is necessary in terms of the kinematical approximation that $t < V_0/\lambda\Phi(H)$ (Vainshtein 1956). In the light of the results of the dynamical theory, this problem will be later treated in more detail.

The dependence of $I(H)$ on δ for a uniform plate thickness has been discussed in detail in the section on the properties of the interference function $D^2(\delta) = D^2(\delta_3/c)$. In a first approximation, $D_2(\delta)$ varies linearly decreasing from $N^2 = (t/c)^2$ for $\delta = 0$ to 0 for $\delta = 1/t$, where c is the repeat distance along the normal to the surface of the plate. If the reflecting sphere were planar, i.e., $\lambda \to 0$, then for an ideal parallel-sided plate and for the incident beam direction coinciding with c, reflections hko would be observed having integrated intensities determined by the expression (2.58). However, the Ewald sphere has a certain curvature even for very fast electrons. Therefore, with the increasing distance from the reciprocal lattice origin the surface of the sphere will become more and more distant from the centers of the nodes hko. This will lead, according to (2.57b), to a consequent decrease in the intensity for reflections with large hk.

Let us find the deviation x of the Ewald sphere from the plane at the distance H from the origin. Figure 12a shows that

$$x = \delta' = PP' = AA' = k_0 - (k_0^2 - H^2)^{1/2} \approx \frac{\lambda H^2}{2}.\qquad(2.59)$$

In the limiting case of $x = 1/t$, the rod-like node will not be in contact with the sphere due to its curvature. Therefore

$$\lambda H^2/2 \leqslant 1/t \quad \text{and} \quad d(hko) > \sqrt{\lambda t/2}.$$

For an accelerating voltage of 100 kV, the electron wave length $\lambda = 0.037$ Å. Diffraction patterns from parallel-sided crystals having the thicknesses of 50, 100, and 500 Å will contain reflections corresponding to $d(hko)$ spacings greater than 0.96, 1.36, and 3.0 Å, respectively. Hence, if a sufficiently thin plate with large repeat distances a and b is used, a relatively large number of reflections will be present in the diffraction pattern, even if the plate is strictly parallel-sided. For example, for $t = 100$ Å, $\lambda = 0.037$ Å and $a = 10$ Å, seven orders of hoo reflections will be observed, but there will be only three if $a = 5$ Å. The increase of the thickness up to 500 Å will lead to a single reflection observed for $a = 5$ Å. Radiation being monochromatic, the integrated intensity should be determined according to (2.57b) assuming that $\delta = \lambda H^2/2 = \lambda/2\, d^2(hko)$ for each hko reflection.

The integrated reflection intensity in a point electron diffraction pattern from a thin, ideally parallel-sided crystal is given, with (2.57a) and (2.59) taken into account, by

$$I(H) = 4\,I_0 \left[\frac{\Phi(H)}{V_0} \right]^2 \frac{\sin^2 \pi A_3 \lambda H^2/2}{\pi^2 H^4}\, S .$$

Thus, to allow for the decrease in the integrated intensity for a reflection with the given H caused by the Ewald sphere curvature, one should know the crystal thickness A_3 or, to be more precise, the thickness of the domain of coherent scattering in the direction parallel to the incident beam.

In a more general case, the incident beam direction does not necessarily coincide with the crystal surface normal. To evaluate the deviation of the Ewald sphere from the plane passing through the reciprocal lattice origin parallel to the crystal surface, consider Fig. 12b. A normal is drawn from the point A to the above plane containing the vector H. The point of intersection of the normal and the plane of interest is specified by the vector r_0^*. Figure 12b shows that

$$\delta = PP' = AA' = [k_0^2 - |(H - r_0^*)|^2]^{1/2} - [k^2 - r_0^{*2}]^{1/2} .$$

Under certain conditions, mosaicity of samples is important for the formation of point patterns, as crystal films of sufficiently large lateral dimensions ($\sim 0.01 - 1$ mm) generally consist of a large set of coherent scattering domains disorientated according to some law. The reciprocal lattice for such a sample may be represented by a set of reciprocal lattices with a common origin and the corresponding orientations. As a result, each node hkl will be "spread" on the surface of the sphere within a certain solid angle according to the distribution of orientations of coherent domains.

Comparatively small angular spread of orientations permits to widen considerably the region where the diffraction conditions are satisfied, which enables one to register quite a large number of reflections. For example, variations in orientation of $\pm 2°$ correspond to such a "spreading" of node centers as permits to observe all reflections down to $d \approx 0.5$ Å, because

$$x = \alpha/d = 2°/57.3° = \lambda/2\, d^2 \quad \text{and} \quad d \approx 0.5 \text{ Å}$$

[see (2.59)].

To estimate the integrated intensity for electrons scattered by a thin mosaic film, the law of the distribution of orientations for each set of reflecting planes (hkl) should be known. However, characteristics of this kind are difficult to determine, and therefore one has to assume that crystallities are equally distributed within a certain solid angle.

Vainshtein (1956) has shown that the kinematical integrated intensity $I_I(hkl)$ of a reflection in a point pattern obtained from a thin mosaic crystal is given by

$$I_I(hkl) = I_0\lambda^2\frac{\Phi^2(hkl)}{V_0}\,nV\frac{d(hkl)}{\alpha} = K[\Phi^2(hkl)]\,d(hkl)\,,\tag{2.60}$$

where V is the coherent domain volume (or that of a mosaic block), n is the number of mosaic blocks in the crystal volume contributing to the pattern, $d(hkl)$ is the d-spacing corresponding to the reflection analyzed, α is the effective angular spread for mosaic blocks disoriented about the axes lying in the basal plane of the crystal. It is assumed that α equal to several degrees ($1° - 3°$) is greater than the half-width of the interference function defined by the coherent domain thickness; K is a constant.

The smearing of the reciprocal lattice nodes in a mosaic microcrystal within the same angular interval leads to the following effect. The increase in the reflection order leads to the increase in the area over which the node $nh\ nk\ nl$ is spread. Consequently, decrease in the intensity for higher orders results from the decrease in the portion of crystals in reflecting positions. Due to this, the factor d_{hkl} appears in (2.60).

If the interaction between the electrons and the object is kinematical, and the object is close to an idealized single crystallite, so that α is very small, then (2.60) becomes

$$I_I(hkl) = I_0\lambda^2[\Phi(hkl)/V_0]^2Vt = K[\Phi(hkl)]^2\,,\tag{2.61}$$

where t is the crystal thickness.

Sometimes it is more convenient to evaluate the mean intensity per spot area in the pattern plane. It is obvious that the integrated intensity of a reflection should be then divided by the corresponding spot area which is inversely proportional to the area of the domain of coherent scattering in the direction normal to the incident beam.

Then the expression for the mean intensity becomes (Gorshkov and Drits 1984)

$$I_m(hkl) = I_0\lambda^2[\Phi(hkl)/V_0]^2nVS\frac{d(hkl)}{\alpha}$$

$$= I_0\lambda^2[\Phi(hkl)/V_0]^2nS^2t\frac{d(hkl)}{\alpha}$$

$$= I_0\lambda^2[\Phi(hkl)N]^2\frac{nd(hkl)}{\alpha t}\,,\tag{2.62}$$

where N is the number of unit cells in a coherent scattering domain, and S and t are the coherent domain area and thickness, respectively.

It is obvious that point patterns containing reflections hko can be obtained only if the orientations of the axes a and b are practically the same in all the microblocks in the crystal. However, in the case of layer crystals, thin particles are often found where microblocks have a certain angular spread of orientations (β) with respect to the axis normal to the particle surface. Each reflection hkl is then spread over an arc of the radius equal to the reciprocal lattice vector $H(hkl)$ projected onto the basal crystal plane passing through the reciprocal lattice origin.

The arc length is determined by β, and the intensity distribution over the arc is defined by the distribution function $f(\beta)$. The integrated intensity does not depend on the form of $f(\beta)$. However, it is sometimes convenient, especially for hko reflections, to measure the local intensity in the central part of the arc defined by the interval Δ. If $f(\beta)$ is a bell-shaped function, with the range of its nonzero values defined by the angular spread β, then the concept of the effective angular width $\beta' = \beta/2$ may be introduced. Within the limits of β', intensity is distributed with approximately equal density. Then it may be assumed that the integrated intensity is uniformly distributed over the effective arc length $H(hko)\beta'$. The the local intensity $I_l(hko)$ corresponding to the distance Δ (which is equal to, e.g., a detector slit) can be found from

$$\frac{I_I(hko)}{\beta' H(hk)} = \frac{I_l(hko)}{\Delta} .$$

Hence

$$I_l(hko) = I_I(hko) \Delta d(hko)/ \beta' .$$

Taking account of (2.60), we have (Gorshkov and Drits 1984)

$$I_l(hko) = I_0 \lambda^2 [\Phi(hk)/V_0]^2 n V \frac{d^2(hko)\Delta}{\alpha\beta'}$$

$$= K[\Phi(hko)/ V_0]^2 n V d^2(hko) . \tag{2.63}$$

Equations (2.60) – (2.63) are valid for all the cases where the periodic distribution of atoms in a crystal may be described by a single crystal lattice. However, this is not the case for some substances.

There have been reported natural hybrid crystal substances having layer structures (Allman and Lohse 1966; Allman et al. 1968; Evans and Allman 1968; Organova et al. 1973a). To describe these, two interpenetrating lattices are needed that differ in unit cell parameters due to differences in layer structure and/or composition. In this case, the diffracting single crystal volume V_{cr} contains different numbers of unit cells belonging to different lattices. This effect can be readily allowed for, since the formula (2.60) contains the unit cell volume. The reflection intensity $I_i(hkl)$ for the ith sublattice will be then related to the corresponding structure factor by the expression

$$I_i(hkl) = K' [\Phi_i(hkl)/V_{0i}]^2 d_i(hkl) . \tag{2.64}$$

The ratio of the integrated intensities for reflections belonging to different sub-lattices is given by

$$\frac{I_i(hkl)}{I_j(hkl)} = \left[\frac{\Phi_i(hkl)V_{0j}}{\Phi_j(hkl)V_{0i}}\right]^2 \frac{P_i}{P_j} \tag{2.65}$$

where P_i/P_j is unity or $d_i(hkl)/d_j(hkl)$ for the absence and the presence of mosaicity resulting from rotations of blocks about axes lying in the basal crystal plane, respectively. Note also that the value of the ratio in (2.65) is independent of whether it is the integrated, or the mean, or the maximum reflection intensities that are measured. If it is the local intensities for the hko reflections that are measured, then $P_i/P_j = d_i^2(hko)/d_j^2(hko)$.

In the recent decade a number of new natural substances has been revealed (Chuhrov et al. 1980a, b, 1982, 1983a, b, 1984; Drits et al. 1985). The structures of these are described by regular alternation of layers differing in composition and/or structure.

The structures involved have an important peculiarity that makes them different from any other known substances, namely, that layers continuous in two dimensions within the coherent scattering domain may coexist with island-like, i.e., defect layers of the other type. Then, if the incident electron beam illuminates the crystal of the area S and the thickness t, the contributions to scattering for the two components will differ due to a difference in the scattering volumes corresponding to the same area S_{cr}. Since in the structures in question layers differing in cationic composition alternate regularly along the axis c, it is sufficient to take into account the differences in the areas of the unit cells and mosaic blocks (Gorshkov and Drits 1984). It is obvious that n in (2.60) will be then equal to the number of coherent domains for each of the components corresponding to the crystal area S_{cr}. Assume that N_i is the number of unit cells corresponding to the coherent domain area S_i for the ith component, i.e., $N_i = S_i/\sigma_i$, where σ_i is the unit cell area in the basal plane for the ith sublattice. Then the expression for the integrated intensity for the reflection hkl belonging to the ith sublattice can be written

$$I_i(hkl) = K[\Phi_i(hkl)/\sigma_i]^2 n_i S_i d_i(hkl) = K[\Phi_i(hkl)]^2 \frac{n_i N_i}{\sigma_i} d(hkl).$$

The ratio of the integrated intensities for reflections belonging to different sub-lattices is given by

$$\frac{I_i(hkl)}{I_j(hkl)} = \left[\frac{\Phi_i(hkl)\sigma_j}{\Phi_j(hkl)\sigma_i}\right]^2 \frac{n_i S_i d_i(hkl)}{n_j S_j d_j(hkl)}$$

$$= \left[\frac{\Phi_i(hkl)}{\Phi_j(hkl)}\right]^2 \frac{n_i N_i \sigma_j d_i(hkl)}{n_j N_j \sigma_i d_j(hkl)}.$$

The ratio of maximum intensities for reflections corresponding to different sub-lattices, as well as that of their mean intensities is given by

$$\frac{I_i^{max}(hkl)}{I_j^{max}(hkl)} = \frac{I_i^m(hkl)}{I_j^m(hkl)} = \left[\frac{\Phi_i(hkl)\,\sigma_j\,S_i}{\Phi_j(hkl)\,\sigma_i\,S_j}\right]^2 \frac{n_i\,d_i(hkl)}{n_j\,d_j(hkl)}$$

$$= \left[\frac{\Phi_i(hkl)N_i}{\Phi_j(hkl)N_j}\right]^2 \frac{n_i\,d_i(hkl)}{n_j\,d_j(hkl)} . \tag{2.66}$$

Thus, for some layer crystals with defect hybrid structures, the difference in intensity for reflections having the same indices hkl results not only from different unit cell sizes and scattering powers for each component, but also from a defect structure of one of the components.

Therefore, to analyze intensities of reflections for different sublattices on the same scale, one should determine both the ratio of coherent domain areas corresponding to different structure components S_i/S_j and the value n_i/n_j, where n_i and n_j are the numbers of coherent domains for the two components in the diffracting crystal volume. These parameters can be determined only in the case where the chemical composition of the mineral as well as the distribution of cations over different layer types are known. and the half-width of hko reflections depends only on the coherent scattering domain area and instrumental factors (Gorshkov and Drits 1984).

CHAPTER 3

Geometrical Analysis of Point Electron-Diffraction Patterns

3.1 Raypath in a Transmission Electron Microscope for Imaging and Selected Area Diffraction

The possibility of obtaining not only a magnified image of the object but also its diffraction pattern is the main advantage of an electron microscope. It is also essential that both the electron microscopic image and the diffraction pattern correspond to the same part of the specimen studied. With $V = 100\,\text{kV}$, the dimensions of the object analyzed may be as small as several thousands of Å, and even several hundreds of Å under special conditions.

Raypaths in the simplest three-lens system in imaging and diffraction modes are shown in Fig. 13. Although electromagnetic lenses, due to the vectorial nature of fields, differ fundamentally from the glass optical lenses, a raypath in an electron microscope may nevertheless be treated in terms of geometrical

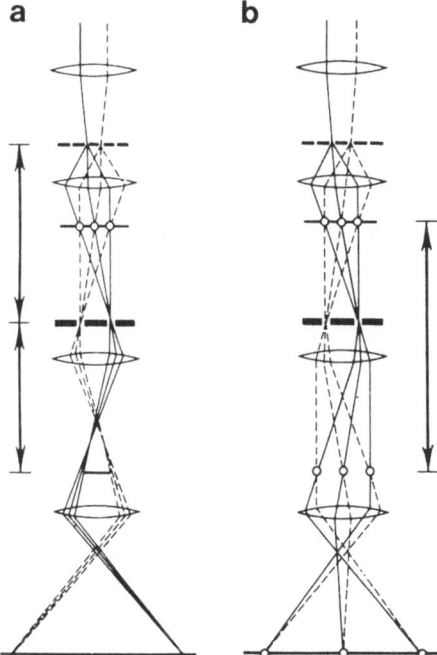

a   **b**

Fig. 13a, b. Ray-path in a transmission electron microscope; **a** imaging; **b** diffraction

optics. Electrons ejected by a point emitter are accelerated by high voltage of the order of 100 kV, which reaches in certain devices from one to several million volts. A system of condenser lenses provides a high illuminance of the sample by a narrow and practically parallel electron beam having a divergence $\leqslant 10^{-6}$ rad. It is only for high magnifications that the divergence is to be increased.

The rays scattered at equal angles with respect to the optical axis of the system converge in the same point in the back focal plane of the objective lens, where the diffraction pattern from the crystal studied is formed. Afterward, electrons reach the plane which is conjugate to the object plane, and the rays scattered by a point in the object converge at a point in the plane of the objective. Thus the image of the object is formed.

The essential advantage of electromagnetic lenses over optical ones is that the focal distance can be easily changed by varying the strength of current. To obtain a diffraction pattern on the screen, the focal distance of the intermediate lens is chosen such as to make the back focal plane of the objective conjugate to the focal plane of the projector lens. Then the projector lens transfers the magnified image of the diffraction pattern of the object to the screen. To obtain on the screen a magnified image of the object, the focal distance of the intermediate lens is decreased, as compared to the previous case, so as to make the focal plane of the projector lens conjugate to the objective plane. In the image plane of the objective there is a selector diaphragm which determines the size of the specimen contributing to the formation of a SAED pattern.

Le Poole (1947) was the first to elaborate this technique. In the back focal plane of the objective lens there is an aperture. The size of the objective aperture limits the number of diffracted beams contributing to the formation of the image. If the aperture removes all the diffracted beams, except the incident one, then the bright-field diffraction contrast is achieved. If the aperture is used to select one of the strong diffracted beams, then dark-field diffraction contrast is obtained. Phase contrast is achieved when the incident beam and several diffracted ones interact in the image plane. The study of the mechanism for the formation of the phase contrast is the central problem in crystal structure investigations by high-resolution electron microscopy.

Most of the modern electron microscopes are provided with goniometric stages allowing to vary the orientation of the object in a sufficiently wide angular range (up to $\pm 60°$). There are two main types of goniometric stage. One of these, *rotation-tilt,* permits to rotate the specimen about an axis normal to the specimen plane and to tilt it about a fixed axis normal to the incident beam. During azimuthal rotation, the object area under study may easily withdraw from the observation field, which is a serious drawback in goniometers of this type. In goniometers of the second type, *double tilt,* the specimen is tilted about two orthogonal axes, one lying in the specimen plane, while the other is normal to the incident beam. A detailed description of the use of both types of stage is given by Gard (1976).

Diffraction patterns are usually calibrated using standard specimens, e.g., thin polycrystal gold or aluminum foils, for which unit cell parameters and d-spacings are known with high accuracy. Since the electron wavelength is small, scattered beams make minor angles with the direction of the incident beam, therefore $\sin \vartheta = \lambda / 2d \approx \vartheta$.

In terms of the small-angle approximation, the equation $rd = L\lambda$ applies exactly enough, where r is the distance between the central spot on the screen or on a photographic plate and the reflection corresponding to the spacing d, and L is the effective distance between the object and the screen. With constant accelerating potential, $L\lambda$ is a constant and may be easily determined by using ring patterns from the corresponding standards.

3.2 Methods for Interpretation of Point Diffraction Patterns: Indexing and Determination of Unit Cells

Vainshtein (1956) was the first to give a thorough and comprehensive treatment of problems concerning the symmetry and the indexing of point electron diffraction patterns and the determination of the unit cell. Methods for the interpretation of point diffraction patterns were later discussed by Hirsch et al. (1965), Gard (1976), Utevsky (1973), and others. However, the most comprehensive and consistent geometrical analysis of electron diffraction patterns for the general case of triclinic lattices has been developed by Zvyagin and his collaborators (Vrublevskaya et al. 1974; Zvyagin 1967, 1968; Zvyagin and Fedotov 1974, 1975; Zvyagin and Gorshkov 1969; Zvyagin and Pinsker 1949; Zvyagin and Vrublevskaya 1972, 1974; Zvyagin et al. 1979). Therefore, only a concise general discussion of the methods for the interpretation of point electron diffraction patterns will be given here. In SAED structure studies, it is first of all necessary to determine the dimensions and the shape of the unit cell, to reveal systematic absences, and to obtain a set of diffraction patterns containing a sufficiently representative array of hkl reflections. Point diffraction patterns obtained with the incident beam parallel to the most symmetrical directions in the crystal, including the basic vectors, are most important.

Since the radius of the Ewald sphere is about a factor of 10^2 greater than the distance between the reciprocal lattice points, a point diffraction pattern may be treated as a plane passing through the reciprocal lattice origin (Fig. 4). If the incident beam is directed along the crystal lattice row $[mnp]$, which may be regarded as a zone axis, then the indices of the reciprocal lattice plane coinciding with the diffraction pattern are given by the same numbers, i.e., $(mnp)^*$. The indices of the reciprocal lattice points lying in this plane satisfy the Eq. (1.11): $hm + kn + lp = 0$. Therefore a point pattern is usually denoted either by the indices of reflections contained in it (e.g., hko) or by the indices of the reciprocal lattice plane coinciding with the Ewald sphere, e.g., $(001)^*$.

In all cases, an electron diffraction pattern contains a two-dimensional net of reflections. To characterize it, two independent variables are sufficient. On the other hand, in the general case a pattern may correspond to an arbitrary plane $(mnp)^*$ and contain reflections with three nonzero indices hkl. The value of each of the indices hkl is determined by the other two, e.g, $l = -(m/p)h - (n/p)k$. Moreover, there is an infinite number of ways to express a node index in terms of the two independent variables (Vainshtein 1956). For example, it may be assumed for reflections in the $(1\bar{2}1)^*$ plane $l = -h'$, $k = k'$. Then $h = h' + 2k'$, where h', k' are any whole numbers.

Indexing of diffraction patterns representing the reciprocal lattice coordinate planes is the simplest case. According to the general rules for choosing the basic vectors in unit cells, two rows are chosen that pass through the center of the pattern and make angles closest to 90° with each other. These lines are assumed to be coordinate directions. In a more general case, where the Ewald sphere does not coincide with a coordinate plane but the indices for two reflections are known, the indexing is also a simple procedure.

Vainshtein (1956) has shown that a three-dimensional reciprocal lattice may be determined by using two diffraction patterns, provided the dihedral angle between the corresponding reciprocal lattice planes is known. These planes intersect along a common direction, which permits to reconstruct the reciprocal lattice. If the angle between the two diffraction patterns analyzed is unknown, then at least three paterns are required to determine the unit cell parameters. Practically, it is important to obtain a series of diffraction patterns having a common lattice row. This is achieved rather easily by using a rotation-tilt goniometric stage. A certain reciprocal lattice axis is assumed to be the coordinate one and it is matched with the direction of the tilt axis of the specimen holder which is normal to the incident beam. Tilting the crystal at different angles with respect to this axis provides a set of diffraction patterns, each of these containing a row of reflections. If all these patterns are brought into coincidence along the common lattice row and are rotated at the corresponding angles with respect to one another, they will represent the space distribution of reflections. Analysis of this distribution enables one to reconstruct the reciprocal lattice of the object under study. Certain difficulties may arise if the reciprocal lattice basic vector parallel to the rotation axis makes oblique angles with the other two basic vectors. This is the case for triclinic crystals or for monoclinic crystals with the $c*$ (or $a*$) rotation axis.

In some cases tilting about a coordinate axis in the direct lattice is more expedient, because this axis is always perpendicular to a coordinate plane in the reciprocal lattice. For example, rotation about the axis a enables one to obtain a set of diffraction patterns representing various reciprocal lattice sections of the type $(onp)*$. If the basic vector direction about which the object is being tilted is each time recorded in diffraction pattern (e.g., by simultaneous imaging of the crystal analyzed), then the reciprocal lattice may be again reconstructed. For example, the distribution of okl lattice points may be obtained as follows. A line normal to the rotation direction of the crystal is drawn passing through the center of each pattern, and the positions of reflections located on this line are marked in. These lines drawn from the same origin at angles corresponding to rotation angles of the crystal will actually represent the distribution of the nodes okl if the crystal is rotated about a.

Examples of the reconstruction of the reciprocal lattice and the determination of unit cell parameters in minerals are discussed in the last four chapters of the present book.

It should be stressed that SAED determinations of dimensions and shape of unit cells in disperse minerals ought to be combined with X-ray diffraction data. The efficiency of this approach is illustrated by numerous examples in the second part of this book.

3.3 Simulation of Diffraction Patterns for Objects with Known Unit Cell and Space Symmetry

The simplest case of interpretation and simulation of diffraction patterns is that of the known direction $[mnp]$ of the incident beam, as the indices $(mnp)^*$ of the reciprocal lattice plane coinciding with the Ewald sphere are specified. Using (1.11) one may determine the indices of reflections (hkl) located in the given plane $(mnp)^*$. The corresponding $d(hkl)$ spacings and the angles between $H_{h_i k_i l_i}$ and $H_{h_j k_j l_j}$ are calculated for the set of lattice points in question. These data are sufficient to reproduce the two-dimensional net of reflections in the reciprocal lattice plane involved.

However, in some cases it proves useful to imagine a priori diffraction patterns that should be expected from crystals oriented in such a way that the indices of the lattice plane (hkl) normal to the incident beam are known. In Chapter 1 the indices of a plane (hkl) and those of the row $[mnp]$ normal to it have been shown to be related by an expression of the type (1.7) or (1.13). Thus, provided the indices of the plane (hkl) are known, the indices of the reciprocal lattice plane $(mnp)^*$ which is parallel to it may be easily found (Zvyagin 1968). The further procedure is exactly similar to the one described above.

Sometimes it is useful to calculate a priori the required tilt angles with respect to the coordinate direction chosen, so as to obtain diffraction patterns containing an informative set of reflections. For example, if the crystal is supposed to be tilted about b^*, the lattice plane (010) normal to b^* is to be constructed. A reference direction, which coincides for the crystal in the initial position with the incident beam should be marked in this plane. The angles between this direction and lattice rows $[mop]$ should be found. Each direction $[mop]$ corresponds to a reciprocal lattice plane $(mop)^*$, and the distribution of lattice points lying in this plane is determined according to the procedure described above. Analysis of the whole set of simulated patterns enables one to choose the crystal orientations corresponding to sections that seem preferable.

If the crystal is intended to tilt about a coordinate axis in the crystal lattice, the coordinate plane in the reciprocal lattice normal to this axis should be constructed and the reference direction should be marked in it. The coordinate axis chosen is supposed to make oblique angles with the other two basic vectors in the crystal lattice. Otherwise the problem would be exactly similar to the one described above. Assume that the tilt axis coincides with the b axis and, therefore, the reciprocal lattice coordinate plane which is normal to it contains reflections hol. The problem may be solved in two steps. First, the angles are determined between the reference direction and the rows $[hol]^*$. Then Eqs. (1.11) and (1.13) are applied to find for each (hol) plane the distribution of lattice points in the reciprocal lattice plane $(mnp)^*$ which is parallel to it. Thus diffraction patterns may be simulated for each given crystal orientation.

3.4 Interpretation and Simulation of Diffraction Patterns for Triclinic Lattices with a Fixed Coordinate Plane

Zvyagin et al. (1979) have elaborated a method for the analysis of triclinic lattices with a fixed coordinate plane and used it for interpretation and simulation of diffraction patterns. This case is important, as very thin plate crystals are often used as objects in SAED studies. At the preliminary stage it is convenient to choose the basal plane of the crystal as the coordinate plane. It is especially expedient for minerals having layer structures.

Any crystal lattice may be represented by a set of parallel two-dimensional nets. If lattice planes parallel to the crystal basal face are selected and the distribution of lattice points in each of these is determined by vectors a and b, the only problem will consist in the reasonable choice of the axis c. In triclinic lattices, there are numerous alternative choices of c. Zvyagin et al. (1979) proposed to compare alternative choices of unit cell in the same lattice in terms of the values x_n, y_n of the components of the projection of c on the ab plane, choosing the unit cell with the minimum normal projection of c.

Regardless of the c direction, the set of planes parallel to ab in the direct lattice specifies unambiguously the set of parallel hk rods normal to ab in the reciprocal lattice, having the repeat distance of c^*. Such a representation of the reciprocal lattice simplifies essentially the analysis of diffraction patterns.

Let us first consider the peculiarities in diffraction patterns to be observed with the incident beam normal to the ab plane chosen. Without any loss of generality, this problem may be considered for layer crystals differing in the degree of three-dimensional ordering. Assume that in the first case, the layers are stacked at random, i.e., the crystal cannot be described by a three-dimensional lattice, although each layer has a two-dimensional periodicity with the basic vectors a and b. The reciprocal space for such a crystal consists of parallel rods perpendicular to the plane ab, the intensity distribution along these being continuous. Therefore each hk rod has a continuous variable l. Their arrangement depends on the unit cell parameters in a layer, a, b, γ, because the two-dimensional distribution of points in the section of the rods by the plane ab is specified by a parallelogram having the sides $1/a \sin \gamma$ and $1/b \sin \gamma$, and the angle $\gamma^* = \pi - \gamma$ between them. The geometry of this section coincides with that of a diffraction pattern from a disordered crystal or from a single layer with the incident beam being parallel to c^*.

Let us now consider a crystal consisting of the same layers regularly stacked. Assume that the c axis in the three-dimensional lattice of this crystal is inclined to the ab plane. The reciprocal space of the crystal analyzed will again consist of a set of lines parallel to c^*, their arrangement being exactly the same as in the previous case. The points of intersection of these lines (for the time being we ignore the specific distribution of lattice points along them) by the ab plane are distributed according to the unit cell parameters $1/a \sin \gamma$, $1/b \sin \gamma$ and γ^*.

To prove this, consider Fig. 14 which shows the scheme of the arrangement of axes in triclinic direct and reciprocal lattices. It may be seen that the projections of a^* and b^* on the ab plane make angles $\gamma - \pi/2$ with a and b, and $\gamma^* = \pi - \gamma$ with each other. According to Fig. 14 we have:

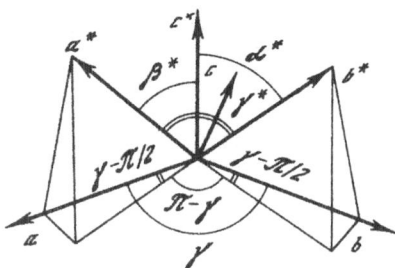

Fig. 14. Relationship between coordinate parameters in the direct lattice and those in the reciprocal one (Zvyagin and Vrublevskaya 1972)

$$\cos(aa^*) = \cos(\pi/2 - \beta^*)\cos(\gamma - \pi/2) = \sin\beta^* \sin\gamma, \qquad (3.1)$$

$$\cos(bb^*) = \sin\alpha^* \sin\gamma$$

$$a^* \sin\beta^* = \sin\beta^*/a\cos(aa^*) = 1/a\sin\gamma, \quad b^* \sin\alpha^* = 1/b\sin\gamma.$$

The substantial difference between the two cases consists in the intensity distribution along each rod. In the ordered structure, each rod is represented by a discrete set of equidistant points having a constant hk, but differing in l. Since the c axis is inclined to the ab plane, the vectors a^* and b^* do not belong to the ab plane. This results in substantial differences in geometry for patterns obtained from ordered and random crystals when the incident beam coincides with c^*. In the case of the reciprocal lattice, the plane of the diffraction pattern may cut the lattice rows into some intermediate points instead of the reciprocal lattice points. This would lead to the absence in the point pattern of some the hk reflections depending on c, α, and β. To avoid misunderstanding, it should be stressed that the pattern geometry will vary for different crystal lattices with different c, α, and β, but it will be preserved if alternative c, α, and β are chosen in the same lattice. In all these cases we are dealing with lattices having the same a and b.

Since the geometry of an electron diffraction pattern depends on specific positions of hkl nodes with respect to the ab plane, it is essential to evaluate the distance between a node and this plane. It is determined by the distance lc^* between the points hkl and hko along the lattice row with the given hk and by the sum of the projections of the vectors ha^* and kb^* on the c^* axis. Expressed in terms of fractions of c, this distance is given by $ha^* \cos\beta^*/c^* + kb^* \cos\alpha^*/c^* + l$. Zvyagin and Vrublevskaya (1974) have shown that the projections of a^* and b^* on c^* are equal in terms of fractions of c^* to the components x_n and y_n of the normal projection of c on the ab plane, taken with an opposite sign, i.e., $a^* \cos\beta^*/c^* = -x_n$; $b^* \cos\alpha^*/c^* = -y_n$.

As an example, consider layer silicates. For a disordered structure, the hk reflections distribution is described by hexagonal symmetry with $h + k = 2n$, in accord with the C-centered orthogonal unit cell in silicate layers.

Many layer silicates with regular structures have monoclinic cells with $x_n = -1/3$, $y_n = 0$ or $x_n = 0$, $y_n = 1/3$. Then $a^* \cos\beta^*/c^* = 1/3$ or $b^* \cos\alpha^*/c = 1/3$, and the reciprocal lattice planes parallel to ab, according to (1.13), have the indices (103)* or (013)*. Therefore, reflections with $h = 3n \pm 1$ should be missing in patterns from crystals having $\beta \neq \pi/2$, and those with $k = 3n + 1$ in patterns from crystals with $\alpha \neq \pi/2$. Thus the geometry of the diffraction pattern

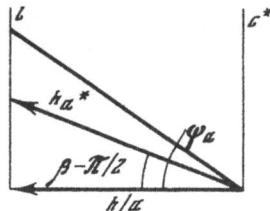

Fig. 15. The relationships between the indices l and h (Zvyagin and Groshkov 1969)

should depend essentially on layer stacking. However, under real experimental conditions the forbidden reflections may be observed in diffraction patterns due to the extension of the reciprocal lattice points along c^* resulting from the small thickness of the coherent scattering domains.

Interpretation of a diffraction patterns from a single crystal rotated by a certain angle with respect to the ab plane requires parameters describing the distribution of lattice points in an arbitrary section of the reciprocal lattice. Any reciprocal lattice plane may be specified by the direction along which it intersects with the plane ab and by the dihedral angle φ between them. This direction may be regarded as the axis of the tilt of the diffraction pattern by the angle φ with respect to the plane ab. For an oblique section, the distance r_{hk} from the origin to the point of the intersection of the hk rod by the pattern plane is by $1/\cos \psi_{hk}$ greater than the r_{hk}^0 distance to the intersection point of the same rod by the plane ab, where ψ_{hk} is the angle between r_{hk} and r_{hk}^0. Therefore, the values of r_{hk} in a diffraction pattern with $\varphi \neq 0$ vary, as compared to r_{hk}^0, with the azimuth with respect to the tilt axis, so that the maximum $r_{hk} - r_{hk}^0$ corresponds to the direction normal to the tilt axis. Comparison of distances between reflections corresponding to different azimuthal directions enables one to establish the direction with the minimum deviation from r_{hk}^0, which coincides with the pattern tilt axis or is close to it (Zvyagin and Gorshkov 1969). The direction of this axis is conveniently designated by the indices $[hk]$.

The relationships between angular parameters defining the space orientation of a fixed tilt axis with respect to various $[hk]$ directions in the plane of the pattern and in the ab plane may be established as follows. Figure 15 shows two planes intersecting at the angle φ, one of these being represented by the ab plane and the other by the plane of the pattern. Directions r_{ho}^0 and r_{ho} coincide with the projection of a^* on the ab plane and on the diffraction pattern plane, respectively, and ψ_h is the angle between these directions. If δ_h and δ_h' are the corresponding angles between the tilt axis and the directions r_{ho}^0 and r_{ho}, then $\operatorname{tg} \psi_h = \operatorname{tg} \varphi \sin \delta_h'$ (Zvyagin and Gorshkov 1969). Similarly, $\operatorname{tg} \psi_k = \operatorname{tg} \varphi \sin \delta_k'$ for angles δ_k' and δ_k between the tilt axis and the directions r_{ok}^0 and r_{ok}.

The angles δ_h' and δ_k' will have opposite signs if the tilt axis lies within the octant formed by the directions $[ho]$ and $[ok]$ in the ab plane, whereas they will be of the same sign if it is within the neighboring octant (Zvyagin et al. 1979). Obviously $\delta_k' - \delta_h' = \gamma^*$.
Hence,

$$\operatorname{tg} \psi_h = \operatorname{tg} \varphi \sin \delta_h' = -\operatorname{tg} \varphi (\sin \gamma \cos \delta_k' + \cos \gamma \sin \delta_k'),$$

$$\operatorname{tg} \psi_k = \operatorname{tg} \varphi \sin \delta_k' = \operatorname{tg} \varphi (\sin \gamma \cos \delta_h' - \cos \gamma \sin \delta_h') .$$

(3.2)

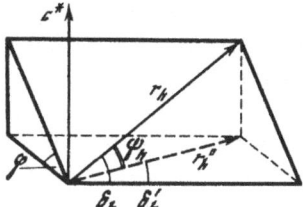

Fig. 16. The relationships between the angles φ, ψ, δ, δ'

Consequently

$$\operatorname{tg}\delta_h' = t/(1+t\cos\gamma), \qquad \operatorname{tg}\delta_k' = -\sin\gamma/(t+\cos\gamma), \tag{3.3}$$

$$\operatorname{tg}^2\varphi = (\operatorname{tg}^2\psi_h + \operatorname{tg}^2\psi_k + 2\operatorname{tg}\psi_h\operatorname{tg}\psi_k\cos\gamma)/\sin^2\gamma,$$

where $t = \operatorname{tg}\psi_h/\operatorname{tg}\psi_k$.

Thus, provided ψ_h and ψ_k are known, the values of δ_h' and δ_k' characterizing the position of the tilt axis as well as the angle φ (the latter — up to a sign) may be easily found according to (3.3).

The intensity of reflections in the diffraction pattern, with the node extension along lattice rows taken into account, depends on the proximity of these to the surface of the Ewald sphere. The closer the hkl node to the intersection point of the hk rod by the diffraction pattern, the higher the intensity of the corresponding reflection, other conditions being equal. To determine the coordinate of a hkl lattice point in the plane of the diffraction pattern, one should know the unit cell parameters and the angles ψ_h and ψ_k. For example, in the particular case of $\gamma = \pi/2$ the axes a^* and b^* are projected on the diffraction pattern plane along the same directions as the axes a and b. Figure 16 shows that the l coordinate for the ho points lying along the $[ho]$ direction in the plane of the diffraction pattern is then given by

$$l = [(h/a)\operatorname{tg}\psi_h - ha^*\cos\beta]/c^* = \rho_h h. \tag{3.4}$$

Similarly, the l coordinate for the points along $[ok]$ is given by

$$l = [(k/b)\operatorname{tg}\psi_k - kb^*\cos\alpha]/c^* = \rho_k k. \tag{3.5}$$

Consequently, an hk point in the plane of the diffraction pattern has the l coordinate

$$l = \rho_h h + \rho_k k, \tag{3.6}$$

where ρ_h and ρ_k are determined by (3.4) and (3.5) (Zvyagin and Gorshkov 1969). In a more general case of lattices having $\gamma \neq \pi/2$ the expressions for ρ_h and ρ_k are more complicated (Zvyagin and Fedotov 1974). With the reciprocal lattice points exactly coinciding with the plane of the diffraction pattern, the l coordinates are integers, and the expression (3.6) may be regarded as the equation for straight lines in the diffraction pattern containing reflections with equal values of l.

In the light of the results obtained above, the sequence of operations for simulating diffraction patterns containing reflections with the given indices hkl is as follows (the unit cell is supposed to be known):

1. The arrangement of the intersection points of hk rods by the ab plane is determined; the directions $[ho]$ and $[ok]$ coinciding with the projections of $a*$ and $b*$ on ab, are established; reciprocal lattice nodes are found that have l integer (the lattice nodes correspond to the strongest reflections).
2. Equation (1.11) is applied to find the indices $(mnp)*$ of the plane coinciding with the diffraction pattern simulated. According to the equation $l = -(m/p)h-(n/p)k$, the values $\rho_h = -m/p$ and $\rho_k = -n/p$ are found. Using ρ_h and ρ_k, (3.4) and (3.5) are applied to find ψ_h and ψ_k.
3. Equation (3.3) is used to determine the maximum tilt angle φ and angles δ'_h and δ'_k characterizing the position of the tilt axis with respect to the directions $[ho]$ and $[ok]$ in the plane ab.

A number of additional operations is needed to determine the geometry of the diffraction pattern. Using δ'_h and δ'_k, the tilt axis is drawn in the scheme containing the intersection points hk of the lattice rows by the ab plane. Then the tilt angles δ'_{hk} between the tilt axis and the radii r^0_{hk} drawn from the origin into all hk points are measured directly or calculated. The values

$$r^0_{hk} = [(ha* \cos \beta*)^2 + (kb* \cos \alpha*)^2 - 2hka*b* \cos \alpha* \cos \beta* \cos \gamma]^{1/2}$$

are also calculated.

The formula $\mathrm{tg}\,\psi_{hk} = \mathrm{tg}\,\varphi \sin \delta'_{hk}$ is applied to determine the angles between the $[hk]$ directions in the plane of the diffraction pattern and in the ab plane. These data are sufficient to calculate the radii $r_{hk} = r^0_{hk}/\cos \psi_{hk}$. Thus the distribution of hk points is reconstructed in the diffraction pattern plane intersecting at oblique angles the lattice rods parallel to $c*$. Finally, the straight lines should be drawn, satisfying the equation $l = -(m/p)h-(n/p)k$, which is used to reveal reciprocal lattice points in the pattern corresponding to $\varphi \neq 0$. To do this, in the equation involved it is consequently assumed that $l = 0, \pm 1, \pm 2$ etc., and the corresponding lines are drawn. The given l is ascribed to all hk points lying along the line $l = \mathrm{const}$.

If a diffraction pattern already obtained is to be interpreted, then the sequence of operations is different.

1. As the crystals under study are supposed to be platy, it is worthwhile to obtain an experimental diffraction pattern parallel to the ab plane. This permits to imagine the peculiarities in the distribution of the strongest reflections and to establish precisely the directions $[ho]$ and $[ok]$.
2. Having compared the diffraction patterns with $\varphi = 0$ and $\varphi \neq 0$, the directions $[ho]$ and $[ok]$ are found in the latter pattern. The values of r^0_h and r^0_k are used to calculate ψ_h and ψ_k according to formulae of the type $\cos \psi_h = r^0_{ho}/r_{ho}$. These values may be determined only up to a sign. If the tilt axis passes through the octant formed by the positive directions of oh and ko, then ψ_h and ψ_k are of the opposite sign, whereas they are of the same sign if it passes through the adjacent octants.

While the determination of the position of the tilt axis in the pattern analyzed may be relatively simple, the establishment of the positive and negative directions of $[ho]$ and $[ok]$ meets with a number of difficulties (Zvyagin et al. 1979).

Therefore, further calculations are carried out for two versions for the signs for ψ_h and ψ_k. For each of these versions, ρ_h and ρ_k are calculated from (3.4) and (3.5) and then (3.3) is applied to find δ'_h, δ'_k and ψ. The version where integer values of l are obtained for all strong reflections is chosen and only these reflections are involved in a system of lines l = const. The values of δ'_h and δ'_k calculated should be in agreement with the true direction of the tilt axis in the diffraction pattern. Since $\rho_h = -m/p$ and $\rho_k = -n/p$ are known, the $(mnp)^*$ indices of the plane of the diffraction pattern are easy to find. Note that at the first steps the values l for strong reflections may not be exactly integer. Their rounding to the nearest integer permits to refine the values of ρ_h, ρ_k, ψ_h, ψ_k. In some cases the choice of $[ho]$ and $[ok]$ directions may turn out to be ambiguous. Then all the calculations are to be carried out for each of the versions possible until the main features of the pattern are explained.

3.5 Determination of the Bravais Cell and the Space Group. Secondary Diffraction Effects

In Chapters 1 and 2 it has been noted that systematic absences of reflections with definite indices may indicate the Bravais lattice type and the space symmetry of the object.

Vainshtein (1956) has emphasized two difficulties that may arise in the determination of the crystal symmetry and that should be taken into account in interpreting the point patterns. If a very strong diffracted beam arises during the propagation of electrons through the crystal, then it may act as an incident one leading to secondary scattered waves. Consequently, the secondary scattering effect will lead to a superposition of two diffraction patterns. The center of the first one coincides with the incident beam, while that of the second one with strong diffraction maximum involved, hkl. With the two centers brought into coincidence, the resultant pattern would reflect the secondary scattering effects.

Figure 17 shows that secondary scattering may lead to weak reflections that should be absent for the given symmetry. Note that this is possible only for reflections having one or two zero indices. Secondary scattering does not affect extinctions defined by the Bravais lattice type, since a diffraction pattern coincides with itself when shifted along any H_{hkl} vector. Weak "forbidden"

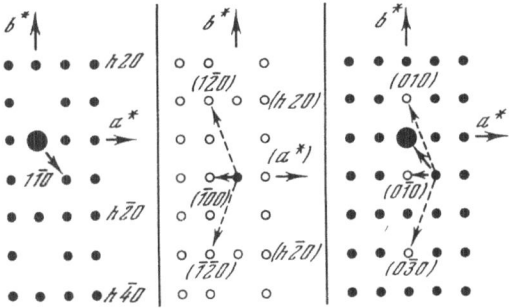

Fig. 17. Forbidden reflections resulting from secondary diffraction (Gard 1976)

reflections should disappear when the crystal is rotated about the axis containing extinctions at a certain angle which makes the "generating" strong reflections weak.

The second difficulty is associated with two-dimensional diffraction effects that occur with very thin or disordered crystals that lead to elongation of the reciprocal lattice points. In the extreme case, rods appear having continuous intensity distribution. For example, with an A-centered unit cell all forbidden hko reflections with k odd may appear.

Secondary diffraction is distinctly observed in the case of the intergrowth of two crystals differing either in orientation or in unit cell parameters, or in some other features. Generally, the combination of two crystals leads to the sum of diffraction patterns from each crystal. However, strong diffracted beams from one of the crystals may act as incident ones propagating through the second crystal. This leads to secondary diffraction effects complicating the diffraction pattern. These effects were thoroughly analyzed by Zvyagin and Gorshkov (1969).

CHAPTER 4

Diffraction Methods in Structure Analysis

4.1 Fourier Series and Integrals: Their Role in the Theory of Diffraction

Any continuous differentiable function $\Psi(x)$ may be expanded by its Fourier series:

$$\Psi(x) = A_0 + 2 \sum_{n=1}^{\infty} [A_n \cos 2\pi n(x/a) + B_n \sin 2\pi n(x/a)]$$

$$= \sum_{-\infty}^{\infty} [A_n \cos 2\pi n(x/a) + B_n \sin 2\pi n(x/a)] \tag{4.1}$$

where a is the period, A_0, A_n and B_n are constants, and $A_n = A_{-n}$ and $B_n = -B_{-n}$. This means that for each given x, the value of $\Psi(x)$ may be obtained by adding the sine and the cosine harmonics, if the appropriate coefficients A_n and B_n are selected for each harmonic.

Thus the problem of expanding a periodic function by a Fourier series amounts to the determination of coefficients for each term of the series. There are some other forms in which a Fourier series may be written. According to the Euler formula

$$A_n \cos 2\pi nx/a + B_n \sin 2\pi nx/a = (1/2)(A_n + B_n) \exp 2\pi i(nx/a)$$
$$+ (1/2)(A_{-n} + B_{-n}) \exp -2\pi i(nx/a) .$$

Assume $C_0 = A_0$, $C_n = A_n + B_n$, $C_{-n} = A_{-n} + B_{-n}$ then

$$\Psi(x) = C_0 + \sum_{n=1}^{\infty} \left(C_n \exp 2\pi i \frac{nx}{a} + C_{-n} \exp \left[-2\pi i \frac{nx}{a} \right] \right)$$

$$= \sum_{-\infty}^{\infty} C_n \exp 2\pi i \frac{nx}{a} . \tag{4.2}$$

A_n and B_n in (4.1) as well as C_n in (4.2) may, in general, be complex.

With n and m integer

$$\int_0^q \exp \left[2\pi i \frac{nx}{a} \right] dx = 0 \quad \text{for all } n,$$

$$\int_0^q \exp \left[2\pi i \frac{(n-m)x}{a} \right] dx = \begin{matrix} a & \text{for} & n = m \\ 0 & \text{for} & n \neq m . \end{matrix} \tag{4.3}$$

To find C_0, both sides in (4.2) are integrated between 0 and a

$$\int_0^a \Psi(x)\,dx = \int_0^a C_0\,dx + \sum_{n-1}^{\infty}\left[C_n\int_0^a \exp\left[2\pi i\frac{nx}{a}\right]dx\right.$$

$$+ \left.C_{-n}\int_0^a \exp\left[2\pi i\frac{-nx}{a}\right]dx\right].$$

Since the integral of each term with $n \neq 0$ are zero

$$C_0 = (1/a)\int_0^a \Psi(x)\,dx. \tag{4.4}$$

Multiplying both sides in (4.2) by $\exp[2\pi i(mx/a)]$ and integrating between 0 and a, we have

$$\int_0^a \Psi(x)\exp\left[-2\pi i\frac{mx}{a}\right]dx = \sum_{-\infty}^{\infty} C_n\int_0^a \exp\left[2\pi i\frac{(n-m)x}{a}\right]dx = aC_m.$$

Hence C_n for each term is given by

$$C_n = (1/a)\int_0^a \Psi(x)\exp\left[-2\pi i(nx/a)\right]dx. \tag{4.5}$$

Quite similarly, a function of two or three variables can be written

$$\Psi(x,y,z) = \sum_{h,k,l=-\infty}^{\infty} C_{hkl}\exp\left[2\pi i\left(\frac{hx}{a} + \frac{ky}{b} + \frac{lz}{c}\right)\right], \tag{4.6}$$

where h, k, l are integers. The coefficients in (4.6) are given by

$$C_{hkl} = \frac{1}{abc}\int_0^a\int_0^b\int_0^c \Psi(xyz)\exp\left[-2\pi i\left(\frac{hx}{a} + \frac{ky}{b} + \frac{lz}{c}\right)\right]dx\,dy\,dz. \tag{4.7}$$

Although each term in (4.6) is complex, the summation leads to real values for $\Psi(x, y, z)$, because there is a corresponding complex conjugate for each term. If the function is not periodic, it may be described by the Cauchy formula:

$$\Psi(x, y, z) = \int_{-\infty}^{\infty} \exp\left[-2\pi i(xx^* + yy^* + zz^*)\right]dx^*\,dy^*\,dz^* \int_{-\infty}^{\infty} \Psi(x', y', z')$$

$$\times \exp[2\pi i(x'x^* + y'y^* + z'z^*)]\,dx'\,dy'\,dz'. \tag{4.8}$$

Having written

$$\Phi(x^*, y^*, z^*) = \int_{-\infty}^{\infty} \Psi(x, y, z)\exp[2\pi i(xx^* + yy^* + zz^*)]\,dx\,dy\,dz, \tag{4.9}$$

we obtain

$$\Psi(x, y, z) = \int_{-\infty}^{\infty} \Phi(x^*, y^*, z^*)$$

$$\times \exp[-2\pi i(xx^* + yy^* + zz^*)]\,dx^*\,dy^*\,dz^*. \tag{4.10}$$

If two functions are related by an expression of the type (4.9) or (4.10), one of them is called the Fourier transform of the other. Thus in (4.10) $\Psi(x, y, z)$ is the Fourier transform of $\Phi(x^*, y^*, z^*)$. The most important property of a Fourier transformation is its reversibility. In other words, if the function $\Phi(x^*, y^*, z^*)$ is known, the formula (4.10) is applied to determine $\Psi(x, y, z)$, while the reverse transformation (4.9) permits to find $\Phi(x^*, y^*, z^*)$ if $\Psi(x, y, z)$ is known.

The reversibility of Fourier transforms is widely used in the theory of structure analysis. If the coordinates x^*, y^*, z^* determine the position of points in the reciprocal space which is specified by vectors a^*, b^*, c^* satisfying Eqs. (1.9) and (1.10), then the function $\Phi(x^*, y^*, z^*)$ describes the diffraction properties of the object having the scattering matter distribution described by $\Psi(x, y, z)$. If $\Phi(x^*, y^*, z^*)$ is known, (4.10) is used to solve the main problem in structure analysis, i.e., the determination of $\Psi(x, y, z)$.

If the positions of points in direct and reciprocal space are given by vectors

$$r = xa + yb + zc \quad \text{and} \quad s = x^*a^* + y^*b^* + z^*c^*, \tag{4.11}$$

then (4.9) and (4.10) may be written

$$\Psi(r) = \int \Phi(s) \exp[-2\pi i sr] dV_s, \tag{4.12}$$

$$\Phi(s) = \int \Psi(r) \exp[2\pi i sr] dV_r. \tag{4.13}$$

Using (4.13), the concept of the reciprocal lattice may be readily introduced. The Fourier transform of the function $R(r)$ [see Eq. (1.2)] describing the position of the crystal lattice points is given by

$$D(s) = \int R(r) \exp[2\pi i sr] dV_r = \sum_m \sum_n \sum_p \int \delta(r - r_{mnp}) \exp[2\pi i rs] dV.$$

Owing to the properties of the δ-function, the integral is zero for all r except $r = r_{mnp}$ and

$$D(s) = \sum_m \sum_n \sum_p \exp 2\pi i sr_{mnp} = \sum_m \sum_n \sum_p \exp 2\pi i(mx^* + ny^* + pz^*). \tag{4.14}$$

The expression (4.14) represents the interference function. Therefore $D(s)$ in (4.14) has nonzero values only for integer h, k, l components of the vectors s in the reciprocal space. Thus the Fourier transform of the function describing the crystal lattice is also a lattice function

$$D^2(H) = \sum_{hkl} \delta(H - H_{hkl}), \tag{4.15}$$

where $s = H$, and $H_{hkl} = ha^* + kb^* + lc^*$ is the reciprocal lattice vector.

A significant role in the diffraction theory is played by the operation of convolution of functions. Convolution is the result of the distribution of a function $\Psi_1(r)$ according to a law specified by another function $\Psi_2(r)$. Mathematically, this is given by

$$\Psi_1 * \Psi_2 = \int \Psi_1(r - r') \Psi_2(r') dr'. \tag{4.16}$$

The symbol $*$ is used to denote a convolution operation.

Here are several examples illustrating the result of a convolution of two functions. Figure 18 shows two functions, $\Psi_1(x')$ and $\Psi_2(x')$, as well as $\Psi_2(x - x')$

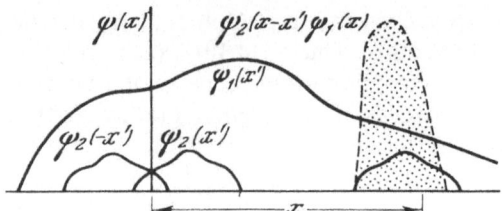

Fig. 18. The convolution of Ψ_1 with Ψ_2

which is shifted by the distance x along the positive direction of the x axis after it has been reflected in the mirror plane m passing through the origin. The dotted region represents the value of the convolution for the given x. Convolution of the two functions should be taken for all possible values of the argument.

Practically, the case where one of the functions is represented by the sum of δ-functions is important, i.e., $\Psi_1 = \sum\limits_{n} \delta(x - x_n)$. It may be easily seen that $\Psi_1 * \Psi_2 = \sum\limits_{n} \Psi_2(x - x_n)$. Thus the convolution leads to distributing the function Ψ_2 according to law defined by the sum of δ-functions.

Let us consider the properties of the convolution operation. Assume

$$f_1(s) = \int \Psi_1(r) \exp[2\pi i sr] dr , \quad f_2(s) = \int \Psi_2(r) \exp[2\pi i sr] dr ;$$

$$\Psi_1(r) = \int f_1(s) \exp[-2\pi i sr] ds , \quad \Psi_2(r) = \int f_2(s) \exp[-2\pi i sr] ds .$$

Let us find

$$f_1(s) f_2(s) = \int [\int \Psi_1(r') \Psi_2(r - r') dr'] \exp[2\pi i sr] dr .$$

Taking account of (4.16), we have

$$f_1(s) f_2(s) = \int [\Psi_1 * \Psi_2] \exp[2\pi i sr] dr . \tag{4.17}$$

In other words, the Fourier transform for the convolution of Ψ_1 with Ψ_2 is the product of the Fourier transforms for each of these functions. The reverse Fourier transform of (4.17) gives

$$\Psi_1 * \Psi_2 = \int f_1(s) f_2(s) \exp[-2\pi i sr] ds . \tag{4.18}$$

Thus the convolution of Ψ_1 with Ψ_2 is equal to the Fourier transform of the product of Fourier transforms for each of these functions.

Consider the product of Ψ_1 and Ψ_2

$$\Psi_1(r) \Psi_2(r) = \int [\int f_1(s') f_2(s - s') ds'] \exp[-2\pi i sr] ds$$
$$= \int [f_1(s) * f_2(s)] \exp[-2\pi i sr] ds .$$

Using the reverse Fourier transform we obtain

$$f_1(s) * f_2(s) = \int \Psi_1(r) \Psi_2(r) \exp[2\pi i sr] dr .$$

Hence the Fourier transform of the product of two functions is the convolution of the transforms for each of the functions.

Convolution may be taken repeatedly by involving new functions regardless of their order. For example

$$\Psi_1(r) * \Psi_2(r) * \Psi_3(r) = [\Psi_1(r) * \Psi_2(r)] * \Psi_3(r)$$
$$= \int f_1(s) f_2(s) f_3(s) \exp[-2\pi i s r] ds .\tag{4.19}$$

To obtain this formula, the expression (4.18) has been employed, where the Fourier transform of the convolution of the first two functions was considered according to (4.17) as $f_1(s) \cdot f_2(s)$.

4.2 Fourier Series: Representation for the Electrostatic Potential and Use in Structure Analysis

Since the distribution of values of the electrostatic potential may be described by a periodic function $\varphi(x, y, z)$ with periods equal to the parameters of the crystal lattice, the potential in a point (x, y, z) may, according to (4.6), be written as

$$\varphi(x, y, z) = \sum_{h,k,l=-\infty}^{\infty} F(hkl) \exp 2\pi i \left(h\frac{x}{a} + k\frac{y}{b} + l\frac{z}{c} \right)\tag{4.20}$$

where

$$F(hkl) = \frac{1}{abc} \int_0^a \int_0^b \int_0^c \varphi(x, y, z) \exp\left[-2\pi i \left(h\frac{x}{a} + k\frac{y}{b} + l\frac{z}{c} \right) \right] dx\, dy\, dz .$$

Since $dx\,dy\,dz/abc = dV_0/V_0$, where V_0 is the unit cell volume, then

$$F(hkl) = \frac{1}{V_0} \int_0^a \int_0^b \int_0^c \varphi(x, z, y) \exp\left[-2\pi i \left(h\frac{x}{a} + k\frac{y}{b} + l\frac{z}{c} \right) \right] dV_0 .\tag{4.21}$$

Comparing (4.21) with (2.41), it can be seen that the Fourier coefficients for the electrostatic potential are related to the kinematical structure amplitudes by a simple expression

$$\Phi(hkl) = \frac{2\pi m e V_0}{h^2} F(hkl) ,\tag{4.22}$$

if atomic scattering factors for electrons are expressed in terms of cm^{-1}. This is one of the most fundamental results which relates the parameters of a diffraction pattern with the Fourier coefficients. These may be used in (4.20) to obtain the distribution of the electrostatic potential in the unit cell volume. The problem of the crystal structure determination is thus reduced to the determination of structure amplitudes including the initial phases.

The zero-order term of the Fourier series for the electrostatic potential is

$$F(000) = \frac{1}{V_0} \int_0^a \int_0^b \int_0^c \varphi(x, y, z) dV_0 = \overline{\varphi(x, y, z)} = \varphi_0 ,$$

representing the mean value for the internal potential of the crystal. We shall hereafter use one symbol $\Phi(hkl)$ to designate both the coeffecient in the Fourier series for the electrostatic potential and the structure amplitude, since they are related by a simple expression (4.22). For the values of the electrostatic potential calculated according to (4.20) to be real, $\Phi(hkl)$ should be equal to $\Phi^*(\bar{h}\bar{k}\bar{l})$.

For a centrosymmetrical crystal, $\alpha_{hkl} = 2\pi$ or π and

$$\Phi(hkl) = \pm |\Phi(hkl)| = \Phi(\bar{h}\bar{k}\bar{l}).$$

Consequently,

$$\varphi(x, y, z) = \sum_{h,k,l=-\infty}^{\infty} \Phi(hkl) \cos 2\pi \left(\frac{hx}{a} + \frac{ky}{b} + \frac{lz}{c} \right). \tag{4.23}$$

The expression (4.20) can be rewritten with (4.22) taken into account and without constant factors

$$\varphi(x, y, z) = \sum_{HKL=-\infty}^{\infty} \Phi(HKL) \exp 2\pi i \left(H\frac{x}{a} + K\frac{y}{b} + L\frac{z}{c} \right). \tag{4.24}$$

Here h, k, l are replaced by H, K, L to emphasize that they are any whole numbers that may have a common factor, whereas for lattice planes symbols hkl will be used, which are, by definition, integers without a common factor. Note that $H:K:L = h:k:l$. Using different symbols to designate the indices for structure amplitudes and for lattice planes will help to avoid confusion in the discussion of the physical meaning of the expansion of the electrostatic potential by a Fourier series.

If the position of an arbitrary point in the unit cell is described by the vector $r = (x/a)a + (y/b)b + (z/c)c$, then the plane passing through this point parallel to the plane (hkl) is represented by the equation

$$rH_{HKL} = H\frac{x}{a} + K\frac{y}{b} + L\frac{z}{c} = \frac{nr\cos\beta}{d(hkl)} = \frac{np}{d(hkl)}, \tag{4.25}$$

where H_{hkl} is the reciprocal lattice vector of the length $n/d(hkl)$, β is the angle between r and H_{hkl}, and $p = r\cos\beta$ is the shortest distance between the origin and the plane parallel to the (hkl) plane, which is passing through the point given by r, or the projection of r on the direction normal to (hkl). The phase factors in (4.24) are, in fact, equations representing the planes passing through points at x/a, y/b, z/c parallel to the planes (hkl). If, in addition, it is taken into account that

$$\Phi(HKL) = |\Phi(HKL)| \exp 2\pi i \alpha_{HKL},$$

the formula (4.24) may be written

$$\varphi(x, y, z) = \sum_{H,K,L=-\infty}^{\infty} |\Phi(HKL)| \exp 2\pi i \left(\frac{np}{d(hkl)} + \alpha_{HKL} \right). \tag{4.26}$$

Summing up the complex conjugates pairwise, we obtain

$$\varphi(x, y, z) = \sum_{H,K,L=-\infty}^{\infty} |\Phi(HKL)| \cos 2\pi (rH_{HKL} + \alpha_{HKL})$$

$$= \sum_{H,K,L=-\infty}^{\infty} |\Phi(HKL)| \cos 2\pi \left(\frac{np}{d(hkl)} + \alpha_{HKL} \right). \tag{4.27}$$

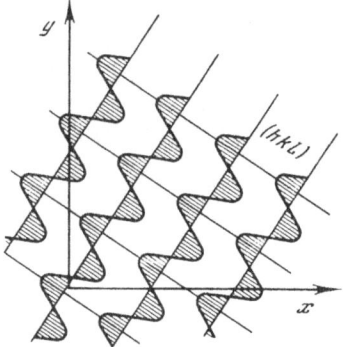

Fig. 19. The electrostatic potential distribution related to one harmonic

Thus the distribution of the electrostatic potential in the object may be presented as a set of harmonics of the type $|\Phi(HKL)|\cos 2\pi[H(x/a) + K(y/b) + L(z/c) + \alpha_{HKL}]$. Each harmonic corresponds to a wave field of the electrostatic potential, which has the same value for all points belonging to the constant phase plane. This plane is represented by the equation

$$H\frac{x}{a} + K\frac{y}{b} + L\frac{z}{c} = \frac{np}{d(hkl)} = \text{const.} \tag{4.28}$$

Since the Fourier coefficients $|\Phi(HKL)|$ and the initial phases for each harmonic are constant, the satisfaction of (4.28) implies the constant values for the potential at all points x/a, y/b, z/c lying in the plane which passes parallel to (hkl) at the distance p from the origin along H_{HKL}. The values of the potential corresponding to this harmonic vary according to a sine law along the normal to the (hkl) set of planes with the period $d(hkl)/n$. The function $\cos 2\pi[(np/d(hkl)) + \alpha_{HKL}]$ will indeed be periodical with the period $p' = p + kd(hkl)/n$, where k is any integer. The initial phase α_{HKL} leads to a shift of the maxima in each harmonic by $-\alpha_{HKL} \cdot d(hkl)/n$ with respect to the origin and the positions of the set of (hkl) planes. The distribution of the electrostatic potential may be presented schematically as a one-dimensional periodic function varying along the normal to the planes (hkl) with the period $d(hkl)/n$, the maximum amplitude $|\Phi(HKL)|$, and the initial phase α_{HKL} (Fig. 19). This function may be pictured by a straight line passing through the origin parallel to the normal to the lattice planes (hkl), so that corresponding values of the electrostatic potential are marked along this line for each given p. Moving this line parallel to itself along the whole volume of the crystal, we can reconstruct the space distribution of the electrostatic potential wave for each given harmonic.

The above discussion elucidates the close connection between the Fourier series representation of the electrostatic potential and the reciprocal lattice. Each harmonic is described by parameters characterizing a definite reciprocal lattice point. The periodic variation of the harmonic $|\Phi(HKL)|\cos 2\pi(rH_{HKL} + \alpha_{HKL})$ proceeds along the vector connecting the origin with the HKL lattice point. The period, or the wavelength, is inversely proportional to the length of the vector, and the maximum amplitude and the initial phase are determined by the "weight" of the lattice point HKL. Since each reciprocal lattice point corresponds to one definite term in the Fourier series, each electrostatic potential wave characterized by definite direction, period, amplitude, and initial phase is responsible for one definite diffraction reflection. The values of the potential determined by a Fourier term are constant within each plane normal to the vector H_{HKL} along which the potential variation takes place.

Assume that there is only the electrostatic potential wave described by the function $\Psi(r)$, i.e.,

$$\Psi(r) = |\Phi(HKL)| \cos 2\pi \left(H\frac{x}{a} + K\frac{y}{b} + L\frac{z}{c} + \alpha_{HKL} \right)$$

$$= \tfrac{1}{2} |\Phi(HKL)| \{\exp 2\pi i (r H_{HKL} + \alpha_{HKL}) + \exp[-2\pi i (r H_{HKL} + \alpha_{HKL})]\} .$$

To calculate the scattering amplitude, the expression (4.13) should be used:

$$A(H) = \int \Psi(r) \exp[-2\pi i r H] dV_r$$

$$= \frac{1}{2} |\Phi(HKL)| \left\{ \exp 2\pi i \alpha_{HKL} \int_{V_0} \exp 2\pi i \left[(H-H')\frac{x}{a} + (K-K')\frac{y}{b} \right. \right.$$

$$\left. + (L-L')\frac{z}{c} \right] dV_0 + \exp(-2\pi i \alpha_{HKL}) \int_{V_0} \exp 2\pi i \left[(H+H')\frac{x}{a} \right.$$

$$\left. \left. + (K+K')\frac{y}{b} + (L+L')\frac{z}{c} \right] dV_0 \right\} .$$

The integral in the first term is zero for all reciprocal lattice points unless $H' = H$, $K = K'$, $L' = L$, and that in the second term is zero unless $H' = -H$, $K' = -K$, $L' = -L$. Therefore,

$$A(H) = \Phi(HKL) .$$

Consequently, the scattering of electrons by the electrostatic potential represented by a sine wave yields only one reflection. The amplitude and the phase for this reflection are unambiguously related to the maximum amplitude and the initial phase of the wave involved, while the direction of the scattered wave is defined by the indices HKL.

If the electrostatic potential distribution in the crystal is specified by a wave exactly similar to the one considered above, but shifted along the positive direction of the vector H_{HKL} with respect to the origin, i.e.,

$$\Psi(r) = |\Phi(HKL)| \cos 2\pi (r H_{HKL} - \alpha_{HKL}) ,$$

then the electron scattering amplitude will be given by $A(H) = \Phi(\bar{H}\bar{K}\bar{L})$. The diffraction intensity is the same in both cases.

Thus the electrostatic potential distribution in the crystal may be regarded as a set of waves characterized by definite periods, amplitudes, initial phases, and "propagation" directions. The interaction between electrons and the scattering potential is selective in the sense that each wave scatters electrons in a definite direction and with a definite amplitude. This representation is, of course, a kind of a mental construction used only to establish the relationships between the properties of the object and the diffraction pattern.

It is clear that one of the basic difficulties in structure analysis is associated with determination of initial phases for all reflections. The shift of the crest of the wave representing the electrostatic potential along H_{HKL} with respect to the origin does not affect the diffraction intensity but influences considerably the electrostatic potential distribution.

In SAED studies, intensities in point patterns contain information on the projection of the structure in the direction coincident with the incident electron beam. Therefore, in practice the electrostatic potential projections are often constructed.

Obviously,

$$\varphi(x, y) = \int_0^c \varphi(x, y, z)\, dz$$

$$= \frac{c}{abc} \int_0^c \Sigma\,\Sigma\,\Sigma\, \Phi(hkl) \exp\left[-2\pi i\left(\frac{hx}{a} + \frac{ky}{b} + \frac{lz}{c}\right)\right] \frac{dz}{c}.$$

Since

$$\int_0^c \exp\left[2\pi il\frac{z}{c}\right] d\frac{z}{c} = \begin{array}{ll} c & \text{for } l = 0 \\ 0 & \text{for } l \neq 0, \end{array}$$

we have

$$\varphi(x, y) = \frac{1}{S} \sum_{h,k=-\infty}^{\infty} \Phi(hko) \exp\left[-2\pi i\left(h\frac{x}{a} + k\frac{y}{b}\right)\right]. \tag{4.29}$$

The construction of a projection of the electrostatic potential is most advantageous in those cases where the repeat distance is small and so is the probability of multiple overlapping of atoms. The projection directions usually coincide with the basic unit cell vectors. However, there are exceptions, and some examples of these are considered in the analysis of the structure of a three-chain silicate (see Chap. 11).

In the general case, a projection of the electrostatic potential on the reciprocal lattice plane $(mnp)^*$ may be obtained if the incident beam is parallel to the direction $[mnp]$ in the direct lattice. Therefore it is appropriate to proceed to a new coordinate system, such as to bring the c' axis into coincidence with the zone axis $[mnp]$. If this direction does not lie in the same plane as vectors a and b of the "old" system, the matrix for the transition from the old system to the new one is given by

$$\begin{vmatrix} 1 & 0 & 0 \\ 0 & 1 & 0 \\ m & n & p \end{vmatrix}.$$

The transition from the old indices to the new ones is described by the same matrix. The index $l' = mh + nk + pl$ being zero by the definition, we may write

$$\varphi(x', y') = \frac{1}{S} \sum_{h'} \sum_{k'} \Phi(h'k'o) \exp\left[2\pi i\left(h'\frac{x'}{a'} + k'\frac{y'}{b'}\right)\right]$$

$$= \frac{1}{S} \sum_{h} \sum_{k} \Phi(hkl_0) \exp\left[-2\pi i\left(h\frac{x}{a} + k\frac{y}{b}\right)\right], \qquad (4.30)$$

where $l_0 = -(1/p)(mh + nk)$.

Various relations for structure amplitudes resulting from crystal symmetry simplify the expressions for double or triple Fourier series. The corresponding "working" formulae for all space groups are given in the International Tables for X-Ray Crystallography (1962).

In a calculation of a Fourier map, the number or experimental structure amplitudes should exceed considerably the number of atomic coordinates to be found. Therefore, in all cases, one should try to obtain as many reflections as possible. However, in the SAED practice, one has to keep to projections of the electrostatic potential.

Another fundamental difficulty in the use of Fourier series is connected with termination errors, as one always has to deal with a finite number of terms. As a result, the so-called termination waves appear that may distort the true pattern in the electrostatic potential distribution, shift the maxima, lead to false peaks, etc. Specifically, the presence of noticeable termination waves hinders revealing light atoms, e.g., of hydrogen. Since, in electron diffraction, the atomic factor falls off sharply, diffraction intensity decreases rapidly with $\sin \vartheta/\lambda$, which reduces the termination effect. At the same time, this leads to a smearing of the maxima of atomic electrostatic potentials, which decreases the accuracy in the localization.

4.3 The Trial-And-Error Method

In terms of the kinematical approximation the measurement of the diffraction intensity permits to determine the values of structure factors $|\Phi(hkl)|^2$ or the moduli of structure amplitudes only, whereas the initial phases remain unknown. Therefore Fourier series cannot be used directly for determining unknown crystal structures. There are various approaches to the phase problem.

One of these is the trial-and-error method which is employed to test the adequacy of structural model to the real structure under study. A structural model is constructed on the basis of the data on chemical composition, unit cell parameters, space symmetry, density, ionic radii, by making use of the structural information on related substances. The structural model includes the arrangement of the atoms in the unit cell, i.e., the atomic coordinates and scattering powers. These data are used in a formula of the type (2.55) to calculate a set of structure amplitudes Φ_c. The values $|\Phi_c|$ are compared with the $|\Phi_0|$ set obtained from experimental intensities. The reliability factor $R = \sum_{hkl}||\Phi_0| - |\Phi_c||/ \sum |\Phi_0|$ serves as a quantitative measure of the convergence of Φ_0 and Φ_c. At the first stage of a structure study one may know the position of only a few atoms, usually those having maximum scattering powers. If this is the case, the positions of the remaining atoms may be determined by the successive approximations method.

First, structure amplitudes $\Phi_c'(hkl)$ are calculated for atoms with the known coordinates and scattering powers. The initial phases for the values $\Phi_c'(hkl)$ are ascribed to the corresponding experimental values of $\Phi_0(hkl)$. The Fourier map is then computed to obtain an approximate distribution of the electrostatic potential.

After revealing new additional maxima which are ascribed to the positions of some new atoms, new values $\Phi_c''(hkl)$ are calculated. The phases of these are again ascribed to the values $\Phi_0(hkl)$ to obtain a revised distribution of the potential, etc. The procedure is repeated until the whole motif of atoms is determined. If there is a three-dimensional array of precisely measured intensities and the nature of the interaction between electrons and the object is known, the Fourier maps reveal not only the precise coordinates and scattering powers for all atoms, but also some important details in the crystal structure, even the nature of interatomic bonding.

On the other hand, analysis of the precision in the determination of the electrostatic potential peak positions shows that Fourier syntheses have a relatively low sensitivity to errors in the estimation of $|\Phi|$. At any rate, the general outline of the structure under study can be obtained even with a 50% error in the estimation of intensity. The accuracy in the determination of atomic coordinates will be then quite low, especially for light atoms.

4.4 Interatomic Vector Space. Patterson Function: Properties and Application to Structure Analysis

Analysis of the interatomic Patterson function is one of the most efficient methods to gain structural information on the only basis of an array of measured structure factors (Buerger 1959). Consider the information contained in a set of structure factors.

The structure amplitude, $\Phi(H)$, is given by

$$\Phi(H_{hkl}) = \sum_{j=1}^{N} f_j \exp 2\pi i r_j H_{hkl} . \tag{4.31}$$

The structure factor is then given by

$$\Phi^2(H) = \sum \sum f_j f_i \exp 2\pi i (r_j - r_i) H_{hkl} = \sum_{j,i=1}^{N} f_j f_i \exp 2\pi i t_{ij} H_{hkl} . \tag{4.32}$$

Figure 20 shows that t_{ij} is an interatomic vector, connecting an atom of the type i and that of the type j. If atoms i and j are located at the points $x_i\, y_i\, z_i$ and $x_j\, y_j\, z_j$ respectively, the components of t_{ji} will be given by $u_{ji} = x_j - x_i$, $v_{ji} = y_j - y_i$, $w_{ji} = z_j - z_i$. Therefore the structure factor is a function of interatomic vector components rather than atomic coordinates. If the total number of atoms in the unit cell is N, it may be seen from (4.32) that for each given j summation is taken over all i values, from 1 to N. On the whole, components of interatomic vectors are summed for all the atoms in the unit cell. Summation over j and i from 1 to N may be replaced by summation over one index q from 1 to N^2. Assuming $f_i f_j = F_q$, we may write

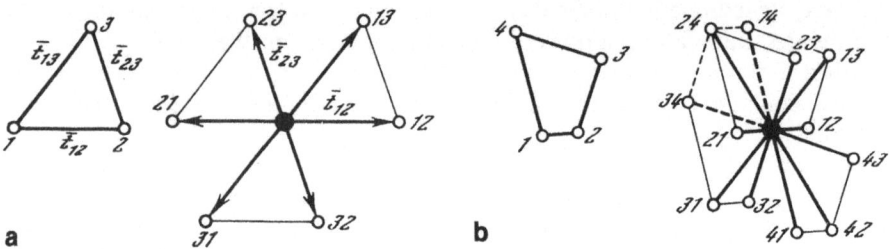

Fig. 20a, b. Relations between the initial atomic pattern and the interatomic vector space; **a** construction of interatomic vectors; **b** interatomic vector space as a superposition of N atomic patterns. *Thin lines* connect the tips of interatomic vectors forming the initial motif. *Dashed lines* show the pattern which is centrosymmetrical to the initial one; $t_{12} = r_1 - r_2$; $t_{13} = r_1 - r_3$; $t_{23} = r_2 - r_3$

$$\Phi^2(H_{hkl}) = \sum_{q=1}^{N^2} F_q \exp 2\pi i t_q H_{hkl} = \sum_{q=1}^{N^2} F_q \exp 2\pi i(hu_q + kv_q + lw_q). \qquad (4.33)$$

This formula is similar to the expression (4.31) for the structure amplitude. To impart a definite meaning to this analogy, the interatomic vector space is introduced which has the same periodicity as the real object. While the scattering power of a unit cell in the crystal is represented by the structure amplitude $\Phi(H)$, that of a unit cell in the space of interatomic vectors is described by the structure factor $\Phi^2(H)$. In other words, the structure factor acts as a structure amplitude of a unit cell whose scattering centers and scattering powers are given by vectors t_q and the values $F_q = f_i f_j$, respectively.

Assume that the arrangement and the scattering powers of atoms in the unit cell are known (Fig. 20a). To construct the unit cell in the interatomic vector space, the following operations should be performed.

The center of atom 1 should be connected by straight lines with the rest of atoms. The set of interatomic vectors obtained should be place into the origin of the unit cell in the interatomic vector space. A scattering power equal to the product of scattering powers for the two atoms connected by a given vector is ascribed to the tip of this vector. The same procedure should be performed using the second atom, the third, etc., as the initial ones. As a result, there will be $N(N-1)$ vectors in the unit cell. The set of points given by interatomic vectors may be obtained by placing the origin into each atom consequently, with the positions of other atoms being marked each time (Fig. 20a). If the tips of interatomic vectors are regarded as point maxima with "weights" F_q, then their arrangement is, in fact, an N-fold superposition of the same atomic pattern. If the atomic positions in the unit cell are given by vectors r_j, it is obvious that the atomic pattern is shifted by $-r_i$ when the origin is placed into the ith atom.

Thus the whole set of shifts of the initial atomic motif is described by the set of vectors r_j taken with an opposite sign. Having marked the ith atom in each of the N patterns we obtain a structural motif which is the inverse image of the initial pattern. If the latter is centrosymmetrical itself, then the procedure described yields the initial distribution of atoms.

Thus the set of interatomic vectors contains all the necessary information on the crystal structure. Structure determinations would be especially simple in the case of point atoms, as the vector space would then be also represented by a set of point maxima. However, in reality the "scattering" matter is the electrostatic potential distribution described by a continuous function $\varphi(x, y, z)$. Consequently, whatever the position of the given vector, it is hardly probable that both of its tips would fall into points where the potential is zero. Therefore the interatomic vector space should also be described by a continuous function $\mathscr{P}(t)$ or $\mathscr{P}(uvw)$, where u, v, w are the components of the vector t. If the structure amplitude is given by

$$\Phi(H) = \int_{V_0} \varphi(r) \exp[2\pi i r H] dV, \tag{4.34}$$

the structure factor may be written

$$\Phi^2(H) = \int\int \varphi(r') \varphi(r'') \exp[2\pi i(r' - r'')H] dV_{r'} dV_{r''}.$$

Putting $t = r' - r''$ and $r = r''$ we have

$$\Phi^2(H) = \int\int \varphi(r) \varphi(r+t) \exp[2\pi it H dV_t] dV_r$$
$$= \int \mathscr{P}(t) \exp[2\pi it H] dV_t, \tag{4.35}$$

where

$$\mathscr{P}(t) = \int \varphi(r) \varphi(r+t) dV_r. \tag{4.36}$$

The expressions (4.34) and (4.35) are again similar, but here it is the continuous interatomic vector function $\mathscr{P}(t)$ that acts as the scattering matter. Comparison of (4.16) and (4.36) shows that the Patterson function is the convolution of $\varphi(r)$ with $\varphi(-r)$

$$\varphi(r) * \varphi(-r) = \int \varphi(r) \varphi(t+r) dr = \mathscr{P}(t). \tag{4.37}$$

$\mathscr{P}(t)$ is centrosymmetrical and periodic with the same periods as $\varphi(r)$.

It follows from (4.36) that the value of $\mathscr{P}(t)$ for each t is determined by the sum of products of electrostatic potential for the points that are connected by t, as the latter is moved parallel to itself over the unit cell. It is obvious that $\mathscr{P}(t)$ has large values for those distances between the points r and $r+t$ which correspond to simultaneously large φ values.

The value of each maximum specified by a vector t_q is proportional to the electrostatic potential values for atoms located at the corresponding distance.

The properties of the convolution operation imply

$$\mathscr{P}(t) = \varphi(r) * \varphi(-r) = \int \Phi^2(H) \exp[-2\pi it H] dV_H. \tag{4.38}$$

Since $\mathscr{P}(t)$ is continuous and periodic, it can be written as a Fourier series. According to (4.7),

$$K(hkl) = \frac{1}{V_0} \int_0^a \int_0^b \int_0^c \mathscr{P}(t) \exp[2\pi it H] dV_0$$

$$= \frac{1}{V_0} \int_0^a \int_0^b \int_0^c \mathscr{P}(uvw) \exp[2\pi i(hu + kv + lw)] dV_0, \tag{4.39}$$

where u, v, w and hkl are components for the vectors t and H, respectively.

Comparing (4.39) and (4.35) one may see that structure factors should be used as Fourier coefficients, as

$$K(hkl) = \Phi^2(hkl)/V_0.$$

Hence

$$\mathscr{P}(uvw) = \frac{1}{V_0} \sum \sum \sum \Phi^2(hkl) \exp 2\pi i(hu + kv + lw).$$

$\mathscr{P}(uvw)$ being centrosymmetrical, we obtain

$$\mathscr{P}(uvw) = \frac{1}{V_0} \sum_{h,k,l=-\infty}^{\infty} \sum \sum \Phi^2(hkl) \cos 2\pi(hu + kv + lw). \tag{4.40}$$

For a projection on the axis w, we have

$$\mathscr{P}(uv) = \frac{1}{S} \sum \sum \Phi^2(hko) \cos 2\pi(hu + kv).$$

Computation of Fourier series with structure factors as coefficients (the Patterson synthesis) permits to construct an interatomic vector function without any a priori assumptions on the structure under study. The problem is to find the generating function $\varphi(xyz)$ by analyzing $\mathscr{P}(uvw)$. The solution is complicated by a number of factors. If there are, e.g., 20 atoms in the unit cell, the Patterson function will contain, theoretically, 380 maxima, which moreover are smeared and may overlap. Maxima corresponding to vectors connecting light atoms are especially difficult to reveal. Since the peak at the origin corresponds to N atoms, it veils all the details in the Patterson function near the origin.

Structures containing one heavy atom are the most simple to determine by the Patterson synthesis. It is evident that the strongest maxima in the interatomic function will represent the atomic pattern.

However, analysis of the Patterson function is, in general, a complex problem, and the present author has to confine himself to the discussion of the basic principles in its interpretation. The properties of the Patterson function and practical methods for its analysis are treated in detail in the literature. Note that a Patterson synthesis is even less sensible to errors in determination of structure factor moduli than a Fourier synthesis.

4.5 Direct Phasing Methods

To determine a crystal structure using Eqs. (4.20) and (4.29), the complex values of structure amplitudes are required. In X-ray crystallography the so-called *direct phasing* methods are in wide use, which are especially convenient for centrosymmetrical crystals. These methods are based on analysis of certain equalities and inequalities relating phases to moduli and indices of structure amplitudes (Hauptman 1972; Woolfson 1961). The data processing is carried out with the help of computers, and for a representative set of initial data the structure determination becomes completely automatized. The same methods are also applicable to electron crystallography provided the influence of dynamical effects on diffraction intensity is eliminated (Dorset and Hauptman 1976). New approaches to phasing that are based on the analysis of high resolution micrographs with the help of optical diffractometers have been lately elaborated. Various methods for the solution of the phase problem with the help of electron microscopy have been recently reported by Miscell (1978). A more detailed treatment of these problems will be given in Chapter 7.

4.6 Refinement of Atomic Coordinates by the Least-Squares Method

Having constructed a structure model one can refine the atomic coordinates by the least-squares method. This method consists in varying atomic coordinates and thermal factors in order to achieve the best agreement between the observed structure amplitudes and those calculated according to (2.52) and (2.55). The value to be minimized is

$$K = \sum_{hkl} (|\Phi_{exp}| - |\Phi_{calc}|)^2 .$$

Atomic coordinates corresponding to the smallest K should be the most adequate to the real values. The factor R is used as a quantitative criterion for convergence of Φ_{exp} and Φ_{calc}.

An essential advantage of the least-squares method is that it does not require the use of the whole set of reflections. In addition, in minimizing discrepancy between Φ_{exp} and Φ_{calc}, the degree of reliability in determination of $|\Phi|_{exp}$ may be taken into account. This is done by introducing weight factors that increase contributions of those reflections whose intensities have been estimated with higher accuracy. Moreover, it is possible to exclude reflections having intensities too distorted by dynamical effects or other factors. Finally, isotropic or anisotropic thermal parameters can be also refined.

CHAPTER 5

Dynamical Theory of Electron Diffraction (Two-Beam Approximation)

5.1 Quantum-Mechanical Solution

The main assumption in the kinematical theory is that amplitudes of diffracted waves are negligible as compared to the amplitude of the incident electron wave. At the same time, electrons interact with matter much more strongly than X-rays. Therefore, multiple scattering effects appear in the crystal under certain conditions, so that amplitudes of diffracted waves might become comparable to the amplitude of the incident beam.

Solution of the problem of electron scattering in a crystal with account taken of multiple scattering, inelastic scattering, and boundary conditions is the subject of the dynamical theory. The behavior of electrons in the crystal potential field was at first analyzed in a classical work by Bethe (1928), and then developed and treated in detail by other authors (Blackman 1939, Heidenreich 1964, Hirsh et al. 1977, Cowley 1967, 1975a, b).

First of all, the two-beam approximation is to be considered, assuming that only two waves, the incident and the diffracted, interact in the crystal. On the one hand, even this approximation reveals all the main physical consequences of the dynamical theory. On the other hand, two-beam calculations produce criteria for transition from the kinematical interaction between electrons and matter to the dynamical, so that the field of applicability for the kinematical theory is defined more strictly. Finally, despite an a priori approximate character of the two-beam theory of Bethe-Blackman, it gives a satisfactory description for various experimental data even when there are a number of strong reflections in a diffraction pattern.

In terms of nonrelativistic energies, the behavior of an electron moving in a potential crystal field $\varphi(r)$ is described by time-independent wave functions satisfying the Schrödinger equation

$$\nabla^2 \Psi(r) + (8\pi^2 m/h^2)[E - U(r)] \Psi(r) = 0 . \tag{5.1}$$

Being continuous and periodic, the electrostatic potential function may be written as a Fourier series:

$$U(r) = -e\varphi(r) = -(h^2/2m) \sum v_H \exp 2\pi i r H . \tag{5.2}$$

Since $\varphi(r)$ is real, then $v_h = v^*_{-H}$. The solution of (5.1) is sought in the form of superposition of waves of the type

$$\Psi(r) = \sum_H C_H(k_0) \exp 2\pi i r(k_0 + H) = \sum_H C_H(k_0) \exp 2\pi i k_H r. \qquad (5.3)$$

Each wave in (5.3) has constant amplitude with a definite wave number. The expression (5.3) will be shown below to be equivalent to a superposition of plane waves. Substituting (5.3) into (5.1) and taking account of (5.2) we have

$$-4\pi^2 \sum_H k_H^2 C_H(k_0) \exp 2\pi i(k_0 + H)r + \frac{8\pi^2 m}{h^2} E \sum_H C_H(k_0) \exp 2\pi i r(k_0 + H)$$

$$+ \frac{8\pi^2 m}{h^2} \frac{h^2}{2m} \sum_{H'} v_{H'} \exp[2\pi i r H'] \sum_{H''} C_{H''}(k_0) \exp 2\pi i(k_0 + H'')r = 0.$$

Canceling out common factors and factoring the zero term in the potential Fourier series outside the sum over H' and then replacing H' by N and $(H' + H'')$ by H, we obtain

$$(-k_H^2 + 2mE/h^2 + v_0) \sum_H C_H(k_0) \exp 2\pi i(k_0 + H)r$$

$$+ \sum_N{}' \sum_H v_N C_{H-N} \exp 2\pi i(k_0 + H)r = 0.$$

The sum of coefficients of exponential terms with the same phase factors should be equal to zero, i.e.,

$$(-k_H^2 + 2mE/h^2 + v_0) C_H(k_0) + \sum_N{}' v_N C_{H-N}(k_0) = 0. \qquad (5.4)$$

The mean inner crystal potential v_0 changes the wavelength, so that the square wave number is given by [see Eq. (2.33)]

$$k^2 = (2mE/h^2) + v_0. \qquad (5.5)$$

In (5.4), the primed sum sign implies that the zero term is excluded.

Since we are interested in the two-beam approximation, (5.4) may be written, taking account of (5.5)

$$(k^2 - k_0^2) C_0(k_0) + v_{-H} C_H(k_0) = 0,$$

$$v_H C_0(k_0) + (k^2 - k_H^2) C_H(k_0) = 0. \qquad (5.6)$$

Two simultaneous equations have a solution if

$$(k^2 - k_0^2)(k^2 - k_H^2) - v_H v_{-H} = 0. \qquad (5.7)$$

As $\varphi(r)$ is real $v_H^2 = v_H v_{-H}$. Therefore

$$k^4 - k^2(k_0^2 + k_H^2) - v_H^2 + k_0^2 k_H^2 = 0. \qquad (5.8)$$

It should be stressed that the full energy of an electron is a constant value and, consequently, k^2 is also constant [see Eq. (5.5)]. On the other hand, constant k^2 corresponds, according to (5.8), to two expressions:

$$k^2 = \tfrac{1}{2}(k_0^2 + k_H^2) \pm \sqrt{\tfrac{1}{4}(k_0^2 - k_H^2)^2 + v_H^2}.$$

After transformations we obtain

$$k^2 - k_0^2 = \tfrac{1}{2}(k_H^2 - k_0^2) \pm \sqrt{\tfrac{1}{4}(k_H^2 - k_0^2)^2 + v_H^2}.$$

Since $E: ev_H \simeq 10^5$, the values k_0, k_H and k are much greater than their differences, so that

$$2(k - k_0) = k_H - k_0 \pm \sqrt{(k_H - k_0)^2 + \left(\frac{v_H}{k}\right)^2}. \tag{5.9}$$

It follows from (5.9) that

$$k_0^{(1)} = k - \frac{1}{2}(k_H^{(1)} - k_0^{(1)}) - \left[\frac{(k_H^{(1)} - k_0^{(1)})^2}{4} + \left(\frac{v_H}{2k}\right)^2\right]^{1/2}, \tag{5.10}$$

$$k_0^{(2)} = k - \frac{1}{2}(k_H^{(2)} - k_0^{(2)}) + \left[\frac{(k_H^{(2)} - k_0^{(2)})^2}{4} + \left(\frac{v_H}{2k}\right)^2\right]^{1/2}. \tag{5.11}$$

Equations (5.10) and (5.11) define acceptable values for wave vectors in a crystal. It may be imagined that two branches of the electron energy arise in the crystal. Each branch corresponds to a pair of wave vectors $k_0^{(1)}$, $k_0^{(2)}$ for the incident wave and a pair $k_H^{(1)}$, $k_H^{(2)}$ for the diffracted one. Thus, an incident electron having a wave vector k passes to two new states owing to a perturbing influence of the crystal potential v_H. The wave vector in one of these states is somewhat greater than k, and in the other it is somewhat smaller than k.

Equation (5.7) relating the energy and the wave vector describes the so-called dispersion surface formed by the tips of wave vectors. The two branches of the dispersion surface [each corresponding to one of the Eq. (5.9)] are shown in Fig. 21. To be more precise, dispersion curves are shown, resulting from the intersection of dispersion surfaces with the plane containing the wave vectors and the reciprocal lattice vector H. The dispersion surface may be obtained by rotating the dispersion curves about the fixed vector H. A plane normal to H and dividing the latter in two is called the *Brillouin zone boundary*. An approximate form of dispersion curves may be found if (5.10) and (5.11) are rewritten as

$$(k - k_0^{(i)})(k - k_H^{(i)}) = (v_H/2k)^2 \quad \text{for} \quad i = 1 \text{ or } 2. \tag{5.12}$$

It may be seen that near the Brillouin zone these curves should have hyperbolic form. The dispersion curve corresponding to (5.12) with $i = 1$ [see also Eq. (5.10)] will be denoted by 1 and with $i = 2$ by 2 [see also Eq. (5.11)]. Thus the vectors $k_0^{(1)}$ and $k_H^{(1)}$ connect the point $\mathcal{P}^{(1)}$ in the curve (1) with the reciprocal lattice points 0 and H_{hkl}, while the vectors $k_0^{(2)}$ and $k_H^{(2)}$ are drawn from the point $\mathcal{P}^{(2)}$ in the curve (2) to the same reciprocal lattice points.

Bragg's Law applies exactly if the points $\mathcal{P}^{(1)}$ and $\mathcal{P}^{(2)}$ lie in the plane of the Brillouin zone. Then $k_0^{(i)} = k_H^{(i)}$, where $i = 1$ or 2 and, following from (5.12), $k_0^{(2)} - k = k - k_0^{(1)} = v_H/2k$. It may be seen in Fig. 21 that the distance Δk between the dispersion curves is also related to $k - k_0^{(1)}$

$$k - k_0^{(1)} = (\Delta k/2) \cos \vartheta_0 = v_H/2k.$$

Hence

$$\Delta k = v_H/k \cos \vartheta. \tag{5.13}$$

Here the perturbing effect of the potential is revealed most vividly. Decrease in v_H leads to the shift of the dispersion curves toward each other (as Δk

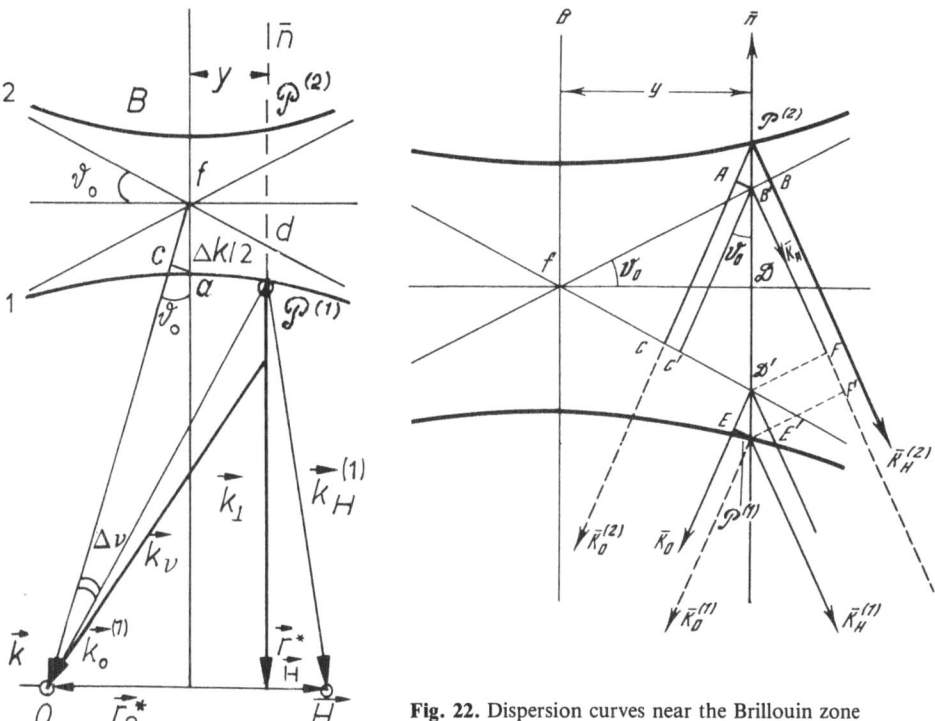

Fig. 22. Dispersion curves near the Brillouin zone

Fig. 21. Dispersion curves formed by the wave vectors $k_0^{(1)}$, $k_0^{(2)}$, $k_H^{(1)}$, $k_H^{(2)}$. B is the trace of the Brillouin zone boundary

decreases). In the limiting case of $v_H = 0$ the curves degenerate into two intersecting spheres. The k value being large, the trace of these spheres is shown in Fig. 21 as two straight lines intersecting at the point f at an angle of $2\vartheta_0$. The behavior of an electron may be described in this case in terms of the kinematical approximation by two vectors, k_0 and k_H, and one spherical dispersion surface. On the contrary, large values of v_H determine substantial redistribution of energy within the limits of the electron energy states permitted.

With the increase in the deviation of the points $\mathscr{P}^{(1)}$ and $\mathscr{P}^{(2)}$ from the Brillouin zone boundary as they are moved along the dispersion curves, the deviation from Bragg's Law increases, and so do the differences between $k_0^{(i)}$ and $|k_0^{(i)} + H|$.

It should be stressed that, among numerous wave vector combinations satisfying (5.12), only those in conformity with the boundary conditions should be chosen. Specifically, the tangential components for all vectors should be the same at $z = 0$. This requirement is each time met by two points, $\mathscr{P}^{(1)}$ and $\mathscr{P}^{(2)}$, lying in the plane normal to the foil plane, or to H, as in the particular case shown in Fig. 21.

Let us show that with the boundary conditions preserved, the following relations are valid

$$|k_0^{(1)} + H| - k_0^{(1)} = |k_0^{(2)} + H| - k_0^{(2)}$$

or

$$k_H^{(1)} - k_0^{(1)} = k_H^{(2)} - k_0^{(2)} .$$

The degree of deviation from Bragg's Law is characterized by the parameter δ, which, as it may be seen in Fig. 8, is equal to

$$\delta = H \operatorname{tg} \Delta \vartheta = \left(\frac{2 \sin \vartheta}{\lambda} \right) \operatorname{tg} \Delta \vartheta = 2k \sin \vartheta \operatorname{tg} \Delta \vartheta .$$

On the other hand, Fig. 21 shows that

$$\operatorname{tg} \Delta \vartheta = df/k = y/k \cos \vartheta_0 = \delta/2k \sin \vartheta_0 .$$

Hence $\delta = 2y \operatorname{tg} \vartheta_0$.

Figure 22 shows a magnified scheme of the arrangement of dispersion branches near the Brillouin zone. It can be seen that $\mathscr{P}^{(2)} B = K_H^{(2)} - K$, $\mathscr{P}^{(2)} C = K_0^{(2)} - K$ and $K_0^{(2)} - K_H^{(2)} = \mathscr{P}^{(2)} C - \mathscr{P}^{(2)} B$. $B' A$ is drawn perpendicular to $\mathscr{P}^{(2)} C$. Right-angled triangles $A B \mathscr{P}^{(2)}$ and $B' \mathscr{P}^{(2)}$ are equal since $\mathscr{P}^{(2)} B'$ is a common side and $<A \mathscr{P}^{(2)} B' = <B \mathscr{P}^{(2)} B'$. Hence $\mathscr{P}^{(2)} B = \mathscr{P}^{(2)} A$ and

$$k_0^{(2)} - k_H^{(2)} = \mathscr{P}^{(2)} C - \mathscr{P}^{(2)} A = A C = B' C' = B' D' \cos \vartheta_0$$
$$= 2y \operatorname{tg} \vartheta_0 \cos \delta_0 = \delta \cos \delta_0 .$$

Similarly,

$$k - k_0^{(1)} = \mathscr{P}(1) E , \qquad k - k_H^{(1)} = B' F' , \qquad k_0^{(1)} - k_H^{(1)} = B' F = \delta \cos \vartheta_0 .$$

Therefore (5.10) may be written as

$$k_0^{(i)} = k + \frac{1}{2} \delta \cos \vartheta_0 \mp \left[\left(\frac{\delta \cos \vartheta_0}{2} \right)^2 + \left(\frac{v_H}{2k} \right)^2 \right]^{1/2} = k + \Delta k_0^{(i)} ,$$

$$k_H^{(i)} = k - \frac{1}{2} \delta \cos \vartheta_0 \mp \left[\left(\frac{\delta \cos \vartheta_0}{2} \right)^2 + \left(\frac{v_H}{2k} \right)^2 \right]^{1/2} = k + \Delta k_H^{(i)} ,$$

(5.14)

where the minus sign corresponds to $i = 1$.

Simultaneous Eqs. (5.6) imply that the coefficients for wave functions are related by

$$C_H^{(i)} = \left[\frac{(k_0^{(i)} - k) 2k}{v_{-H}} \right] C_0^{(i)} .$$

(5.15)

If we neglect the reflected amplitude, the boundary conditions lead to

$$C_0^{(1)} + C_0^{(2)} = 1 , \qquad C_H^{(1)} + C_H^{(2)} = 0 .$$

(5.16)

Solving simultaneous Eqs. (5.15) and (5.16) we obtain

$$C_0^{(1)} = \frac{\delta \cos \vartheta_0 + \left[\delta^2 \cos^2 \vartheta_0 + \left(\frac{v_H}{k} \right)^2 \right]^{1/2}}{2 \left[\delta^2 \cos^2 \vartheta_0 + \left(\frac{v_H}{k} \right)^2 \right]^{1/2}} = \frac{1}{2} \left(1 + \frac{\delta}{\omega} \right) ,$$

$$C_0^{(2)} = \frac{1}{2}\left(1 - \frac{\delta}{\omega}\right), \tag{5.17}$$

$$C_H^{(1)} = \frac{-v_H}{2k\left[\delta^2 \cos^2 \vartheta_0 + \left(\frac{v_H}{k}\right)^2\right]^{1/2}} = \frac{1}{2t_e\omega} = -C_H^{(2)}. \tag{5.18}$$

To simplify the above formulae, we put

$$t_e = \frac{k \cos \vartheta_0}{v_H}, \quad \omega = \left[\delta^2 + \left(\frac{v_H}{k \cos \vartheta_0}\right)^2\right]^{1/2} = (\delta^2 + t_e^{-2})^{1/2}.$$

t_e is called the extinction distance. Thus

$$\Psi(r) = C_0^{(1)}(k_0^{(1)}) \exp 2\pi i k_0^{(1)} r + C_0^{(2)}(k_0^{(2)}) \exp 2\pi i k_0^{(2)} r$$
$$+ C_H^{(1)}(k_0^{(1)}) \exp 2\pi i (k_0^{(1)} + H)r + C_H^{(2)}(k_0^{(2)}) \exp 2\pi i (k_0^{(2)} + H)r \tag{5.19}$$

and $k_0^{(i)}$, $k_H^{(i)}$, $C_0^{(i)}$ and $C_H^{(i)}$ are expressed by (5.14) and (5.17).

The relationships (5.14) determining the exponential factors in $\Psi(r)$ imply that the factor $\exp 2\pi i k r$ is common for all the terms in (5.19). Therefore

$$\Psi(r) = A(r) \exp 2\pi i k r,$$

where $A(r)$ ist the sum of amplitudes of plane waves, i.e.,

$$A(kr) = \frac{\delta + \omega}{2\omega} \exp 2\pi i \Delta k_0^{(1)} r + \frac{\omega - \delta}{\omega} \exp 2\pi i \Delta k_0^{(2)} r$$

$$- \frac{1}{\omega t_e} \exp 2\pi i (\Delta k_0^{(1)} + H)r + \frac{1}{\omega t_e} \exp 2\pi i (\Delta k_0^{(2)} + H)r. \tag{5.20}$$

Thus $\Psi(r)$ for an electron is indeed a superposition of plane waves, although it is more convenient to present it in the form (5.3) corresponding to a superposition of the so-called Bloch waves.

Amplitudes of electron waves with vectors $k_0^{(1)}$ and $k_0^{(2)}$ corresponding to different branches of the dispersion surface are described by

$$A^{(1)}(k_0^{(1)}, r) = \frac{1}{2}\left[\frac{\omega + \delta}{\omega} - \frac{\exp 2\pi i r H}{\omega t_e}\right] \exp 2\pi i (k_0^{(1)} - k)r,$$

$$A^{(2)}(k_0^{(1)}, r) = \frac{1}{2}\left[\frac{\omega - \delta}{\omega} + \frac{\exp 2\pi i r H}{\omega t_e}\right] \exp 2\pi i (k_0^{(2)} - k)r. \tag{5.21}$$

δ may take two values depending on the crystal orientation. If the angle of incidence for the initial beam exceeds the Braggs's angle ϑ_0, then $\delta < 0$; otherwise, $\delta > 0$ (Fig. 1b). Therefore, according to (5.21), $A^{(1)}(r) > A^{(2)}(r)$ for $\delta > 0$, and vice versa.

To find the incident and the diffracted amplitudes at the distance z inside the crystal, one should take into account that, according to boundary conditions, k, $k_0^{(1)}$ and $k_0^{(2)}$ have equal values in a projection on xy, but different projections on z. It can be seen in Fig. 22 that

$$\mathscr{P}_2 \mathscr{D}' = k_{0z}^{(2)} - k_z = \frac{k_0^{(2)} - k}{\cos \vartheta_0} = \frac{\delta + \sqrt{\delta^2 + t_e^{-2}}}{2} = \frac{\delta + \omega}{2} = \Delta k_{0z}^{(2)}.$$

Similarly,

$$\Delta k_{0z}^{(1)} = k_{0z}^{(1)} - k_z = \frac{k_0^{(1)} - k}{\cos \vartheta_0} = \frac{\delta - \omega}{2}.$$

Therefore

$$A_0(z) = \left[\frac{(\delta + \omega)}{2\omega} + \frac{(\omega - \delta)}{2\omega} \exp 2\pi i \omega z \right] \exp 2\pi i \Delta k_{0z}^{(1)} z,$$

(5.22)

$$A_H(z) = [\exp 2\pi i \omega z - 1] \exp (2\pi i \Delta k_0^{(1)} z)/2\omega t_e.$$

The full wave function is then given by

$$\Psi(r) = [A_0(z) + A_H(z) \exp 2\pi i r H] \exp 2\pi i k r.$$

Intensities of the incident and the diffracted beams at the exit surface of a crystal having the thickness t are

$$I_0(\delta, t) = \frac{\omega^2 \cos^2 \pi \omega t + \delta^2 \sin^2 \pi \omega t}{\omega^2} = 1 - \frac{\sin^2 \pi \omega t}{\omega^2 t_e^2},$$

(5.23)

$$I_H(\delta, t) = \frac{\sin^2 \pi \omega t}{\omega^2 t_e^2} = \frac{v_H^2 \sin^2 \pi \omega t}{k^2 \cos^2 \vartheta_0 \omega^2} = \frac{\Phi_H^2 \sin^2 \pi \omega t}{\pi^2 k^2 V_0^2 \cos^2 \vartheta_0 \omega^2},$$

where Φ_H^2 is the structure factor for the diffracted wave and V_0 is the unit cell volume.

Note that the sum of the intensities of the incident and diffracted beams is always equal to unity for any given t (or z) since $|C_0|^2 = 1$, where C_0 is the amplitude of the initial wave.

Let us analyze the expression for the diffracted intensity in greater detail. It can be seen in (5.23) that I_H vary periodically both with t and δ. Consider first the case of $\delta = 0$, implying that the Ewald sphere cuts the reciprocal lattice node h exactly at the center, and $|k_0| = |k_0 + h|$. Then

$$I_0 = \cos^2 \pi(t/t_e), \quad I_H = \sin^2 \pi(t/t_e).$$

(5.24)

It is obvious that the incident beam intensity and the diffracted one oscillate, as they are propagating through the crystal, with the period $t_e = k \cos \vartheta_0 / v_H = \pi V_0 \cos \vartheta_0 / \lambda \Phi_H$.

Using a wedge-like crystal as a model elucidates the pattern in the diffracted and incident intensity distribution. This enables one to observe the dynamics in the variation of electron intensities at the lower surface of a crystal having a continuously varying thickness. Figure 11 shows that $I_0 = 0$ and $I_H = 1$ for the wedge thickness $t_e/2$. This is also the case for all thicknesses that satisfy the condition $t = (n + 1/2) t_e$.

Thus at the exit surface of the wedge there will be straight lines corresponding to $I_0 = 0$ and $I_H = 1$. On the contrary, with $t = nt_e$, I_H will be zero, while I_0

will be unity at the exit surface. For wedge thicknesses $t = [(2n + 1)/4] t_e$, $I_0 = I_H$ at the exit surface.

If the diffracted beam is removed by means of the objective aperture, then the so-called thickness fringes appear in the image of a wedge-like crystal, i.e., the alternation of dark and light fringes parallel to the wedge edge with the period $t_e \operatorname{ctg} \alpha$, where α is the wedge angle. The physical meaning of t_e is thus elucidated. The extinction distance $t_e = \pi V_0 / \lambda \Phi_H$ depends not only on λ and V_0, which are constant irrespective of the reciprocal lattice node involved, but also on the structure amplitude which is determined by the behavior of scattered electrons in the crystal. Therefore, analysis of extinction contours is also of practical interest since the distance $t_e \operatorname{ctg} \alpha$ between the neighboring fringes can be used to evaluate t_e, so that $\Phi(hkl)$ may be then calculated if the unit cell parameters and α are known.

If $t < t_e$, the sine in (5.24) may be replaced by its argument

$$I_H \simeq \pi^2 t^2 / t_e^2 = \lambda^2 (\Phi / V_0)^2 t^2 ,$$

which is similar to (2.58) derived in the kinematical theory. On the other hand, for very thick crystals where $t > t_e$ and $\delta = 0$ the kinematical approximation leads to $I_H / I_0 S > 1$, which is impossible physically. The dynamical treatment of this problem eliminates this contradiction.

If a wedge-like crystal is orientated so that $\delta \neq 0$, the incident and the diffracted waves will still oscillate as they propagate through the crystal. However, the oscillation period will be equal to an effective extinction distance $t_{ef} = t_e (1 + \delta^2 t_e^2)^{-1/2}$. Figure 11 shows that the incident beam intensity does not decrease to zero. Oscillations of I_0 and I_H in the crystal are directly connected with the fact that each of the waves involved is a superposition of two waves having slightly different values of wave vectors.

Assume that a thin plate crystal is uniformly bent to form a semi-cylinder, and Bragg's Law applies exactly ($\delta = 0$) in the plane containing the incident beam and the cylinder axis. It is evident that δ increases with the distance from this plane, so that $\delta > 0$ at one side of the crystal and $\delta < 0$ at the other. To satisfy the two-beam approximation, assume that the left half of the crystal does not contribute to diffraction. The formula (5.23) shows that for constant plate thickness, I_H is a periodic function of δ. If the crystal thickness is $t = p t_e$ where p is a fixed number (not necessarily integer), then

$$I_H = \frac{\sin^2 \pi p (1 + \delta^2 t_e^2)^{1/2}}{1 + \delta^2 t_e^2} . \tag{5.25}$$

If $\delta = 0$, $I_H = \sin^2 \pi p$ and $I_0 = 1 - \sin^2 \pi p$. Extreme values will appear under the following conditions: for I_H maxima

$$p(1 + \delta^2 t_e^2)^{1/2} = \frac{2n+1}{2} , \quad \delta_{max} = \pm \frac{1}{p t_e} \left[\left(\frac{2n+1}{2} \right)^2 - p^2 \right]^{1/2} , \quad n \geqslant \frac{2p-1}{2}$$

for any integer n;

for zero I_H

$$p(1 + \delta^2 t_e^2)^{1/2} = n, \qquad \delta_{\min} = \pm \frac{1}{p t_e}(n^2 - p^2)^{1/2}, \qquad n \geqslant p \qquad (5.25a)$$

for any integer n.

Distances between extreme intensity positions vary, so that with the increase in n the period tends to $1/\delta$, i.e., to the case of kinematical oscillations. A similar result is obtained for small t where p may be neglected. The maximum values of I_H depend on n as

$$I_H^m = 4(t/t_e)^2/(2n + 1)^2. \qquad (5.26)$$

It is of interest that for $n \geqslant 1$ the values of I_H^m coincide exactly with the maxima of the function (2.57) defining the integrated diffracted intensity in terms of the kinematical approximation (Vainshtein 1956).

Figure 23 shows $I_{rel} = I_I/I_0 S$ plotted versus δ for the $\bar{3}31$ reflection from $2M_1$ muscovite ($\Phi = 61.5$; $V_0 = 930$ Å^3 $\lambda = 0.037$ Å; $t_e = 400$ Å). One can clearly see all the main peculiarities in the diffracted intensity distribution in terms of the kinematical and dynamical approximations (I^K and I^D, respectively). For very thin crystals, ($t = 0.5 \cdot t_e = 200$ Å), I^K curves are close to I^D curves (Fig. 23). For relatively thick crystals ($t = 3 \cdot t_e = 1200$ Å) a strong maximum in I^K at $\delta = 0$ corresponds to $I^D = 0$. Although additional maxima in I^K and I^D are shifted with respect to one another, the nearest I^K and I^D maxima are of the same height. If the objective aperture is used to select the strongest diffracted beam from a uniformly bent crystal, then in accordance with (5.25) a series of alternating light and dark fringes will be observed on the dark-field image. These are the so-called rocking curves. They are highly sensitive to the sample orientation, so that they move over the image as the orientation changes. If the distances $\delta_{\min}^{(n-1)}$, $\delta_{\min}^{(n)}$ and $\delta_{\min}^{(n+1)}$ between three successive dark fringes are measured, the values of n, t, and t_e can be found by solving three simultaneous equations of the type (5.25a) (Amelinckx 1964). The value of t_e may be used to evaluate the structure amplitude without direct estimation of the intensity of the corresponding reflection (Goodman and Lehmpfuhl 1967).

A similar effect may be also observed for a parallel-sided crystal by use of the convergent beam diffraction technique, where a cone-shaped convergent electron beam is focused into a point in the middle between the upper and the lower surfaces of the plate. It is obvious that a continuous set of orientations of k_0 is equivalent to a continuous set of crystal orientations with a fixed k_0, as in the case of a uniformly bent crystal. The fine structure of a reflection will then consist of a series of maxima and minima distributed as prescribed by (5.25) and (5.25a). Extinction distances and structure amplitudes may be again determined without direct measurement of reflection intensity.

Experimental study of rocking curves and thickness fringes has shown that the periodicity of fringes is in agreement with the theory, while their number and intensity distribution are not (Hirsh et al. 1977). For example, even for not very thick crystals ($t = 5 t_e$), only about five extinction fringes are observed, although beams with high values of I_0 and I_H pass through twice thicker crystals. In addition, in the case of rocking curves, an asymmetrical distribution of the transmitted intensity is observed on different sides of the central fringe corresponding to

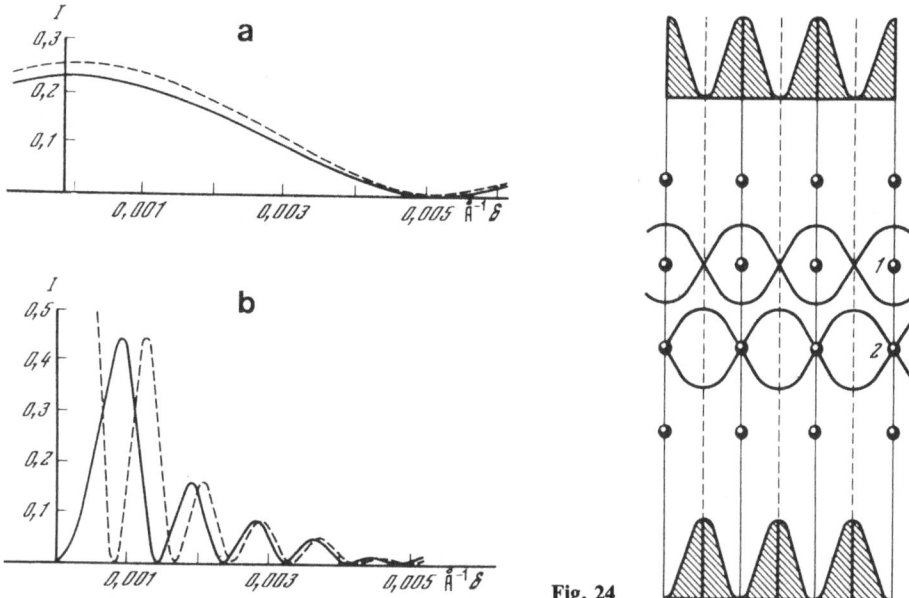

Fig. 24

Fig. 23 a, b. The integrated intensity of the $\bar{3}31$ reflection for 2 M_1 muscovite plotted against δ for different thicknesses t; **a** $t = t_e/2$; **b** $t = 3\ t_e$. *Solid curves* are calculated from the dynamical theory and the *dashed curves* from the kinematical theory

Fig. 24. Differences in propagation of waves with vectors $k\,_0^{(1)}$ and $k\,_0^{(2)}$ along atomic planes in a cubic crystal

$\delta = 0$. For instance, the diffracted intensity for $\delta > 0$ was lower than that for $\delta < 0$.

It has been noted above [see Eq. (5.21)] that the relationship between amplitudes $A^{(1)}$ and $A^{(2)}$ depends on the δ sign. Therefore the processes of propagation of waves with different wave vectors $k\,_0^{(1)}$ and $k\,_0^{(2)}$ in the crystal can be a priori assumed to be not quite identical. We shall discuss this for the case $\delta = 0$ where

$$I^{(1)} = \sin \pi r H, \qquad I^{(2)} = \cos \pi r H. \tag{5.27}$$

The product $r \cdot H_{hkl} = n$, since it is an equation for the scattering planes. In any structure the scattering lattice planes may be brought into coincidence with atomic planes, i.e., they can be assumed to contain scattering atoms arranged in patterns defined by the specific structure. Imagine a structure consisting of "vacant" planes and those occupied by atoms, alternating regularly with a spacing of $d(hkl)/2$. Integer values of $r \cdot H$ will then correspond to atomic planes, while $r \cdot H = (2n + 1)/2$ will correspond to vacant planes. Then, according to (5.27), waves associated with $k\,_0^{(1)}$ would be localized along "vacant" planes, whereas the corresponding intensity would be zero along atomic planes. On the contrary, waves associated with $k\,_0^{(2)}$ would concentrate near atomic planes and have zero intensity between these.

Figure 24 illustrates the propagation of waves corresponding to different wave vectors. In this case the interacting incident and scattered waves have the same wave number. The waves with the wave vector $k_0^{(1)}$ are not affected by the scattering centers, while those with the wave vector $k_0^{(2)}$ should be affected most. It may seem that the strong attenuation of some waves and the anomalous transmission of others explain, at least qualitatively, the observed asymmetrical intensity distribution in extinction contours. It is indeed clear from (5.21) that $A^{(1)}$ exceeds $A^{(2)}$ for $\delta > 0$. Besides, the waves with the vector $k_0^{(1)}$ are located in the minimum potential regions and, therefore, are least affected by scattering centers. Hence, the intensity $|A^{(1)}|^2$ will be greater than $|A^{(2)}|^2$ at the exit surface of the crystal.

For $\delta < 0$, $A^{(2)}$ should be greater than $A^{(1)}$. On the other hand, owing to anomalous absorption, $A^{(2)}$ decreases rapidly as the waves with $k_0^{(2)}$ propagate through the crystal. Thus the resultant transmitted intensity at the exit surface will be low for $\delta < 0$.

However, the situation is much more complicated due to inelastic electron scattering, plasmon oscillation (Heidenreich 1964) and other phenomena that are accompanied by energy losses. These effects are strong for all waves irrespective of their wave vectors, so that the true nature of anomalous absorption is difficult to reveal. Nevertheless, anomalous absorption can be approximately described in terms of a phenomenological theory where an imaginary potential $i\varphi'(r)$ is added to the real $\varphi(r)$ values (Hirsh et al. 1977, Howie and Whelan 1961).

The function $\varphi(r) + i\varphi'(r)$ may be written as a Fourier series with coefficients $v_H + iv_H'$. Instead of $v_H = k \cos \vartheta_0 / t_e$ and $v_0 = k \cos \vartheta_0 / t_0$ we have

$$v_H + iv_H' = k \cos \vartheta_0 (1/t_e + i/t_e'),$$

$$v_0 + iv_0' = k \cos \vartheta_0 (1/t_0 + i/t_0').$$

If $v_H \gg v_H'$ and $v_0 \gg v_0'$ the incident and the diffracted amplitudes will be given by

$$A_0(z) = \left\{ \frac{1}{2} \left(\frac{\omega + \delta}{\omega} \right) \exp\left[-\pi i \left(\omega + \frac{i}{t_e' t_e \omega} \right) z \right] \right.$$

$$\left. + \frac{1}{2} \left(\frac{\omega - \delta}{\omega} \right) \exp\left[\pi i \left(\omega + \frac{i}{t_e' t_e \omega} \right) z \right] \right\} \exp\left[-\frac{\pi z}{t_0'} \right], \qquad (5.28)$$

$$A_H(z) = -\frac{i}{t_e \omega} \left\{ \exp\left[-\pi i \left(\omega + \frac{i}{t_e t_e' \omega} \right) z \right] - \exp \pi i \left(\omega + \frac{i}{t_e t_e' \omega} \right) z \right\}$$

$$\times \exp\left(-\frac{\pi z}{t_0'} \right).$$

The only difference between (5.28) and (5.22) is that additional exponentials appear in (5.28) while $C_0^{(i)}$ and $C_H^{(i)}$ remain the same. The normal absorption which is the same for the transmitted and the diffracted waves is taken into account by the exponential with the absorption factor $\mu = \pi/t_0' = \pi k \cos \vartheta_0 / v_0'$ which is inversely proportional to the imaginary component of the mean internal crystal potential.

The second terms, in both expressions (5.28), contain the exponential $\exp[-\pi(1/t_0' + 1/t_e t_e' \omega)z]$ and decrease with z more rapidly than the first terms containing the exponential $\exp[-\pi(1/t_0' - 1/t_e t_e' \omega)z]$. Therefore, the waves associated with the $\mathbf{k}_0^{(1)}$ would undergo greater energy losses than those corresponding to the other dispersion surface branch. It is also clear from (5.28) that $I_0(\delta) \neq I_0(-\delta)$ and $I_H(\delta) = I_H(-\delta)$. Therefore, the transmitted intensity distribution for $\delta > 0$ should differ from that for $\delta < 0$. Altogether, the introduction of the imaginary component of the crystal potential and, as a consequence, of the absorption distance t_e' allows more adequately for anomalous absorption of electrons.

In trying to estimate t_e and t_e' quantitatively, one should take account of relativistic effects (Fujiwara 1961). The use of high-energy electrons leads to a substantial change in electron mass and wavelength. Since the extinction distance is proportional to the electron velocity, dynamic effects might be expected to be weaker with higher voltage. This problem remains disputable and requires special investigation. Avilov et al. (1973) reported a weakening in dynamical effects with the increase of the voltage from 100 to 400 kV in studying the reflection intensity from polycrystalline Ge and $CuSbS_3$ foils.

5.2 Integrated Diffraction Intensity in Terms of the Two-Beam Dynamical Theory

If the incident electron beam intensity in vacuum is assumed $I_0 = C_0^2$ and the illuminated crystal area is S, then the expression (5.23) for diffraction intensity becomes

$$I_I = I_0 S \frac{\sin^2 A(1 + \delta^2 t_e^2)^{1/2}}{1 + \delta^2 t_e^2}, \tag{5.29}$$

where

$$A = \pi \frac{t}{t_e} = \lambda \frac{\Phi_H}{V_0 \cos \vartheta_0} t.$$

The integrated diffracted intensity from a mosaic single crystal film formed by a set of sufficiently large blocks disoriented according to the law $f(\alpha)$ may be obtained by integrating (5.29) over $d\alpha = d(hkl) d\delta$ (Blackman 1939; Vainshtein 1956) so that

$$I_{\text{rel}} = \frac{I_I}{I_0 S} = \int_{-\infty}^{\infty} \frac{\sin^2 A(1 + \delta^2 t_e^2)^{1/2}}{1 + \delta^2 t_e^2} d\alpha$$

$$\approx \frac{d(hkl)}{t_e} \int_{-\infty}^{\infty} \frac{\sin^2 A(1 + \delta^2 t_e^2)^{1/2}}{1 + \delta^2 t_e^2} d(\delta t_e) = \frac{d(hkl)}{t_e} I(A). \tag{5.30}$$

The function $I(A)$ may be written

$$I(A) = \int_0^A J_0(2x) dx,$$

where $J_0(2x)$ is the Bessel function.

For small A, the Bessel function is unity and $I(A) = A$. Therefore

$$I_{rel} = \lambda^2 \frac{\Phi_H^2}{V_0^2} \frac{d(hkl)}{\pi} t, \tag{5.31}$$

which is equivalent to the expression (2.60) in the kinematical theory. For very large A when $t \gg t_e$, $I(A) = 1/2$ and

$$I_{rel} = \frac{d(hkl)}{2t_e} = \frac{\lambda \Phi_H}{V_0} \frac{d(hkl)}{2\pi}. \tag{5.32}$$

Thus the dynamical integrated diffraction intensity is proportional not to the structure factor, but to the structure amplitude, i.e., Φ_H.

Blackman (1939) has calculated the variation of $I(A)$ with A for sufficiently small A. He suggested to use $I(A)$ for introducing dynamical corrections for observed intensities when these are used to evaluate the structure amplitudes. Since the two-beam dynamical theory involves all the main feature of the kinematical theory, one might suppose that the interaction between electrons and matter in thin films will be of an intermediate nature. Therefore Blackman proposed to make use of the two-beam dynamical theory for interpreting the experimental data.

It is obvious that this theory in the form described above has a number of serious drawbacks and, in fact, lacks strict theoretical grounds. Dealing with only two waves propagating through the crystal is indeed unrealistic; it is not appropriate to treat reciprocal lattice nodes as points when the dimensions of the object are finite, etc. Paradoxical as it is, it is the two-beam approximation that has been up to now used most successfully in electron crystallography (Imamov 1977).

A convenient method to apply Blackman corrections has been suggested by Vainshtein and Lobachev (1961). Using (5.29), the formula (5.30) may be modified as

$$I^{\mathscr{D}} = I^K (I(A)/A) = I^K \mathscr{D}(A), \tag{5.33}$$

where I^K defines, according to (5.31), the relative kinematical integrated intensity for a reflection hkl and

$$\mathscr{D}(A) = \frac{1}{A} \int_0^A J_0(2x) \, dx.$$

It may be assumed that

$$\left(\frac{I^{\mathscr{D}}}{I^K}\right)^{1/2} = \frac{|\Phi(hkl)|^{\mathscr{D}}}{|\Phi(hkl)|^K} = \sqrt{\mathscr{D}(A)} = K(A). \tag{5.34}$$

The functions $\mathscr{D}(A)$ and $K(A)$ are sketched in Fig. 25. In practice, they are used as follows. The strongest reflections concentrated in the region where $\sin \vartheta / \lambda < 0.3$ are most affected by dynamical effects. Therefore the relative values $|\Phi|$ are calculated for these reflections on the basis of measured intensities with account of the relationship between $I_{exp}^{\mathscr{D}}$ and $|\Phi|_{exp}^{\mathscr{D}}$ given by $I \sim \Phi^2$ or

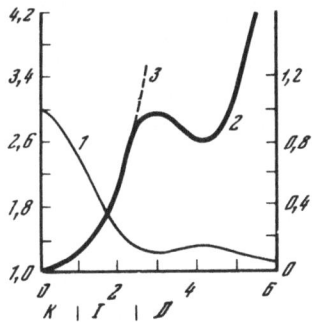

Fig. 25. The functions $\mathscr{D}(A)$ and $K(A)$ (*1* and *2*) used for dynamical corrections; *3* is the part of $K(A)$ that can be approximated by an exponential; K, D, and I are the regions where the interaction of electrons with matter is of kinematical, dynamical or intermediate nature (Vainshtein and Lobachev 1961)

$I \sim \Phi^2/d(hkl)$. Since a preliminary structure model is already established, the values $\Phi^K_{(hkl)}$ are calculated. For each strong reflection the ratio $|\Phi(hkl)|^{\mathscr{D}}_{\text{exp}}/|\Phi(hkl)|^K_{\text{cal.}} = K(A)$ is found. Using the curve $K(A)$, the corresponding values of $A(hkl)$ are determined for each K, and the crystal thickness $t = A(hkl)\,V_0/\lambda\,\Phi(hkl)$ is evaluated. The mean crystal thickness in the sample t_{mean} is obtained by averaging the set of t values. Then the $A(hkl)$ values are calculated for all the reflections observed, the $K(A)$ set is determined and the values $|\Phi(hkl)|^K_{\text{exp}}$ are found according to (5.34).

5.3 Criteria for the Range of Validity of the Kinematical Approximation

Despite the obvious limitations, the two-beam dynamical theory reveals the general features of the interaction between electron waves and crystals. Specifically, analysis of these features produces the conditions for the validity of the kinematical approximation. At the same time, situations implying high probability for dynamical effects become also clear.

Putting $\omega = \delta_{ef}$ we may rewrite (5.23) and (2.57) as

$$I^{\mathscr{D}} = (\lambda\,\Phi_H/V_0)^2 \sin \pi\delta_{ef}t/(\pi\delta_{ef})^2 , \tag{5.35}$$

$$I^K = (\lambda\,\Phi_H/V_0)^2 \sin^2 \pi\delta t/(\pi\delta)^2 . \tag{5.36}$$

Equations (5.35) and (5.36) are similar in appearance. They become identical for $\delta = \delta_{ef} = (\delta^2 + t_e^{-2})^{1/2}$. This is possible when

$$1/t_e = \lambda\,\Phi_H/\pi V_0 \to 0 . \tag{5.37}$$

In the limiting case the Eq. (5.37) is equivalent to the requirement of zero intensity for all reflections. It is obvious that (5.35) and (5.36) would lead to close results for small $1/t_e$ values. It follows from (5.37) that the kinematical approximation is favored by small structure amplitudes, high accelerating voltage and large unit cell dimensions. To clarify the range of validity of the kinematical theory, let us consider the Eq. (5.35) with $\delta = 0$. The minimum crystal thickness t_c, implying equal intensities for the incident and diffracted waves at the exit surface, may be chosen as a parameter defining the boundary between the kinematical and the dynamical scattering.

$$\pi(t_c/t_e) = \pi/4 , \qquad t_c = \pi V_0/4\lambda \Phi_H = t_e/4 . \tag{5.38}$$

On the other hand, the condition $I_I = I_0 S$ may be also assumed as a boundary between the kinematical theory and the dynamical theory. Using (2.58) we have

$$(\lambda \Phi_H/V_0) t_c = 1 , \qquad t_c = V_0/\lambda \Phi_H . \tag{5.39}$$

In both cases, close results are obtained.

This problem has been treated in detail by Vainshtein (1956). It is seen from (5.38) that the critical crystal thickness is inversely proportional to the Φ_H values which are directly connected with the factors f. Since the values of f-curves generally increase with the number in the periodic Table, Z, compounds with light elements are the most favorable objects for electron crystallography. The decrease in f with $\sin \vartheta/\lambda$ for any z leads to the fact that the strongest reflections are concentrated in the region of small scattering angles irrespective of the object structure. Therefore a set of different Φ_H values is present in a diffraction pattern from any object and, consequently, the conditions (5.38) – (5.39) may be satisfied to a different extent for different reflections for the given plate thickness. Intensities of the strongest reflections may be affected by dynamical effects, whereas the kinematical approximation may be valid for the rest of reflections. Thus it may appear necessary to use different formulae to proceed to structure amplitudes from intensities for different reflections and, specifically, to apply Blackman corrections.

On the whole, SAED structure analysis is difficult in the case of simple structures having small unit cells and containing heavy atoms. The situation is reversed for complex structures containing a relatively large number of atoms in large unit cells.

Since it is crystals of this kind that generally serve as objects for electron diffraction, a serious problem is to prepare crystallites with the minimum possible thickness which should, in all cases, be less then extinction distances for the strongest reflections.

CHAPTER 6

Dynamical n-Beam Scattering of Electrons

6.1 The "Physical Optics" Approach

Limitations of the two-beam dynamical theory become obvious if we take into account that, under experimental conditions, the object is of finite dimensions that are very small at least in one direction. Thus dozens and even hundreds of diffracted beams, including very strong ones, may appear simultaneously. The interaction of all the scattered waves should lead to new effects that cannot be predicted in terms of the two-beam theory. Therefore, since the classical work by Bethe (1928), there have been numerous attempts to find a more general solution allowing for multiple interaction of diffracted beams and finite dimensions of the object, as well as for various energy losses, etc.

There are several approaches to the n-beam scattering problem (Allpress et al. 1972; Cowley 1967, 1975a, b, 1978; Cowley and Iijima 1972, 1976; Cowley and Moody 1957, 1960; Cowley and Pogany 1968; Cowley et al. 1979; Fujimoto 1959; Fujiwara 1961; Howie and Whelan 1961; Sturkey 1957, 1962; Vainshtein 1956). Some of them are based on quantum mechanics, others proceed from wave optics. The former methods use the Schrödinger equation as the starting point. It would be natural to solve, as in the case of the two-beam approximation, the simultaneous dispersion Eqs. (5.4) involving scattering amplitudes, Fourier coefficients for the potential $\varphi(r)$, and electron energy, i.e., all the parameters that determine the process of scattering.

However, a general solution for this set of equations has not been found so far. The two-beam approximation is, in fact, the only exception. Nevertheless, Howie and Whelan (1961) have shown that the simultaneous dispersion Eqs. (5.4) may be solved using matrix formulation with the help of numerical computing methods. To solve the Schrödinger equation, Sturkey (1957, 1962) has introduced the scattering matrix defining the changes in amplitude for diffracted beams transmitted through a thin parallel-sided crystal. He has shown that raising a scattering matrix to successively increasing powers with the help of numerical methods one can find the intensity distribution for diffracted beams at the exit surface for any crystal thickness. Fujiwara (1961) has obtained a general solution by the successive Born approximations. Cowley and Moodie (1960) have developed another approach to the n-beam diffraction problem. Basing themselves on the physical optics, they succeeded in obtaining an analytical solution.

Since the quantum-mechanical approach based on the Schrödinger equation has been already considered in the discussion of the two-beam approximation, the *n*-beam electron scattering will be treated in terms of the physical optics based on the results obtained by Cowley and Moodie (1960), Cowley (1975b).

In the formulation of this theory, a plate crystal of the thickness t is represented by a set of N planes separated by distances Δz. The nth plane at z_n is described by a two-dimensional modulated potential distribution, which is a planar section of $\varphi(x, y, z_n)$. This representation is possible if Δz is chosen so small that the potential distribution is constant within the interval $z_n + \Delta z$ in a slice, so that a slice can be replaced by a plane. Then it may be assumed that there is vacuum between the neighboring planes. The problem is thus reduced to analysis of changes in amplitude and phase for the electron wave on transmission through a set of scattering planes chosen as described.

For fast electrons having a very small wavelength, the scattering process can be treated in terms of the small-angle approximation. Thus the calculations are simplified, since, with small angles of incidence, reflected waves may be ignored and boundary conditions may be imposed independently for each surface of the thin crystal. Moreover, for small ϑ angles, the propagation of waves between closely spaced scattering planes is adequately described by the Fresnel diffraction theory.

Let us first consider the changes in a monochromatic coherent electron wave passing through a thin slice with a mean internal potential φ_0. Assume that a plane wave corresponding to a wave vector k which is equal in vacuum to

$$k = \left(\frac{2\, me}{h^2} V \right)^{1/2}$$

is incident on a slice of the thickness t (V is the accelerating voltage).

The angle between the incident beam and the specimen surface is assumed to be 90° for convenience. As the reflected waves may be ignored, the boundary conditions are relaxed, the only requirement being that of continuity for the incident and the transmitted waves at the interface. Since the mean inner crystal potential φ_0 leads to a change in the wavelength, the wave vector k' for the transmitted wave is given by

$$k' = \left[\frac{2\, me(V + \varphi_0)}{h^2} \right]^{1/2} = \left[\frac{2\, me V}{h^2} \left(1 + \frac{\varphi_0}{V} \right) \right]^{1/2} = k \left(1 + \frac{\varphi_0}{2V} \right). \qquad (6.1)$$

The formula (6.1) is derived under the assumption that $\varphi_0 \ll V$.

Outside the crystal the wave vector becomes again k. Then the wave function in the crystal is given by

$$\Psi_k = C \exp 2\pi i (kR + k'z) = C \exp 2\pi i k \left(R + z + \frac{\varphi_0}{2V} z \right),$$

where R is the distance from the source of the initial wave to the external surface of the sample, and z varies from 0 to t. At $z = 0$, the requirement of continuity for the incident and the transmitted waves at the entrance face of the crystal is satisfied. The wave function at the exit surface is described by the equation

$$\Psi'_k = C \exp 2\pi i k \left(R + t + \frac{\varphi_0}{2V} t \right), \qquad (6.2)$$

and at the distance R' from the exit surface it is given by

$$\Psi' = C_1 \exp 2 \pi i k (R + t + R') .$$

According to the boundary conditions for $R' = 0$

$$\Psi' = \Psi'_k ,$$

$$C_1 \exp 2 \pi i k (R + t) = C \exp 2 \pi i k \left(R + t + \frac{\varphi_0}{2 V} t \right) .$$

Hence

$$C_1 = C \exp 2 \pi i k \frac{\varphi_0}{2 V} t = C \exp i \sigma \varphi_0 t , \tag{6.3}$$

where σ is the electron interaction constant for the given accelerating voltage and is expressed in terms of the relativistic wavelength λ:

$$\sigma = \frac{\pi}{V\lambda} \frac{2}{1 + \left[1 - \dfrac{u^2}{c^2} \right]^{1/2}} . \tag{6.4}$$

Equation (6.3) implies that the amplitude of the incident wave does not change on transmission through a thin slice having the potential φ_0. However, the transmitted wave acquires an additional phase $\sigma \varphi_0 \Delta z$ (with respect to a wave in vacuum).

A two-dimensional object causing a phase change in the transmitted wave without a change in amplitude is called a *phase-object*. In terms of physical optics the phase change depends on the thickness and the refractive index of the specimen. In the case of fast electrons, the phase-object is a very thin plate-shaped crystal of the thickness Δz, which is described by a two-dimensional potential distribution $\varphi(xy)$, i.e., by the projection of the potential on the plate surface

$$\varphi(xy) = \int_0^{\Delta z} \varphi(xyz) dz .$$

Therefore, the constant value φ_0 in (6.3) should be replaced by $\varphi(xy)$. Then the amplitude of the transmitted wave will also be the function of the coordinates xy of the phase-object, so that

$$C_1(xy) = C \exp i \sigma \varphi(xy) \Delta z .$$

This substitution is justified since each potential value averaged over the thickness of a thin crystal for given x, y may be regarded as the mean inner potential at the point x, y. In other words, the potential distribution $\varphi(xy)$ modulates the corresponding distribution of wave-vector values in the phase-object plane.

Assume that a plane coherent incident wave has unity amplitude and zero phase at the entrance surface at $z = 0$. Then the amplitude distribution of the transmitted wave is given by

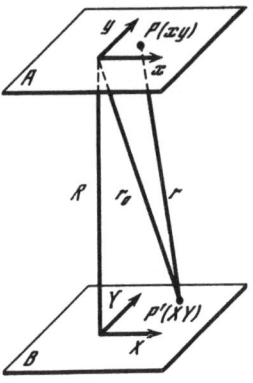

Fig. 26. The wave propagation between the upper surface of one object and the lower surface of the other

$$q(xy) = \exp i\sigma\varphi(xy)\Delta z . \tag{6.5}$$

The function $q(xy)$ is called the *transmission function*.

Let us now consider changes in wave functions on propagation in vacuum from the lower surface of one phase-object to the upper surface of the other. Figure 26 shows two planes, A and B, separated by the distance R. The plane A is a set of point sources of spherical waves. The positions of these scattering sources are given by the coordinates x, y, and the corresponding scattering power is described by the transmission function $q(xy)$. The amplitude of a spherical wave which has propagated from the point x, y in the plane A to the point X, Y in the plane B is given by

$$q(xy)\frac{\exp(2\pi i k r)}{r}\frac{\exp\frac{1}{2}\pi i}{\lambda} , \tag{6.6}$$

where r is the distance between the two points involved.

The second factor in (6.6) allows for the phase difference of $\pi/2$ between the scattered and the initial wave. Since in the small-angle approximation r is close to R we have

$$r = [R^2 + (X-x)^2 + (Y-y)^2]^{1/2} \approx R + \frac{(X-x)^2 + (Y-y)^2}{2R} . \tag{6.7}$$

The resultant amplitude at the point XY is given by superposition of waves of the type (6.6), since waves from all the sources in the plane A contribute to it. Replacing summation by integration over all possible x, y we obtain

$$\Psi(XY) = \frac{i\exp(2\pi i k R)}{\lambda R}\int\int q(xy)$$

$$\times \exp\left[2\pi i k\frac{(X-x)^2 + (Y-y)^2}{2R}\right]dx\,dy . \tag{6.8}$$

The expression (6.8) describes the small-angle Fresnel diffraction where $(x^2 + y^2)^{1/2}$ is of the same order of magnitude as R. According to (4.16) Eq. (6.8) for $\Psi(XY)$ is a convolution of the transmission function $q(x, y)$ with the *prop-*

agation function $p(xy) = (i/\lambda R) \exp[-\pi i k(x^2 + y^2)/R]$. The latter function takes into account the phase difference for interfering waves determined by the path-length difference for beams propagating into the point of observation from different scattering sources.

Thus the amplitude of the spherical wave at the entrance surface of the second phase-object may be written as

$$\Psi(XY) = q_1(xy) * p(xy)$$
$$= \iint q_1(xy) \exp\left[2\pi i k \frac{(X-x)^2 + (Y-y)^2}{2R}\right] dx\,dy. \tag{6.9}$$

Multiplying (6.9) by the exponential $\exp \sigma \varphi_2(xy) \Delta z$, which allows for the phase change on transmission through a phase-object with the potential $\varphi_2(xy)$, we obtain the amplitude distribution for the wave at the exit surface of the second thin crystal plate, i.e.,

$$\Psi_2(XY) = \exp i\sigma \varphi_2(xy) \Delta z[q_1(xy) * p_1(xy)]$$
$$= q_2(xy) \cdot [q_1(xy) * p_1(xy)]. \tag{6.10}$$

To obtain the wave function at the entrance surface of the third plate separated from the second one by the distance R_2, we should take the convolution of (6.10) with the propagation function

$$p_2 = \frac{i}{\lambda R_2} \exp\left[-2\pi i k \frac{x^2 + y^2}{2R_2}\right],$$

so that

$$q_2(xy)[q_1(xy) * p_1(xy)] * p_2(xy).$$

Similarly, the amplitude distribution at the nth phase-object may be expressed in terms of the transmission functions and the propagation functions describing the preceding $n-1$ slices.

Considering again electron scattering in a plate-shaped crystal of a finite thickness t, let us divide the crystal into $N-1$ thin slices of the thickness Δz and replace the slices by a system of equidistant planes. The planes are numbered 1 to N and the z_n coordinate of the nth slice measured from the entrance face of the crystal along the normal is $z_n = (n-1)\Delta z$. The potential distribution in the nth plane is defined by the function $\varphi(xyz_n)$ with a fixed z_n. Therefore, the transmission function for the nth plane is given by

$$q_n(xy) = \exp(i\sigma \varphi(xyz_n)\Delta z). \tag{6.11}$$

If Δz is infinitesimal, (6.11) becomes

$$q_n(xy) = 1 + i\sigma \varphi(xyz_n)\Delta z. \tag{6.12}$$

The phase distribution within each plane for different values of x, y determined by the propagation of spherical waves from one plane to another is taken into account by the propagation function $p_n(xy)$. Since the planes are equidistant, all the functions $p_n(xy)$ are equal, in the small-angle approximation, to

$$p(xy) = \frac{ik}{\Delta z} \exp -2\pi ik \frac{x^2 + y^2}{2\Delta z}. \tag{6.13}$$

Expressing the amplitude distribution in a plane by that in the preceding one, taking account of $p(xy)$, the amplitudes at the exit surface of the crystal may be readily found (Cowley and Moodie 1960):

$$\Psi(xy) = q_N(xy)[\underset{N}{\ldots}[q_2(xy)\underset{2}{[q_1(xy)}\underset{1}{[q_0(xy)} * p_0(xy)] * \underset{1}{p_1(xy)}]\underset{2}{\ldots}]\underset{N}{]}$$

$$* \, p_N(xy). \tag{6.14}$$

Where $q_n(x)$ multiplied by the bracket numbered n represents the wave leaving the slice n. Note that $p_N(xy)$ is the function of propagation from the exit surface of the crystal to the point of observation. If we confine our attention to the amplitude distribution at the exit face of a crystal, $p_N(xy)$ may be neglected.

We are interested mainly in the diffraction pattern which it is convenient to describe by the amplitude distribution for diffracted beams in terms of the reciprocal space coordinates. For this purpose, the function $\Psi(xy)$ in (6.14) should be Fourier transformed. The Fourier transform of $q_n(xy)$ is given by

$$Q_n(x^*y^*) = \int\int q_n(xy) \exp[2\pi i(xx^* + yy^*)] \, dx \, dy$$

$$= \int\int [1 + i\sigma\varphi(xyz_n)\Delta z] \exp[2\pi i(xx^* + yy^*)] \, dx \, dy$$

$$= \delta(x^*y^*) + i\sigma \Phi_n(x^*y^*) \Delta z, \tag{6.15}$$

where

$$\Phi_n(x^*y^*) = \int\int \varphi(xyz_n) \exp[2\pi i(xx^* + yy^*)] \, dx \, dy. \tag{6.16}$$

Since

$$\varphi(xyz_n) = \int\int\int \Phi(x^*y^*z^*)$$

$$\times \exp[-2\pi i(xx^* + yy^* + z_n z^*)] \, dx^* \, dy^* \, dz^*. \tag{6.17}$$

Substituting (6.17) into (6.16) we have

$$\Phi_n(x^*y^*) = \int \Phi(x^*y^*z^*) \exp[-2\pi iz_n z^*] \, dz^*. \tag{6.18}$$

Thus the coefficients $\Phi_n(x^*y^*)$ for the slice n depend on the function $\Phi(x^*y^*z^*)$, which is the Fourier transform of $\varphi(xyz)$. Note that at the origin

$$Q_n(00) = 1 + i\sigma \Phi_n(00) \Delta z, \tag{6.19}$$

but for other values of x^*, y^*

$$Q_n(x^*y^*) = i\sigma \Phi_n(x^*y^*) \Delta z.$$

The Fourier transform of $p(xy)$ gives the corresponding propagation function $\mathscr{P}(x^*y^*)$ in reciprocal space, that is

$$\mathscr{P}(x^*y^*) = \int\int p(xy) \exp[2\pi i(xx^* + yy^*)] \, dx \, dy$$

$$= \frac{ik}{\Delta z} \exp \frac{\pi i\Delta z}{k}(x^{*2} + y^{*2})$$

$$\times \iint \exp\left[(Ax + Bx^*)^2 + (Ay + By^*)^2\right] dx\, dy$$

$$= \exp\frac{\pi i \Delta z}{k}(x^{*2} + y^{*2}), \tag{6.20}$$

where $A = -i\pi k/\Delta z$, $B = \pi \Delta z/ik$ and the double integral is equal to $\Delta z/ik$.
  The Fourier transform of $\Psi(xyz)$ will be hereafter denoted as $\mathcal{F}$, so that

$$\Phi(x^* y^* z^*) = \mathcal{F}\{\Psi(xyz)\}$$

$$= \iiint \Psi(xyz) \exp\left[2\pi i(xx^* + yy^* + zz^*)\right] dx\, dy\, dz.$$

The Fourier transform of the product of two functions, according to (4.18), is

$$\mathcal{F}\{q_n(xy)p_n(xy)\} = \mathcal{F}\{q_n(xy)\} * \mathcal{F}\{p_n(xy)\} = Q_n(x^* y^*) * \mathcal{P}(x^* y^*).$$

On the other hand, according to (4.17)

$$\mathcal{F}\{q_n(xy) * p_n(xy)\} = \mathcal{F}\{q_n(xy)\}\,\mathcal{F}\{p_n(xy)\} = Q_n(x^* y^*)\,\mathcal{P}_n(x^* y^*).$$

According to (6.14), the amplitude of a wave having reached the second plane is
given by

$$q_1(xy)[\underset{1}{q_0(xy)} * \underset{1}{p_0(xy)}]. \tag{6.21}$$

  The Fourier transform of (6.21) gives the amplitude distribution for diffracted waves leaving the first plane. Applying consequently two properties of the
convolution mentioned we have

$$\mathcal{F}\{q_1(q_0 * p_0)\} = \mathcal{F}(q_1) * \mathcal{F}(q_0 * p_0) = Q_1(x^* y^*) * Q_0(x^* y^*)\,\mathcal{P}_0(x^* y^*).$$

Similarly,

$$\mathcal{F}\{q_2(q_1(q_0 * p_0) * p_1)\} = \mathcal{F}(q_2) * \mathcal{F}\{q_1(q_0 * p_0 * p_1)\}$$

$$= \mathcal{F}(q_2) * \mathcal{F}(q_1) * \mathcal{F}(q_0 * p_0 * p_1)$$

$$= Q_2(x^* y^*) * Q_1(x^* y^*) * Q_0(x^* y^*)\,\mathcal{F}(p_0 * p_1)$$

$$= [\underset{2}{Q_2(x^* y^*)} * [\underset{1}{Q_1(x^* y^*)} * \underset{1}{Q_0(x^* y^*)\,\mathcal{P}_0(x^* y^*)}]$$

$$\times \underset{2}{\mathcal{P}_1(x^* y^*)}].$$

  Thus the general reciprocal space expression for amplitudes of waves scattered by a plate-shaped object of the thickness t is (Cowley 1975b)

$$\Psi(x^* y^*) = [\underset{N}{Q_N(x^* y^*)} * \ldots [\underset{2}{Q_2(x^* y^*)} * [\underset{1}{Q_1(x^* y^*)} * \underset{1}{Q_0(x^* y^*)\,\mathcal{P}_0(x^* y^*)}]$$

$$\times \underset{2}{\mathcal{P}_1(x^* y^*)}]\ldots]\underset{N}{\mathcal{P}_N(x^* y^*)}. \tag{6.22}$$

Within each bracket n, as in (6.14), multiplication precedes the operation of convolution.
  The expression (6.22) is universal in respect of its being applicable to an object with an arbitrary distribution of scattering matter on condition that the func-

tion $\varphi(xyz)$ is already known. Let us now take account of the lattice state of crystals. Each plane n in the crystal may be described by the section of the electrostatic potential $\varphi(xyz_n)$ which is given by the Fourier series as

$$\varphi(xyz_n) = \sum_h \sum_k \sum_l \Phi(hkl) \exp\left[-2\pi i\left(h\frac{x}{a} + k\frac{y}{b} + l\frac{z_n}{c}\right)\right]. \tag{6.23}$$

Since z_n is specified, $\varphi(xyz_n)$ may be written as a double Fourier series:

$$\varphi(xyz_n) = \sum_h \sum_k \Phi_n(hk) \exp\left[-2\pi i\left(h\frac{x}{a} + \frac{y}{b}\right)\right]. \tag{6.24}$$

The Fourier coefficients

$$\Phi_n(hk) = \sum_l \Phi(hkl) \exp[-2\pi i l z_n/c] \tag{6.25}$$

may be calculated *a priori*.

Equation (6.24) implies that the potential distribution in each plane n may be described by a two-dimensional lattice function. All the lattice points are of the same weight defined by the potential distribution within one plane unit cell. The Fourier transform of the lattice function in the direct space corresponds to the lattice function in the reciprocal space and the weights of the reciprocal lattice points are equal to the structure amplitudes. Thus

$$\mathscr{F}\{\varphi(xyz_n)\} = \sum_h \sum_k \Phi_n(hk)\,\delta\left(x^* - \frac{h}{a}, y^* - \frac{k}{b}\right).$$

Then the Fourier transform of $q_n(xy)$ which is equal to $1 + i\sigma\varphi(xyz_n)\,\Delta z$ for $\Delta z \to 0$ gives

$$Q_n(x^*y^*) = \mathscr{F}(q_n) = \delta(x^*y^*) + i\sigma\Delta z \sum_h \sum_k \Phi(hk)\,\delta\left(x^* - \frac{h}{a}, y^* - \frac{k}{b}\right)$$

$$= \delta(x^*y^*) + \sum_h \sum_{k=1} Q_n(hk)\,\delta\left(x^* - \frac{h}{a}, y^* - \frac{k}{b}\right)$$

$$= \sum_h \sum_{k=0} Q_n(hk)\,\delta\left(x^* - \frac{h}{a}, y^* - \frac{k}{b}\right). \tag{6.26}$$

For $x^* = y^* = h = k = 0$ Q_n takes the maximum value

$$Q_n(00) = 1 + i\sigma\Delta z\,\Phi_n(00) = -i\Delta z\left[\frac{i}{\Delta z} - \sigma\Phi_n(00)\right].$$

Whereas for $h, k \neq 0$

$$Q_n(hk) = i\sigma\Delta z\,\Phi_n(hk).$$

To simplify the notation we shall hereafter include the constant value σ into $\Phi_n(hk)$ by putting $\Phi_n(hk) = \sigma\Phi_n(hk)$. Assuming $\Phi_n(00) = i/\Delta z - \sigma\Phi_n(00)$ we have $Q_n(00) = -i\Delta z\,\Phi_n(00)$. Substituting (6.20) and (6.26) into (6.22), the amplitude $\Psi(hk)$ of a beam diffracted into the direction given by the indices hk may be found.

For convenience, let us calculate first the one-dimensional function $\Psi(h)$ (Cowley 1975b). If the initial wave is incident normal to the object surface, and has unity amplitude and zero phase when reaching the first plane, then the expression in brackets 1 in (6.22) corresponds to the amplitude distribution for waves diffracted by the first plane and will be given by

$$Q_1(x^*) = i\Delta z \sum_{h_1=0}^{h} \Phi_1(h_1)\,\delta(x^* - h_1/a)\,.$$

The expression in the brackets 2 in (6.22) defining the amplitudes scattered by the first two planes is

$$\left[Q_1(x^*) * Q_2(x^*)\exp 2\pi i\frac{\Delta z}{2k}x^{*2}\right]_2 = (i\Delta z)^2 \sum_{h_1}\sum_{h_2} \Phi_1(h_1)\,\Phi_2(h_2)$$

$$\times \int\delta\left(x^* - x^{*\prime} - \frac{h_2}{a}\right)\delta\left(x^{*\prime} - \frac{h_1}{a}\right)$$

$$\times \exp\left[\pi i\frac{\Delta z}{k}x^{*\prime 2}\right]dx^*\,.$$

The integrand will become zero for all $x^{*\prime}$ except $x^{*\prime} = x^* - h_2/a$.
Therefore

$$\left[\dots\right]_2^2 = (i\Delta z)^2 \sum_{h_1}\sum_{h_2} \Phi_1(h_1)\Phi_2(h_2)$$

$$\times \exp\left[\pi i\frac{\Delta z}{k}\left(x^* - \frac{h_2}{a}\right)^2\right]\delta(x^* - h_1/a - h_2/a)\,. \tag{6.27}$$

The expression (6.27) represents the effect of double diffraction from two successive planes, or, to be more exact, from lattice rows, as we are considering the one-dimensional case. A linear set of beams with indices h_1 and the weight $\Phi_1(h_1)$ corresponds to single diffraction. For double diffraction one should assume that the structure amplitude with the index h_1 may appear to be the initial one generating a diffracted beam with the index h_2. If the treatment is limited to diffraction from two successive planes (lattice rows), the direction of doubly diffracted beams will be given by $x^* = h/a = h_1/a + h_2/a$, as implied by (6.27). Therefore (6.27) becomes

$$(i\Delta z)^2 \sum_{h_1}\sum_{h_2} \Phi_1(h_1)\,\Phi_2(h_2)\exp \pi i\frac{\Delta z}{k}\frac{h_1^2}{a^2}\,. \tag{6.28}$$

Let us choose a certain diffraction direction with a fixed index $h = h_1 + h_2$. It is obvious that $h_1 = 0, 1, 2 \dots h$ corresponds to $h_2 = h, h-1, h-2, \dots 0$. It is assumed the structure amplitude of the lattice point h_1 is initial with respect to the lattice point $h_2 = h - h_1$. Moreover, one has to assume that a phase factor $\exp \pi i(\Delta z/k)(h_1/a)^2$ is added to the structure amplitude of the lattice point h_1. Since the amplitude of secondary waves is proportional to that of the incident ones, the result of the double diffraction in the direction $h = h_1 + h_2$ for the given h_1 and h_2 is defined by

$$\Phi_1(h_1) \exp \pi i \frac{\Delta z}{k} (h_1/a)^2 \Phi_2(h-h_1) . \tag{6.29}$$

All the possible contributions of the structure amplitudes to diffraction in the direction h are obtained by summing (6.29) over h_1 from 0 to h, i.e.,

$$\sum_{h_1=0}^{h} \Phi_1(h_1) \Phi_2(h-h_1) \exp \pi i \frac{\Delta z}{k}(h_1/a)^2$$

which is equivalent to (6.28). Here the phase factor depends only on h_1 although the diffraction direction is specified by the sum $h_1 + h_2$.

The amplitudes of waves diffracted by three successive planes (or lattice rows in the one-dimensional case) are given by

$$[\ldots]_3 = (i\Delta z)^3 \exp\left[\pi i \frac{\Delta z}{k} x^{*2}\right] \sum_{h_1} \sum_{h_2} \sum_{h_3} \Phi_1(h_1)\,\Phi_2(h_2)\,\Phi_3(h_3)$$

$$\times \int \delta(x^* - x^{*\prime} - h_3/a)\,\delta(x^{*\prime} - (h_1 + h_2)/a) \exp \pi i \frac{\Delta z}{k}(x^{*\prime})^2$$

$$\times \exp\left[\pi i \frac{\Delta z}{k}(x^{*\prime} - (h_2/a))\right] dx^{*\prime} .$$

The integrand is zero for all $x^{*\prime}$ except $x^{*\prime} = x^* + h_3/a$. Therefore

$$[\ldots]_3 = (i\Delta z)^3 \exp \pi i \frac{z_3}{k} x^{*2} \sum_{h_1} \sum_{h_2} \sum_{h_3} \Phi_1(h_1)\,\Phi_2(h_2)\,\Phi_3(h_3)$$

$$\times \exp\left\{2\pi i \frac{\Delta z}{k}[(h_3^2/a^2 + h_2 h_3/a^2 + h_2^2/2\,a^2) - x^*(h_2/a + 2 h_3/a)]\right\}$$

$$\times \delta(x^* - (h_1 + h_2 + h_3)/a)$$

$$= (i\Delta z)^3 \exp\left[\pi i \frac{z_3}{k} x^{*2}\right] \sum_{h_1} \sum_{h_2} \sum_{h_3} \Phi_1(h_1)\,\Phi_2(h_2)\,\Phi_3(h_3)$$

$$\times \exp\left[\frac{2\pi i}{k} \sum_{n=1}^{3} z_n(-h_n x^*/a + h_n^2/2\,a^2 + \sum_{m=n+1}^{3} h_n h_m/a^2)\right]$$

$$\times \delta(x^* - (h_1 + h_2 + h_3)/a) . \tag{6.30}$$

Let us treat triple diffraction as consequent transmission through lattice planes (or rows). Directions for nonzero structure amplitudes are given by $x^* = h/a = h_1/a + h_2/a + h_3/a$. Let us fix a diffraction direction with a certain h. It may be assumed that definite structure amplitudes, h_1 and h_2, generate the diffracted beam with the index $h_3 = h - h_1 - h_2$. It should be noted that the contribution of the lattice point h_2 depends both on the structure amplitude $\Phi_2(h_2)$ and the phase factor depending on h_2, as well as on h_1. Similarly, the contribution of the lattice point h_3 is also defined not only by $\Phi_3(h_3)$ but also by the phase factor depending on h_1, h_2, and h_3.

Thus, while the contribution of a plane n to the resultant diffracted amplitude is determined by the value $\Phi_n(h_n)$ for the given plane, the phase relationships depend on the whole scattering process preceding the diffraction from the plane involved. If, for the given h, the value of h_1 is fixed, the index h_2 may go from 0 to $h - h_1$. The corresponding contribution to diffraction in the direction h is given by

$$\Phi_1(h_1) \sum_{h_2=0}^{h_2=h-h_1} \Phi_2(h_2)\,\Phi_3(h-h_1-h_2)$$

$$\times \exp\frac{\pi i \Delta z}{k}\left[\frac{h_1^2}{a^2} + \frac{(h_1+h_2)^2}{a^2} - \frac{2(h_1+h_2+h_3)^2}{a^2}\right].$$

Summing this over h_1 from 0 to h, we obtain the total contribution from waves diffracted by three successive planes in the direction specified by h.

If we continue the procedure described and expand the results to two dimensions, the expression for diffracted amplitude in the direction x^*y^* will be

$$\Psi(x^*y^*) = (i\Delta z)^N \exp\left[\pi i\frac{t(x^{*2}+y^{*2})}{k}\right.$$

$$\times \sum_{h_1}\sum_{k_1}\sum_{h_2}\sum_{k_2}\cdots\sum_{h_N}\sum_{k_N}\Phi_1(h_1 k_1)\,\Phi_2(h_2 k_2)\ldots\Phi_N(h_N k_N)$$

$$\times \exp\left\{2\pi i\lambda \sum_{n=1}^{N} z_n\left[\left(\frac{h_n^2}{2a^2}+\frac{k_n^2}{2b^2}+\sum_{m=n+1}^{N}\right)\left(\frac{h_n h_m}{a^2}+\frac{k_n k_m}{b^2}\right)\right.\right.$$

$$\left.\left.-x^*\left(\frac{h_n}{a}\right)-y^*\left(\frac{k_n}{b}\right)\right]\right\}\delta\left(x^* - \sum_{n=1}^{N}\frac{h_n}{a},\, y^* - \sum_{n=1}^{N}\frac{k_n}{b}\right) \qquad (6.31)$$

where the δ-function defines diffraction directions specified by $h = \sum_{n=1}^{N} h_n$ and $k = \sum_{n=1}^{N} k_n$. All the possible combinations $h_n k_n$ leading to the given h, k are taken into account. Thus the function $\Psi(x^*y^*)$ represents the amplitude of a diffracted wave with the indices $x^* = h/a$ and $y^* = k/b$. Inserting the relevant values of x^* and y^* into (6.31) we have

$$\Psi(hk) = (i\Delta z)^N \exp\left[\pi i\frac{t(x^{*2}+y^{*2})}{k}\right]$$

$$\times \sum_{h_1}\sum_{k_1}\cdots\sum_{h_N}\sum_{k_N}\Phi_1(h_1 k_1)\ldots\Phi_N(h_N k_N)$$

$$\times \exp\left\{-\frac{\pi i}{k}\sum_{n=1}^{N} z_n\left[\frac{h_n^2}{a^2}+\frac{k_n^2}{b^2}+2\sum_{m=1}^{n-1}\left(\frac{h_n h_m}{a^2}+\frac{k_n k_m}{b^2}\right)\right]\right\}. \qquad (6.32)$$

It has been shown that with the incident beam normal to the surface of a crystal with an orthogonal unit cell the Ewald sphere misses the center of the reciprocal lattice node given by the vector $H = ha^* + kb^*$ by the value of the excitation error measured along c^*, which is

$$\delta = -H^2\lambda/2 = -\frac{\lambda}{2}(h^2/a^2 + k^2/b^2) . \tag{6.33}$$

The sign for δ depends on whether the lattice point is inside ($\delta > 0$) or outside ($\delta < 0$) the Ewald sphere. Putting

$$\delta_n = -\frac{\lambda}{2}\left[\left(\frac{h^{(n)}}{a}\right)^2 + \left(\frac{k^{(n)}}{b}\right)^2\right], \tag{6.34}$$

where $h^{(n)} = \sum\limits_{i=1}^{n} h_i, \; k^{(n)} = \sum\limits_{i=1}^{n} k_i$.

We obtain the power of the exponential in (6.32)

$$-\pi i\lambda \sum\limits_{n=1}^{N} z_n\left[h_n^2/a^2 + k_n^2/b^2 + 2\sum\limits_{m=1}^{n-1}(h_n h_m/a^2 + k_n k_m/b^2)\right]$$

$$= 2\pi i \sum\limits_{n=1}^{N} z_n(\delta_n - \delta_{n-1}) = 2\pi i\left[t\delta - \Delta z \sum\limits_{n=1}^{N-1} \delta_n\right] \tag{6.35}$$

where δ is the excitation error for the reflection analyzed. The final expression for the amplitude of a diffracted wave with the indices hk, which was obtained by Cowley and Moodie (1960), Cowley (1975b), is given by

$$\Psi(hk) = (-i\Delta z)^N \sum\limits_{h_1}\sum\limits_{k_1}\dots\sum\limits_{h_N}\sum\limits_{k_N}\Phi_1(h_1 k_1)\dots\Phi_N(h_N k_N)$$

$$\times \exp\left\{2\pi i\left[\delta t - \Delta z \sum\limits_{n=1}^{N-1} \delta_n\right]\right\}. \tag{6.36}$$

As it has been shown, for the given h, k summation over h_2, k_2 depends on h_1, k_2, summation over h_3, k_3 depends on h_1, k_1 and h_2, k_2, etc.

6.2 Numerical Methods for Calculation of Diffraction Patterns

The dynamical theory has provided a basis for numerical calculation of electron diffraction effects. Various techniques for computation of diffracted amplitudes and phases make use of different theoretical approximations.

Goodman and Moodie (1974) and later Self et al. (1983) have analyzed in detail the relative merits of the different methods available, revealed the fields of the most effective use for each technique, and formulated guidelines for accurate computation. They have noted that the modern computational techniques may be divided into two groups. The first group involves matrix operations and consists of the eigenstate formulation developed by Bethe (1928) and the scattering matrix method proposed by Sturkey (1957, 1962). The second group consists of methods involving a mathematical slicing of the crystal. Below is given a general outline of computational procedures in the two most widely used methods based on the theories developed by Cowley and Moodie (1957) and Bethe (1928).

The Multi-Slice Computation. The analytical solution for N-beam dynamical electron scattering obtained by Cowley and Moodie cannot be used for direct calculations since the formulae have been derived for the limiting case of $\Delta z \rightarrow 0$. However, using this theory as a basis, Goodman and Moodie (1974) have worked out a method for calculating diffraction patterns and high-resolution electron-microscopic images of crystals which they called the multi-slice method. Although in terms of this theory the object is represented by a set of equidistant planes separated by vacuum, the numbers of the planes and the spacings have finite values.

Each plane n at $z = z_n$ corresponds to a slice localized within the interval $z_n = \Delta z$. The two-dimensional potential distribution $\varphi_n(xy)\Delta z$ is considered constant within the slice and is given by

$$\varphi_n(xy)\Delta z = \int_{z_n}^{z_n + \Delta z} \varphi(xyz)dz .$$

The transmission function is then given by

$$q_n = \exp[i\sigma\varphi_n(xy)\Delta z] . \tag{6.37}$$

Since Δz is finite ($1 - 3$ Å and more), the approximation (6.12) is no longer possible. Therefore, the Fourier transform of q_n is carried out directly according to the formula

$$Q_n(x^*y^*) = \mathscr{F}\{\cos[\sigma\varphi_n(xy)\Delta z]\} + \mathscr{F}\{i\sin[\sigma\varphi_n(xy)\Delta z]\} . \tag{6.38}$$

The propagation function in terms of reciprocal lattice coordinates for the incident beam parallel to the z axis of the orthogonal unit cell is given by

$$\mathscr{P}_n(hk) = \exp[2\pi i\delta_n(hk)\Delta z] \tag{6.39}$$

where $\delta_n(hk)$ is the excitation error defined by (6.34).

The problem is again to express consequently $N - 1$ times the amplitude distribution for a slice in terms of that for the preceding one, i.e., to calculate the function

$$\Psi_n(hk) = \sum_{h'}\sum_{k'} \Psi_{n-1}(h'k')\mathscr{P}_{n-1}(h'k')Q_n(h-h', k-k') . \tag{6.40}$$

The calculations for the given number of slices N yield the amplitude and the phase for a diffracted beam with the indices hk. Repeating similar calculations for different values of hk, one may obtain any set of calculated diffracted amplitudes or intensities.

It is a computational advantage if slices of equal thickness Δz are used, although this is not a necessity. If the unit cell parameter c is less than 5 Å, iteration can use not only equal Δz but also the same $q(xy)$ and $p(xy)$ for all slices (Self et al. 1983). $\varphi(xy)$ is then given by (6.24) and Δz is chosen such as to ensure accurate evaluation of $Q(hk)$. This simplification is possible since, for small c values, the contribution of reciprocal lattice points located out of the zeroth Laue zone to diffraction is negligible. If this condition does not hold, the slice thickness should be chosen so that $c = m\Delta z$ where m is an integer. The computation should then include m different functions $q_m(xy)$ and $Q_m(hk)$.

Errors in multi-slice computations are generally associated with the choice of the slice thickness and the number of diffracted beams. The following two tests are commonly used to minimize these errors.

The first is based on the fact that $|q(x, y)|^2 = 1$ and therefore

$$Q(hk) * Q^*(-h, -k) = \sum_{h'k'} Q(h'k')Q^*(h+h', k+k') = \begin{cases} 1, & \text{if } h, k = 0 \\ 0, & \text{if } h, k \neq 0 . \end{cases}$$

The number of beams and the slice thickness are chosen adequately if the sum varies from unity by less than 10^{-6}.

The second test consists in the calculation of the sum of the intensity in all the beams included in the iteration. The sum should equal unity after the first slice and then decrease slowly with increasing number of slices.

If the above requirements are obeyed, electron energy losses may be included. It becomes necessary to introduce the absorption potential $\mu(xy)$ which, as has been shown, is equivalent to the introduction of the imaginary potential component. The transmission function then may be written

$$q_n(xy) = \exp\{i\sigma\varphi(xy)\Delta z - \mu(xy)\Delta z\}, \qquad (6.41)$$

where

$$\mu(xy) = \sum\sum V^i(hk)\exp\left\{-2\pi i\left(h\frac{x}{a} + k\frac{y}{b}\right)\right\}.$$

If the incident beam is not parallel to the surface normal, the corresponding corrections are readily introduced into the formula for the excitation error (Goodman and Moodie 1974).

The multi-slice method has been widely applied to numerous problems (Allpress et al. 1972; Cowley and Iijima 1976; Goodman and Moodie 1974; Lynch and O'Keefe 1972). This approach permits to involve a great number of diffraction reflections and to allow for absorption and effects resulting from crystal defects in a relatively simple way.

Following the properties of a Fourier transform, the formula (6.40) is written as

$$\Psi_n(hk) = \mathscr{F}\{\mathscr{F}^{-1}[\Psi_{n-1}(hk)\,\mathscr{P}(hk)]\,\mathscr{F}^{-1}[Q_n(hk)]\}$$

where $\mathscr{F}$ and $\mathscr{F}^{-1}$ stand for direct and reverse Fourier transforming, respectively, i.e., Fourier transforming from real space to reciprocal space and vice versa.

Ishizuka and Uyeda (1977) have shown that computation of structure amplitudes and phases making use of the fast Fourier transform algorithm and consequent Fourier transforms in the above form can be more rapid than that based on (6.40), especially if a large number of diffracted beams is included.

The authors propose to determine the number of diffracted beams by the number of points in the unit cell that are sufficient to describe adequately the potential distribution. For this purpose, the unit cell edges are usually divided into intervals of the length $0.2 - 0.3$ Å. Therefore to calculate precisely the image of, e.g., Cu-phthalocyanine having $a = 19.64$ Å and $b = 24.04$ Å, about 8200 reflections were required. The reflections involved are those occurring on consequent transmission through N slices. The number of diffracted beams having left the exit face and registered in the diffraction pattern is roughly a factor of 10^2 less.

The upper limit of the slice thickness is given by the expression $\Delta z \leqslant 2\pi d^2/\lambda$, where d is the distance within which the potential does not change appreciably. Generally, $d \sim 0.2$ Å and does not depend on the sample except for structures with very heavy atoms. Consequently, for $V = 100$ kV the upper limit of the slice thickness is about $5-6$ Å.

Finally, it was found that there is a certain critical crystal thickness, so that for greater thicknesses multiple scattering effects camouflage the real structural features of the object. The relationship between the structure of the object and the contrast distribution in the image holds for $H(\pi\alpha^4/2\lambda) \ll 1$ where α is the scattering angle for each reflection. According to the estimates of Ishizuki and Uyeda (1977), the crystal thickness critical value is 150 Å and 100 Å for accelerating voltages of 500 and 100 kV, respectively.

Direct Solution of the Schrödinger Equation. The direct solution of the Schrödinger equation in the two-beam approximation has been considered in Chapter 5. If this approximation is extended to N reciprocal lattice points, i.e., to N beams, and the back-scattering is ignored, it is again appropriate to seek the solution by writing the wave-function within the crystal as a sum of N Bloch waves. An ith Bloch wave is defined by the parameters $C_H^{(i)}$ and $k_H^{(i)} = k_0^{(i)} + H$ corresponding to the N reciprocal lattice points included in the calculation, i.e.,

$$\Psi^{(i)}(r) = \sum_H C_H^{(i)} \exp(2\pi i k_H r) \, .$$

By analogy with the results of Chapter 5, it may be inferred that N values for $C_H^{(i)}$ as well as for $k_H^{(i)}$ will be associated with each reciprocal lattice point H.

A set of simultaneous equations can be obtained from (5.4) and may be written as

$$\begin{bmatrix} k^2-k_0^2 \dots & v_{-H} & \dots & v_{-G} \\ \vdots & \vdots & & \vdots \\ v_H & \dots k^2-k_H^2 \dots & v_{H-G} \\ \vdots & \vdots & & \vdots \\ v_G & \dots & v_{G-H} & \dots k^2-k_G^2 \end{bmatrix} \begin{bmatrix} C_0^{(i)} \\ \vdots \\ C_H^{(i)} \\ \vdots \\ C_G^{(i)} \end{bmatrix} = 0 \, . \tag{6.42}$$

Obviously, these equations have eigensolutions only if the determinant of the matrix is zero. In order to determine the particular values of $C^{(i)}$ and $k_0^{(i)}$, the boundary conditions should be imposed similar to those used in the two-beam theory. In particular, vectors $k_0^{(i)}$ and k_v projected on the crystal surface should be equal (where k_v corresponds to the incident beam direction in vacuum). This requirement is obeyed if the end-points of these vectors lie on the same surface normal (Fig. 22). The computation is greatly simplified if the surface normal n coincides with the zeroth Laue zone normal. In this case, the normal n passing through the end points of the set of vectors $k_0^{(i)}$ cuts the Laue plane at the point given by r_0^*.

Figure 21 shows that the vector $k_0^{(i)}$ can be divided into components parallel and perpendicular to n, i.e.,

$$k_0^{(i)} = k_\perp^{(i)} n + r_0^* .$$

Then,

$$k_H^{(i)} = k_0^{(i)} + H = k_\perp^{(i)} n + (H - r_0^*)$$

and

$$k^2 - k_H^{(i)2} = k^2 - k_\perp^{(i)2} - (H - r_0^*)^2 = k^2 - k_\perp^{(i)2} - (r_H^*)^2 .$$

For the known experimental conditions and the unit cell parameters for the crystal under study the values for $(r_H^*)^2$ can be readily calculated beforehand.

The expression (6.42) becomes $MC = (k_\perp^{(i)2} - k^2)C$ (Self et al. 1983), where

$$M = \begin{bmatrix} -r_0^{*\,2} \cdots & v_{-H} & \cdots & v_{-G} \\ \vdots & \vdots & & \vdots \\ v_H & \cdots & -r_H^{*\,2} \cdots & v_{H-G} \\ \vdots & \vdots & & \vdots \\ v_G & \cdots v_{G-H} \cdots & -r_G^{*\,2} \end{bmatrix} ; \quad C = \begin{bmatrix} C_0 \\ \vdots \\ C_H \\ \vdots \\ C_G \end{bmatrix} .$$

The elements of M are independent of (i) and, moreover, the on-diagonal elements are of the same order of magnitude as the off-diagonal elements. Thus the problem is to diagonalize the matrix M and to find N eigensolutions which would define N values for $k_0^{(i)}$ and $C^{(i)}$.

In order to ensure a continuous transition of the incident-beam wave-function into the in-crystal wave-function at the top surface of the crystal, boundary conditions are to be applied for amplitude values of Bloch waves.

As in the case of the two-beam approximation, it is necessary that the sum of amplitudes for all the diffracted beams (except the incident beam) at the top surface should be zero. This is achieved by introducing the excitation coefficients α_i that specify such an energy distribution over the Bloch waves that under experimental conditions would correspond to the energy distribution of the diffracted beams (Self et al. 1983).

The electron wavefunction within the crystal is given by

$$\Psi(r) = \sum_i \alpha_i \sum_H C_H^{(i)} \exp(2 \pi i k_H^{(i)} r) .$$

The amplitude of the wave within the crystal corresponding to one reciprocal lattice point H is given by

$$\Psi_H = \sum_i \alpha_i C_H^{(i)} \exp[2 \pi i k_H^{(i)} r] .$$

Following the boundary conditions, all the diffracted amplitudes included in the calculation are zero at the entrance surface, i.e.,

$$\sum_i \alpha_i C_H^{(i)} = \psi_H(0) = 0 \quad \text{for all} \quad h, k, l \neq 0 .$$

If the normalization $\sum_i \alpha_i C_0^{(i)} = 1$ is used for the zeroth-order beam, then

$$\sum_i \alpha_i C_H^{(i)} = \delta_H$$

for all the beams included, where δ_H is the Kronecker delta function.

In the matrix form, the boundary conditions may be written as

$$\psi(0) = C_f \alpha = \Phi$$

where $\psi(0)$ is the electron wavefunction at the top surface,

$$C_f = \begin{bmatrix} C_0^{(1)} & C_0^{(2)} & \cdots & C_0^{(N)} \\ \vdots & \vdots & & \vdots \\ C_H^{(1)} & C_H^{(2)} & \cdots & C_H^{(N)} \\ \vdots & \vdots & & \vdots \\ C_G^{(1)} & C_G^{(2)} & \cdots & C_G^{(N)} \end{bmatrix} ; \quad \alpha = \begin{bmatrix} \alpha_0 \\ \alpha_1 \\ \vdots \\ \alpha_N \end{bmatrix} \quad \text{and} \quad \Phi = \begin{bmatrix} \psi_0(0) \\ \vdots \\ \psi_H(0) \\ \vdots \\ \psi_G(0) \end{bmatrix} .$$

Hence $\alpha = C_f^{-1} \Phi$.

Absorption can be included by introducing complex terms $v_H - i v_H'$ so that the matrix M becomes complex.

The mathematical formulation becomes more complicated if the out-of-zone reflections are to be included and the incident beam forms oblique angles with the surface. In this case, the approximation described by Hirsch et al. (1977) can be used.

CHAPTER 7

Electron Diffraction and High-Resolution Electron Microscopy

7.1 Diffraction Effects and Formation of High-Resolution Electron-Microscopic Images

Improvements in the construction of electron microscopes have ensured that it is possible to obtain, under certain conditions, direct images of crystal structures or at least the main details in the structure motif. The resolution of modern microscopes is high enough for visualizing various fine details in the structure within the unit cell, if not the atoms themselves. It might seem that this is an excellent opportunity to determine structures by direct visualization of the structural motif.

However, interpretation of high-resolution images is rather complicated. It is not always the case that an image observed even with the highest magnification is directly associated with the structure analyzed. In some cases, the interpretation of high-resolution data that had seemed at first indisputable was eventually revised (Allpress and Sanders 1973). On the other hand, there are numerous examples where images obtained under certain conditions represented the crystal structure of the sample under study (Bursill 1979; Bursill and Wilson 1977; Iijima 1971, 1973, 1975a, b, 1978; Iijima and Allpress 1974; Iijima et al. 1973, 1974; van Landuyt and Amelinckx 1975; Veblen 1981; Veblen and Buseck 1979a and many others). The utmost importance of the results obtained is beyond doubt. Not only was it proved that crystal structures may be visualized directly, but also the possibility to determine details, such as separate point defects that cannot be directly revealed by any other method, was vividly demonstrated.

Nevertheless, there is, at present, no sufficiently simple and reliable criterion ensuring an unambiguous interpretation of the phase contrast observed with substances with unknown structures containing relatively heavy atoms. The difficulties result from the fact that the formation of an electron-microscopic image depends on a great number of factors, each of which should be taken into account in analyzing the experimental data. Therefore, reliable interpretation of the results requires an understanding of the mechanism of the formation of an image with diffraction effects, and that the properties of optical systems in electron microscopes be taken into account.

Electron diffraction is the most essential point in the imaging of crystal structures, since diffraction effects contain information on the object which is not only revealed indirectly in a diffraction pattern, but is also displayed in the very possibility to visualize the crystal structure directly. Therefore, it is natural to treat diffraction and imaging in terms of a uniform approach that permits to reveal relationships between them. For this purpose, use is made of profound analogies between electron diffraction and that of light waves, that is, between electron and light optics (Cowley 1975b). Cowley and co-workers (Cowley 1975a, b, 1978; Cowley and Iijima 1972, 1976; Cowley and Moodie 1957, 1960; Cowley and Pogany 1968) have made a valuable contribution to the development of this theory.

7.2 Fraunhofer Diffraction: An Intermediate Stage in the Transfer of Information Between the Object and the Image

Electron-microscopic images and electron diffraction are usually described in terms of the Fraunhofer diffraction theory. Assume that a plane monochromatic electron wave of zero phase and unity amplitude is incident on a thin plate-shaped crystal which is a phase-object. The amplitude distribution at the exit face is defined by the function $q(xy)$. Let us consider the diffraction effect at the point of observation XY. The distance R between the object plane and the plane of observation is very much greater than the overall dimensions of the object.

Each point x, y in the object plane generates a spherical wave with the amplitude

$$q(xy)\frac{\exp(2\pi ikr)}{r}\frac{\exp(\pi i/2)}{\lambda}. \tag{7.1}$$

Figure 26 shows that

$$r^2 = R^2 + (X-x)^2 + (Y-y)^2, \quad r_0^2 = R^2 + X^2 + Y^2.$$

For Fraunhofer diffraction, $r \approx r_0$, $(x^2 + y^2)^{1/2}/r_0 \to 0$, therefore

$$r = r_0 - (Xx + Yy)/r_0. \tag{7.2}$$

Diffraction effects result from the interference of waves generated by different (x, y) points in the object that have reached the point XY in the plane of observation. The resultant effect is given by summing the expression (7.1) over all x, y. Inserting (7.2) and omitting constant factors we have

$$Q(XY) = \iint q(xy) \exp\left[-2\pi i\frac{Xx + Yy}{\lambda r_0}\right] dx\, dy. \tag{7.3}$$

If the angle between the diffracted and the incident beams is designated by β (Fig. 27), then in terms of the small-angle approximation, which is valid for fast electrons, we may introduce new variables

$$x^* = X/r_0\lambda = \beta_x/\lambda, \quad y^* = Y/r_0\lambda = \beta_y/\lambda. \tag{7.4}$$

Then (7.3) becomes

$$Q(x^*y^*) = \iint q(xy) \exp[-2\pi i(xx^* + yy^*)]\, dx\, dy. \tag{7.5}$$

It may be easily seen that the functions $Q(x^*y^*)$ and $q(xy)$ are a pair of Fourier transforms, and x, y and x^*, y^* are coordinates in direct and reciprocal

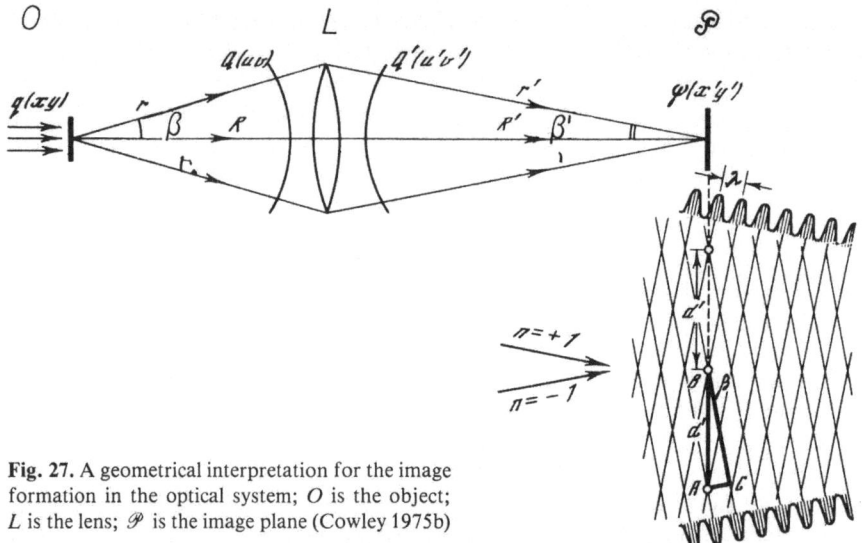

Fig. 27. A geometrical interpretation for the image formation in the optical system; O is the object; L is the lens; $\mathscr{P}$ is the image plane (Cowley 1975b)

space, respectively. If we approximate $q(xy)$ by the projection of the electro-static potential, $Q(x^*y^*)$ will represent a two-dimensional reciprocal lattice, so that the weights of the lattice points are given by kinematical structure ampli-tudes $\Phi(hk)$. Note that if the distance between the object plane and the plane of observation were of the same order of magnitude as dimensions of the object, the electron scattering would be described by the Fresnel diffraction, which has been already discussed. Analysis of the formation of images of crystal structures makes use of both approximations studied.

Proceeding to optical analogies, we consider the simplest case where a narrow parallel light beam is incident on a one-dimensional diffraction grating with a grating constant d (Fig. 27). Scattered light is localized along diffraction direc-tions β_n defined by

$$d \sin \beta_n = n\lambda \,,$$

where n is the order of diffraction.

Consider two diffraction beams with $n = \pm 1$. After being refracted by a lens, they converge in the image plane. The result of their interaction is shown in Fig. 27. The crest of the resultant wave in the image plane will be represented by a set of parallel lines with a spacing

$$d' = \lambda / \sin \beta' = d \sin \beta / \sin \beta' = Md \,,$$

where M is the magnification.

Note that the interference of diffracted waves of the first order gives the in-tensity distribution which has nothing in common with real features of the ob-ject. However, including higher orders of interfering waves should lead, in the idealized case, to the formation of an M times magnified object image that would be more adequate to the object.

The above example shows that the image is formed by the addition of dif-fracted waves that are generated in the object under the action of the incident radiation. The role of the optical system in this idealized case is simply to transform the divergent diffracted beam into one convergent in the image plane. Then, if we disengage ourselves from the actual beam directions, the object and the image may be regarded equivalent with respect to the lens plane, since the incident light beam directed to the image will produce an identical copy of the initial object in the plane of observation. The diffraction direction will be defined by the same basic equation $n\lambda = (Md)\sin\beta'_n = d\sin\beta_n$.

Analysis for a general case gives but a more rigorous treatment of the image formation mechanism which, in fact, has been already considered. If monochromatic radiation is incident on a parallel-sided crystal, the scattered amplitude distribution at the exit face will be given by the transmission function $q(xy)$. The propagation of diffracted waves in space is described in terms of Fraunhofer diffraction, that is, the amplitude distribution at the surface of a spherical wave is represented by the function $Q(x^*y^*)$, whose coordinates depend on β in the small-angle approximation [see Eq. (7.4)] (Fig. 27). The spherical wave formed in the object propagates to the optical system. In the idealized case, the latter transforms the amplitude distribution $Q(x^*y^*)$ into $Q'(x^{*\prime}y^{*\prime})$ distributed over a spherical surface centered on a point in the image plane (Cowley 1975b). For an ideal lens, $Q(XY) = Q'(XY)$. On the other hand, it may be seen in Fig. 27 that

$$X = x^*r = x^{*\prime}r', \quad Y = y^*r = y^{*\prime}r', \quad r'/r = R'/R, \quad x^* = x^{*\prime}R'/R,$$

$$y^* = y^{*\prime}R'/R.$$

Therefore

$$Q(x^*y^*) = Q'(x^{*\prime}y^{*\prime}) = Q(x^{*\prime}R'/R, y^{*\prime}R'/R).$$

The image function $\Psi'(x'y')$ is the Fourier transform of $Q'(x^{*\prime}y^{*\prime})$ owing to the reversibility of the Fourier integral. Hence the function to be determined for, e.g., a one-dimensional case, that is

$$\Psi'(x') = \int Q'(x^{*\prime})\exp[2\pi i x^{*\prime}x']dx^{*\prime}$$

$$= \int\int q(x)\exp\left[2\pi i\left(\frac{x^{*\prime}R'}{R}x + x^{*\prime}x'\right)\right]dx\,dx^{*\prime}$$

$$= \int q(x)\delta\left(x + \frac{R}{R'}x'\right)dx = q\left(-\frac{R}{R'}x'\right) = q(-Mx')$$

is the transmission function inverted and magnified by a factor M (Cowley 1975b).

Thus the imaging is described, mathematically, by the Fourier transform of the transmission function $q(xy)$, which is transformed into the function $Q'(x^{*\prime}, y^{*\prime})$ by the optical system, so that the Fourier transform of $Q'(x^{*\prime}, y^{*\prime})$ is the image function. Physically, this implies that the optical system "synthesizes" diffracted waves.

The relationship between diffraction effects and images is more explicitly demonstrated in terms of the Abbe wave theory. Figure 28 shows a parallel monochromatic light beam incident on an object. The geometry of the diagram implies that the image is inverted and magnified by a factor $M = R_2/R_1$.

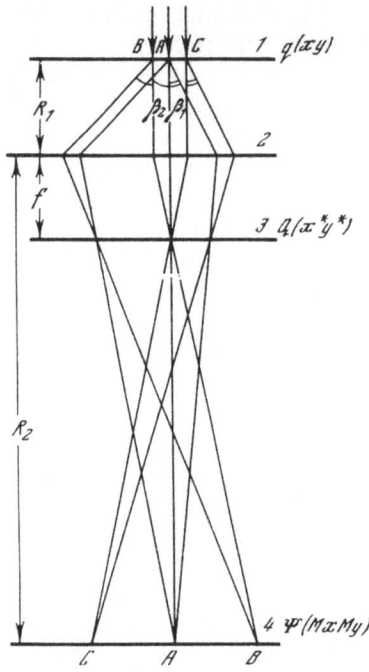

Fig. 28. The relationship between diffraction effects and the image according to the Abbe theory: *1* the object plane; *2* the lens plane; *3* the back-focal plane; *4* the image plane

One of the essential points in the Abbe theory is the demonstration of the possibility to obtain a diffraction pattern using the lens system. It is seen in Fig. 28 that all rays scattered through the same angle with respect to the incident beam direction are brought to a focus at one point in the back-focal plane of the objective lens. Since this is equivalent to observation at infinity, the amplitude distribution at the back-focal plane is that of the Fraunhofer diffraction and is given by the Fourier transform of $q(xy)$. For crystals, $q(xy)$ is a periodic function. Hence its Fourier transform is a lattice function $Q(x^*y^*) = Q(hk)$. The coordinates of diffraction maxima in the back-focal plane are given, in the small-angle approximation, by $X = hf\lambda$ and $Y = kf\lambda$, where f is the focal distance and λ is the wavelength.

In the general case, each point in the back-focal plane may be considered as a source of secondary spherical waves propagating toward the image plane. The formation of an image proceeds in terms of the Fraunhofer diffraction with superposition of all waves from the back-focal plane. The resultant amplitude distribution describing the image function $\Psi(xy)$ is given by the Fourier transform of the amplitude distribution $Q(x^*, y^*)$ at the back-focal plane. Thus imaging is closely associated with Fraunhofer diffraction and is described by two consequent Fourier transforms.

The most important consequence of the Abbe theory is that there is a theoretical possibility to observe and register both the diffraction pattern and the image. To conclude, it should be stressed that idealized lenses have been hitherto considered. This implies equal optical raypaths for rays having passed through the lenses and scattered through equal angles, from the points at the object to the corresponding points at the conjugate image plane. Consequently, diffracted rays for each direction reach the image plane without a phase change.

7.3 Factors Defining the Contrast in Electron-Microscopic Images

Very thin samples are generally used in electron-microscopic studies. A decrease in the sample thickness leads to a decrease in the three-dimensional n-beam interaction, diminishes energy losses of scattered electrons, and improves the resolution of fine details in the images. As it will be shown later, samples of the thickness of about $50 - 100$ Å are the most favorable objects for direct visualization of the atomic motif.

If the Fresnel diffraction within the sample is neglected, the transmission function at the exit face of a very thin crystal is $q(xy) = \exp i\sigma\varphi(xy)$. Ignoring Fresnel diffraction effects implies, in fact, neglecting the lateral scattering of electrons in their propagation through the crystal. Therefore, the distribution of scattering matter in the plane of a phase-object being periodic or aperiodic should not have a perturbing influence on the distribution of $q(xy)$. It is obvious that this approximation is valid either for very thin crystals or for extra-high voltages. At any rate, approximation of a thin crystal containing relatively heavy atoms by a phase-object is more adequate than the kinematical approximation applied to it.

If we ignore the scale and the negative sign in the function $\Psi(-Mx, -My)$, then $q(xy) = \Psi(xy)$. Contrast in the image is represented by an intensity distribution $J(xy) = \Psi\Psi^*$. It is obvious that intensity should be constant for any phase-object since $qq^* = 1$.

Thus, for a perfect lens, there should be no contrast in an image of a phase object. Contrast in images results from changes in phase relations between diffracted beams owing, on the one hand, to various, mainly instrumental, imperfections in optical systems, and, on the other hand, to the defocus of the objective lens. The main instrumental effects are spherical and chromatic aberrations and astigmatism. In modern microscopes the influence of astigmatism as well as that of mechanical vibration are minimized. Therefore, we shall confine the discussion to the influence of aperture limitation, defocus, spherical and chromatic aberrations, and incident beam divergence.

Aperture Limitation. An aperture in the back-focal plane of the objective lens provides contrast owing to a deficiency in scattering, since the rays scattered at angles greater than the limiting angle do not contribute to the image. If, for instance, the aperture removes all the rays except one, only those parts of the object which have made the main contribution into scattering in the given diffraction direction will appear in the image. This is the so-called *dark-field amplitude (or diffraction) contrast*. The bright-field contrast, which is achieved when the aperture blocks out all the rays except the incident one, is more difficult to interpret. Both bright- and dark-field amplitude contrasts are widely used in studying rocking curves, thickness fringes, dislocations, stacking faults, and other features of real crystal structure (Hirsh et al. 1977).

If the incident beam and one diffracted beam, e.g., hoo, pass through the aperture, then it is two waves that reach the image plane:

$$\Psi = \Psi_0 + \Psi_h \exp 2\pi i h(x/a).$$

The path-length difference for these, $\lambda rH = \lambda h(x/a)$ determines the phase factor in the second term (Fig. 28). Interference of these waves in the image plane leads to a set of fringes with the spacing $M(a/h)$ since

$$I = |\Psi|^2 = \Psi_0^2 + \Psi_h^2 + 2\,\Psi_0 \Psi_h \cos 2\pi i h(x'/aM),$$

where $x' = xM$ is the current coordinate in the image plane.

Menter (1956) was the first to obtain images from crystals of Cu- and Pt-phthalocyanine containing sets of dark and bright straight lines alternating regularly with a period equal to the corresponding parameter in the crystal lattice. It seemed at first that fringe imaging was a simple and reliable means for direct determination of various structural imperfections (stress fields, dislocations, etc.). However, Cowley (1975b) and later Hashimoto et al. (1961) (in a more generalized form) have proved that fringe contrast should be treated with care. Cowley has shown that for the two-beam dynamical situation the intensity distribution in the image plane for the symmetrical case $\vartheta_0 = \vartheta_h$ and zero excitation error is given by

$$I = \Psi\Psi^* = 1 - \sin 4\pi\sigma\Phi_h t \, \cos 2\pi(hx/a - \alpha), \qquad (7.6)$$

where $\alpha = \dfrac{1}{2\pi} \mathrm{tg}^{-1}\left(\dfrac{\sigma\Phi_h}{\mu_h}\right)$ and μ_h is the absorption amplitude for the reflection hoo.

Since $\sigma\Phi_h \gg \mu_h$, $2\pi\alpha \simeq \pi/2$. Therefore, the sinusoidal variation of intensity (7.6) is shifted by $a/4$ with respect to the arrangement of the planes. In addition, the contrast is determined by the sample thickness t so that it may vanish ($4\sigma\Phi_h t = n$) or reverse the sign ($t = (2n+1)t_e/2$).

Hashimoto et al. (1961) have shown that the fringe direction may differ from that of reflecting atomic planes if Bragg's Law does not apply strictly. Fringe images often contain terminated fringes. This was usually attributed to the presence of edge dislocations. The calculations of Cockayane et al. (1971) testify that in this case there may be also no relationship between the fringe contrast and the structure. For example, in the case of inclined dislocations the number of terminated fringes varies depending on the geometry of the experiment.

The increase in the number of systematic reflections contributing to imaging makes the interpretation for the contrast obtained even more difficult. On the whole, spacings between fringes, the mode in their alternation, and the intensity distribution depend on a great number of parameters affecting the phase relationships between interfering beams. Under such conditions, fringe images are widely and successfully used in testing the stability and the quality of the work of electron microscopes, since they permit evaluating the instrumental resolution. Possibilities to use fringe images for studying structural features have been recently analyzed by Sinclair (1978).

A fundamentally different situation arises with a large number of diffracted beams passing through the aperture, since this is the case where a direct image of the object structure may be obtained. To allow for the effect of the aperture, the amplitude distribution function in the focal plane of the objective lens should be multiplied by the aperture function, which is unity inside the aperture and zero outside. Thus

$$Q(x^*y^*)A(x^*y^*).$$ (7.7)

For a circular aperture with the radius r_a, we have

$$A(x^*y^*) = \begin{cases} 1 & \text{for} \quad [(x^*)^2 + (y^*)^2]^{1/2} \leqslant \rho, \\ 0 & \text{for} \quad [(x^*)^2 + (y^*)^2]^{1/2} > \rho, \end{cases}$$

where ρ is the aperture radius in terms of reciprocal space coordinates and $r_a = \rho f \lambda$.

Applying (4.18) to (7.7) we have

$$\Psi(xy) * a(xy).$$ (7.8)

In other words, the aperture perturbs the image function. The function $a(xy)$ is a kind of shape-factor for each point in the image, which is smeared over an area whose dimensions and shape depend on the dimensions and shape of the aperture. For a circular aperture,

$$a(xy) = a(r) = \frac{J_1(2\pi\rho r)}{(2\pi\rho r)},$$ (7.9)

where $J_1(2\pi\rho r)$ is the first-order Bessel function.

The intensity distribution in the plane of observation is described by the function

$$\left[\Psi(xy) * \frac{J_1(2\pi\rho r)}{(2\pi\rho r)} \right]^2.$$ (7.10)

Note that the dependence of $J_1(2\pi\rho r)/(2\pi\rho r)$ on $2\pi\rho r$ is almost the same as that for the interference function $\sin x/x$. The difference is that the Bessel function decreases more rapidly than trigonometric functions, so that about 90% of intensity is concentrated within the region restricted by the second maximum. The aperture dimensions are directly connected with the resolution that may be achieved. It is seen in (7.8) that the least resolvable distance will be equal to the width of $a(xy)$. The first minimum in the function $|a(r)|^2$ is achieved for

$$2\pi\rho r_{min} = 1.22\pi \quad \text{and} \quad r_{min} = 0.61/\rho = (0.61/r_a)f\lambda.$$ (7.11a)

According to the Rayleigh criterion, the value r_{min} equal to the distance between the basic maximum and the nearest minimum describes the resolution. The values of the latter are inversely proportional to the aperture radius. To characterize the resolution, the value of the aperture angle is commonly used, equal to half the opening of a cone with the aperture as the base and the apex at the point of intersection of the sample by the optical system axis. We have

$$\beta = r_a/f \quad \text{and} \quad r_{min} = 0.61\lambda/\beta.$$ (7.11b)

Thus the resolution becomes worse with the decrease of β. Since the aperture blocks out electron beams with scattering angles greater than β, the aperture limitation could be regarded as a factor increasing the absorption of electrons in propagation through the object. Naturally, absorption exists by itself, resulting from, e.g., inelastic scattering and other processes causing energy losses and failures of coherence of scattered electrons. Although the nature of these effects is not altogether clear, phenomenological description making use of the imaginary component of the electrostatic potential, or the absorption function $\mu(x)$ is quite adequate. The transmission function becomes

$$q(xy) = \exp[-i\sigma\varphi(xy) - \mu(xy)] . \tag{7.12}$$

Generally, $\mu(xy)$ must take account of both the effect of preventing the electrons from contributing to the image (regardless of the reasons for this) and the influence of inelastic scattering causing deterioration of the image due to oscillating background.

Cowley (1975b) has shown that $\mu(xy)$ is proportional to the mean square deviation of $\varphi(xy)$ from the value of $\varphi(xy)$ averaged over the resolution distance. This is natural since the aperture leads to averaging the intensity in the image plane within the limits defined by the resolution. If we consider now the function (7.8) as equivalent to the image function, then, with (7.12) taken into account, the intensity distribution in the image plane will have contrast, since

$$|q(xy)|^2 = \exp[-2\mu(xy)] . \tag{7.13}$$

This is an explicit implication of the fact that the contrast resulting from the aperture limitation is the amplitude contrast. Since $\mu(xy)$ is periodic with maxima attributed to the scattering centers, the amplitude contrast in (7.13) is also associated with the structure of the object.

Defocus of the Objective Lens. A change in the focal distance of the objective lens for fixed positions of the object, the lens, and the plane of observation causes a change in phase relationships for beams scattered through different angles, as well as a decrease in resolution of details in the plane of observation. It follows from the formula $1/f = 1/R_1 + 1/R_2$ that, for a fixed distance between object and the lens, a change in f corresponds to a change in the distance between the lens and the conjugate image plane by the value

$$\Delta R - (R/f)^2\Delta f - M^2\Delta f , \tag{7.14}$$

where M is the magnification.

For a decrease in the focal distance of Δf, the conjugate image plane will shift upward with respect to the plane of observation by the distance $M^2\Delta f$. Figure 29 shows that the ray OB scattered through the angle β reaches the plane of observation at the point P' separated from the optical axis by the distance $O'P' = \Delta R\beta' = M^2\Delta f(\beta/M) = M\Delta f\beta$. The rays coming from the point O at scattering angles ranging from O to β form a disc of the radius $M\Delta f\beta$ in the plane of observation, which deteriorates resolution in the image. On the other hand, it is seen in Fig. 29 that paraxial rays whose conjugate image plane coincides with the plane

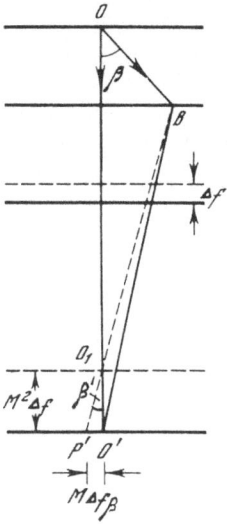

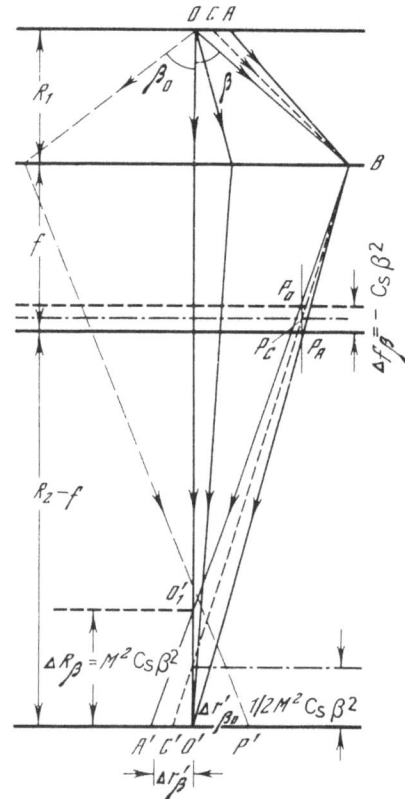

Fig. 29. The raypath in a microscope with defocus

Fig. 30. The raypath for spherical aberration ▶

of observation lag in phase behind those whose conjugate image plane is shifted with respect to the plane of observation, so that the path-length difference

$$\delta r_\beta = O_1 P' - O_1 O' = O' P' (\beta'/2) = \tfrac{1}{2} \Delta f \beta^2 . \tag{7.15}$$

If the focal distance increases by Δf, then the conjugate image plane will shift downward with respect to the plane of observation, and the corresponding phase difference will be $-(\pi/\lambda)\Delta f \beta^2$.

Spherical Aberration. With the increase of the angle of scattering the refraction of the waves by the lens increases and this leads to smearing of points in the image plane. Figure 30 shows the raypath in the case of spherical aberration. Paraxial rays scattered at the point O through minor angles β reach the conjugate point O' in the Gaussian image plane. These rays are practically unaffected by spherical aberration. With increasing β spherical aberration becomes noticeable. Figure 30 shows that a ray scattered at the point O through the maximum angle, β_0, with respect to the optical axis, defined by the aperture radius reaches, after being refracted by the lens, the plane of observation at the point P'. The latter is shifted with respect to O' by the distance $\Delta r'_{\beta_0}$. The rays scattered through angles ranging from O to β_0 fill the region between the points O' and P'.

Thus, due to spherical aberration, each point of the object in the image plane is smeared into a disc of the radius $\Delta r'_{\beta_0}$. On the other hand, owing to spherical aberration the rays scattered through the angle β are focused above the image plane (see Fig. 30). For example, the ray OBA' cuts the optical axis at the point O'_1 separated from the Gaussian image plane by the distance ΔR. Since the distance between the object and the lens plane, R_1 is assumed constant, the decrease in R_2 by ΔR_β should lead to a decrease in the focal distance by Δf_β, so that

$$1/R_1 + 1/(R_2 - \Delta R_\beta) = 1/(f - \Delta f_\beta) \, .$$

The focal distance is known to decrease with β as

$$f_\beta = f - \Delta f_\beta = f - C_s \beta^2 \, , \tag{7.16}$$

where C_s is the spherical aberration coefficient. Taking account of (7.16) we may rewrite (7.14) as

$$\Delta R_\beta = -(R_2^2/f^2)\Delta f_\beta = -M^2 C_s \beta^2 \, .$$

Each cone of scattered rays having an opening of 2β is represented by a smeared disc of the radius $\Delta r'_\beta$ in the image plane (Fig. 30). The latter value is given by the following expressions

$$\beta' = \beta(R_1/(R_2 - \Delta R_\beta)), \qquad \Delta r'_\beta = \Delta R_\beta \beta' = M C_s \beta^3 \, . \tag{7.17}$$

It is often convenient to ascribe the effect of spherical aberration to the influence of the object, i.e., to assume that each point of the object is smeared over the circular area of the radius $\Delta r_\beta = C_s \beta^3$. Figure 30 shows the path for rays scattered through the angle β in the case where the scattering point O is replaced by a set of scattering centers occupying the area of a disc of the radius Δr_β. The ray AB from a boundary point in the disc cuts the focal plane at the point P_A and falls into the point O' of intersection of the optical axis and the plane of observation. The ray OB cuts the optical axis at the point O'_1, which is higher than the plane of observation by $M^2 C_s \beta^2$. Therefore, the focal plane for this ray may be considered shifted by $C_s \beta^2$ with respect to that for paraxial rays.

Let us fix the point P_0 which lies in the focal plane of the ray OB and is separated from the point P_A by the distance $P_A P_0 = C_s \beta^2$ (Fig. 30).

It appears then that all the points of the line $P_A P_0$ are intersected by rays from scattering centers lying along the interval AO, that converge at the point B. In other words, each point in $P_A P_0$ corresponds to a definite ray within the angle OBA. For example, the ray CB cuts the relevant focal plane at the point P_c. Hence, each cone of beams scattered through the angle β corresponds to a set of focal distances varying continuously from f_0 to $f_0 - C_s \beta^2$. In the small-angle approximation it is appropriate to replace this f set by the average focal distance for the given β

$$f_\beta = f_0 + \Delta f_s = f_0 - \tfrac{1}{2} C_s \beta^2. \tag{7.18}$$

Thus the effect of spherical aberration may be considered to be equivalent, to a certain extent, to the effect of defocus of the objective lens. The sign of Δf_β is always negative, i.e., the spherical aberration, in fact, diminishes focal distances, so that each given scattering direction corresponds to a definite defocus Δf_β. On the other hand, the rays affected by spherical aberration lag in phase behind the paraxial rays.

This is clearly seen in Fig. 30. For example, the path traveled by the ray OBA' is longer than that of the ray OBO' by the distance $OB - AB = \beta r_\beta = C_s\beta^4 (OB + BO' = AB + BA'$ by the conditions of the problem and $BA' - BO' = OB - AB)$. Since a set of beams generated by the disc of the radius $\Delta r'_\beta$ may be replaced by a single beam having the effective focal distance of f_β [see Eq. (7.18)], the mean path-length difference for paraxial and nonparaxial rays may be readily obtained. It is equal to the difference between the paths traveled by the rays OBC' and OBO' or $O'_2C' - O'_2O$ since $OB + BO' = OO'_2 + O'_2O'$ (O_2 is the intersection point of OO' and BC' in Fig. 30). Hence

$$O'_2C' - O'_2O = C'O'\frac{\beta'}{2} = \left(\frac{1}{2}MC_s\beta^3\right)\left(\frac{\beta}{2M}\right) = \frac{1}{4}C_s\beta^4. \qquad (7.19)$$

Thus spherical aberration causes changes in phase relationships between waves interfering in the image plane, which affects the contrast observed. Since these phase changes are defined, for each given β, by the effective focal distance depending on Δf_s, the contrast can be controlled by varying the objective lens power. Decrease in f resulting from spherical aberration always leads to a phase lag for rays affected by the spherical aberration with respect to paraxial rays, whereas by increasing or decreasing the lens power it is possible to ensure that paraxial rays would either lead or lag in phase with respect to axial rays.

The resultant focal distance may be written

$$f_\beta = f_0 + \Delta f_s + \Delta f = f_0 - \tfrac{1}{2}C_s\beta^2 + \Delta f, \qquad (7.20)$$

where f_0 is the focal distance of the objective lens for the object plane conjugate to the plane of observation, Δf_s is the defocus resulting from the spherical aberration, and Δf is that resulting from the change in the lens power.

Any change in the focal distance may be compensated by varying the distance from the lens plane either to the plane of observation or to the object plane, assuming that these planes were conjugate before changing f. The resultant variation in the focal distance for each cone of scattered rays of the opening 2β is denoted by

$$\Delta f_\beta = \Delta f_s + \Delta f = \Delta f - \tfrac{1}{2}C_s\beta^2.$$

If the rays reaching the plane of observation are considered focused, the change in the focal distance by Δf_β should lead to a shift of the object plane for the given β by the distance

$$\Delta R_\beta = (R_1^2/f^2)\Delta f_\beta \approx \Delta f_s + \Delta f = \Delta f_\beta.$$

In spite of the position of the object being fixed, we may assume that its initial position is conjugate to the plane of observation for the given f_β. The object in this position is characterized by the transmission function $q(xy)$. The problem is to determine the amplitude distribution in the true object plane. Since the distance $\Delta R_\beta = \Delta f_\beta$ between the planes in question is small, the propagation of spherical waves from one of the planes to the other may be described in terms of Fresnel diffraction with the propagation function $p(xy)$. In other words, the amplitude distribution in the object plane for the defocus Δf_β is defined by the convolution of the transmission function with the propagation function, i.e.,

$$q(xy) * \exp[-\pi ik(x^2 + y^2)/\Delta f_\beta]. \tag{7.21}$$

Taking the Fourier transform of this expression we obtain the amplitude distribution in the back-focal plane of the objective lens which is given by [see Eq. (6.20)]

$$Q(x^*y^*) \exp \pi i\lambda \Delta f_\beta[(x^*)^2 + (y^*)^2] = Q(x^*y^*) \exp i\kappa. \tag{7.22}$$

The value Δf_β is the function of the reciprocal space coordinates, since $\beta^2 = \lambda^2[(x^*)^2 + (y^*)^2]$. Therefore

$$\begin{aligned}\kappa &= \pi\lambda[\Delta f_s + \Delta f][(x^*)^2 + (y^*)^2] \\ &= \pi\{\Delta f\lambda[(x^*)^2 + (y^*)^2] - \tfrac{1}{2}C_s\lambda^3[(x^*)^2 + (y^*)^2]^2\}.\end{aligned} \tag{7.23}$$

Rewriting (7.23) with account taken of the relationship between the phase difference and the path-length difference in terms of the angular aperture coordinates we have [cf. (7.15) and (7.19)]

$$\kappa = \frac{2\pi}{\lambda}\left[\frac{1}{2}\Delta f\beta^2 - \frac{1}{4}C_s\beta^4\right]. \tag{7.24}$$

The value $(1/2)\Delta f\beta^2$ is equal to the path-length difference between the paraxial and the non-paraxial rays resulting from the defocus with negative Δf. As paraxial rays lead in phase with respect to those affected by aberration, the corresponding path-length difference equal to $(1/4)C_s\beta^4$ in (7.24) is negative.

Thus the amplitude distribution in the back-focal plane in the presence of an aperture is given by [see also Eq. (7.7)]

$$Q = Q(x^*y^*)A(x^*y^*) \exp i\kappa, \tag{7.25}$$

and that in the image plane is given by

$$\Psi(xy) = q(xy) * \frac{J_1(2\pi\rho r)}{2\pi\rho r} * \mathscr{F}\{\exp i\kappa\}, \tag{7.26}$$

where the symbol $\mathscr{F}$ implies the Fourier transform.

Before proceeding to analyze (7.25) let us consider in greater detail the influence of spherical aberration and aperture on resolution. The aperture dimensions limit the angle of scattering to a certain value β_{lim}, and resolution improves with the increase of β_{lim} [see Eq. (7.11b)]. On the other hand, spherical aberration leads to smearing of the image points, so that the smeared disc radius defining the resolution is proportional to β_{lim}^3.

It is obvious that there should be an optimum β_{lim} ensuring the highest resolution (Scherzer 1949). It should be also taken into account that for the spherical aberration, the radius of the smeared disc decreases with moving upward from the plane of observation (see Fig. 30). Therefore the disc radius involved may be decreased to $KC_s\beta_{\text{lim}}^3$ by increasing the objective lens power (Zworikin et al. 1945). The value of K depends on the distribution of electrons over the angles of scattering. Therefore the total radius of a smeared disc is given by [see Eq. (7.11b)].

$$r = KC_s\beta_{\text{lim}}^3 + 0.61\,\lambda/\beta_{\text{lim}}.$$

The minimum value for r is that for $dr/d\beta = 0$, i.e., for

$$\beta_{\text{opt}} = (0.2\lambda)^{1/4}(KC_s)^{-1/4}.$$

Therefore

$$r_{\text{min}} = K'\lambda^{3/4}C_s^{1/4}. \tag{7.27}$$

It is supposed that $K' = 0.5$ at the best (Zworikin et al. 1945). If $C_s = 0.7$ mm, then for $V = 100$ kV and $\lambda = 0.037$ Å, $r_{\text{min}} = 2.0$ Å. Thus, under ideal conditions details as small as separate atoms could be resolved in images. It should be noted that, under ideal conditions, the aperture should remove diffracted beams corresponding to d less than or equal to the resolution, since the contribution from these reflections increases the background in the image.

We have been discussing resolution of separate points in the image. However, the resolution, the magnification, and the stability of a modern electron microscope is usually controlled by observing fringe images. In this case, a resolution of about several tenths of 1 Å is achieved, which is not equivalent to resolution that can be obtained in point images. It should be taken into account that, in general, the objects analyzed are neither sets of isolated atoms nor ideally perfect crystals. Therefore, resolution estimates obtained by analyzing either point images or lattice fringes do not reflect to the full measure the real resolution achieved in the images observed.

Chromatic Aberration. Instability in high voltage and current in the objective lens leads to aberration resulting from electron wavelength variations. In other words, each scattering center turns into a smeared disc of the radius

$$\Delta r_c = C_c\beta[(\Delta V/V)^2 + (2\Delta i/i)^2]^{1/2},$$

where C_c is the chromatic aberration coefficient, β is the angle of scattering, $\Delta V/V$ and $\Delta i/i$ are the relative fluctuations in the voltage and the objective lens current (Heidenreich 1964).

It has been noted above that defocus leads to smeared discs of the radius $\Delta r = \Delta f\beta$. Therefore the effect of chromatic aberration may be attributed to variation in the focal distance of the objective lens by the value

$$\Delta f_c = C_c\left[\left(\frac{\Delta V}{V}\right)^2 + \left(\frac{2\Delta i}{i}\right)^2\right]^{1/2}.$$

In modern microscopes $\Delta V/V$ and $\Delta i/i$ are about 10^{-6}, so that, for $C_c = 1.5$ mm, the value $\Delta f_c = 35$ Å resulting from variations in V and i has no effect on the quality of images. However, electrons generated by the emitter are characterized by an energy spread which also leads to a certain pulsational variation in the focal distance. Therefore it is worthwhile to discuss the influence of chromatic aberration on the quality of images in a more general form. This problem was solved by Fejes (1977). According to (7.21) the amplitude distribution for a wave having left the object plane is described, for a defocus Δf, by the convolution of the transmission function $q_0(r)$ ($\Delta f = 0$) with the propagation function, i.e.,

$$q_{\Delta f}(r) = q_0(r) * \exp[-i\pi r^2/\lambda \Delta f] . \tag{7.28}$$

The intensity distribution in the image is then given by

$$J_{\Delta f}(r) = q_{\Delta f}(r) q^*_{\Delta f}(r) . \tag{7.29}$$

In the case of chromatic aberration, the energy spread is represented by a set of various Δf values. The probability of occurrence for these may be described by, e.g., the Gaussian function $\mathcal{D}(\Delta f)$. Consequently the values of $J_{\Delta f}(r)$ ought to be averaged over all possible Δf, so that

$$J(r) = \int J_{\Delta f}(r) \mathcal{D}(\Delta f) d(\Delta f) = \int q_{\Delta f}(r) q^*_{\Delta f}(r) \mathcal{D}(\Delta f) d(\Delta f) , \tag{7.30}$$

where

$$\mathcal{D}(\Delta f) = \frac{1}{\sigma \sqrt{2\pi}} \exp\left[-\frac{(\Delta f - \Delta f_0)^2}{2\sigma^2} \right] . \tag{7.31}$$

In other words, $\mathcal{D}(\Delta f)$ describes the Gaussian distribution of Δf with a standard deviation σ from the mean defocus Δf_0.

According to the expression (7.22), the diffracted amplitude distribution in the back-focal plane of the objective lens for a given defocus Δf is given by

$$Q_{\Delta f}(r^*) = \mathcal{F}\{q_{\Delta f}(r)\} = Q_0(r^*) \exp i\pi \lambda \Delta f(r^*)^2 . \tag{7.32}$$

The intensity distribution averaged over Δf is the Fourier transform of $J(r)$, i.e.

$$I(r^*) = \mathcal{F}\{J(r)\} = \int Q_{\Delta f}(r^*) * Q^*_{\Delta f}(-r^*) \mathcal{D}(\Delta f) d(\Delta f) . \tag{7.33}$$

Taking account of (7.31), (7.32) and the properties of convolution, (7.33) may be written

$$I(r^*) = \int Q_{\Delta f_0}(r^{*\prime}) Q^*_{\Delta f_0}(r^{*\prime} - r^*) \exp\{-\tfrac{1}{2}\pi^2\lambda^2\sigma^2[(r^{*\prime})^2 - (r^{*\prime} - r^*)^2]^2$$
$$\times dr^{*\prime}$$
$$= \int Q_\sigma(r^{*\prime}) Q_\sigma(r^{*\prime} - r^*)$$
$$\times \exp\{\pi^2\lambda^2\sigma^2(r^{*\prime})^2(r^{*\prime} - r^*)^2\} dr^{*\prime} , \tag{7.34}$$

where

$$Q_\sigma(r^*) = Q_{\Delta f_0}(r^*) \exp[-\tfrac{1}{2}\pi^2\lambda^2\sigma^2(r^*)^4] . \tag{7.35}$$

The exponential in (7.35) is called the *envelope function* and σ is the width of the defocus spread.

Fejes (1977) has noted that if all the values of $Q(r^*)$ are relatively small as compared with the strong peak at the origin, the expression (7.34) has a simple and clear physical meaning. The exponential in (7.34) will then be unity since, for strong reflections, either $r^{*\prime} = 0$, or $r^{*\prime} - r^* = 0$, therefore

$$I(r^*) = Q_\sigma(r^*) * Q_\sigma^*(-r^*),$$ (7.36a)

$$J(r) = q_\sigma(r)q_\sigma^*(r).$$ (7.36b)

Consequently, chromatic aberration decreases the diffracted amplitude $Q_{\Delta f_0}(r^*)$, for the mean defocus Δf_0. The exponential $\exp -\frac{1}{2}\pi^2\lambda^2\sigma^2(r^*)^4$ determining this decrease depends on the standard deviation σ which reflects the spread of pulsations of the focal distance. The increase in σ leads to a more rapid decrease in the diffracted amplitudes with r^* and their contribution to the image.

In this respect the chromatic aberration is essentially analogous to the aperture limitation, i.e., it leads to smearing of the points in the image and lowers the resolution due to a decrease in the contribution of "external" reflections to the image. Thus, the chromatic aberration is displayed in an effective decrease in the diameter of the real aperture. Therefore, under the experimental conditions, the resolution of separate details in an image may appear worse than was expected for the given aperture dimensions. If it is supposed that the reflections contributing to the image are attenuated by chromatic aberration by a factor of 10 or less, then the maximum r^* for reflections of the highest orders according to the formula for the envelope function, will be defined by the expression

$$\ln 10 = \pi^2\lambda^2\sigma^2(r^*)^4/2$$

i.e., $r^* = 0.83/(\lambda\sigma)^{1/2}$ or $d = 1.2(\lambda\sigma)^{1/2}$ (7.37a). If the aperture radius is, e.g., 0.33 Å^{-1} but $\sigma = 200 \text{ Å}$, then $r^* = 0.29 \text{ Å}^{-1}$. Hence, a wider aperture used in experiments does not provide any gain in resolution of details in the image. The decrease of the role of the chromatic aberration is associated, first of all, with the greatest possible decrease in σ.

Incident Beam Divergence. It is obvious that, within the cone formed by directions that are defined by the incident beam divergence, each specific direction forms an image differing in details from those formed by rays propagating along other directions. The image observed is an incoherent superposition of the images produced by a set of rays forming the divergent incident beam.

It is evident that the smaller the incident beam divergence, the higher the resolution. However, the decrease of the half-opening of the focused incident beam cone from 10^{-3} to $5 \cdot 10^{-5}$ rad leads to a rapid decrease in the sample illumination, so that the observation and the registration of high-resolution images becomes hardly feasible. Therefore, compromise solutions are important that are based on analysis of the main aspects in the influence of the incident beam divergence on the images observed.

It has been mentioned above that minor variations in the angle of incidence of the initial electron beam change the excitation error, which affects the amplitude-and-phase distribution of diffracted waves. However, for a very thin crystal and a comparatively small (about 10^{-3} rad) incident beam divergence, this effect is negligible. This is equivalent to a superposition of crystal structure projections along directions within the angular spread for the initial beam.

The spherical aberration and out-of-focus imaging increase the effect of the incident beam divergence on the quality of images. For example, in out-of-focus images each non-paraxial ray corresponds to an image shifted in the image plane with respect to that formed by a paraxial ray. Thus, details would be smeared to a greater extent with greater divergencies. O'Keefe and Sanders (1975) reported that in modern microscopes this effect influences appreciably an image with a resolution of about 3 Å only for $\Delta f > 1000$ Å.

A greater perturbing influence of the incident beam divergence on the image is associated with spherical aberration. O'Keefe and Sanders (1975) have shown that, for $C_s = 1.8$ mm, the aperture radius $\rho = 0.33$ Å^{-1} and the half-opening of the focused incident beam $\alpha = 10^{-3}$ rad, the effective resolution is 3.8 Å instead of 3 Å. This is due to the fact that the external diffraction beams practically do not contribute to the image in spite of having passed through the aperture.

By decreasing the incident beam divergence to $2 \cdot 10^{-4}$ rad, Bursill and Wilson (1977) succeeded in resolving details at distances less than 3 Å. Fejes (1977) has obtained an expression reflecting the modulation of the diffracted amplitude distribution resulting from the incident beam divergence

$$\frac{2J_1\{2\pi\alpha[\Delta f r^* + C_s\lambda^2(r^*)^3]\}}{2\pi\alpha[\Delta f r^* + C_s\lambda^2(r^*)^3]} = \frac{2J_1(x)}{x}, \tag{7.37}$$

where α is the half-opening of the cone describing the focused incident beam divergence. Increase in α leads to a more rapid decrease in diffracted amplitudes with r^*. This results in a more significant decrease in the contribution of "external" diffracted beams with relatively large r^* to the image. Consequently, decrease in the resolution due to the incident beam divergence is similar to that resulting from aperture limitation or chromatic aberration.

Thus, to achieve resolution of finest details, it is necessary to minimize C_s and the incident-beam divergence without a loss in illumination, and to improve the electron gun so that the energy spread for emitted electrons are minimized.

7.4 Contrast in Electron-Microscopic Images of Thin Crystals

According to the n-beam dynamical theory, interpretation of high-resolution images of crystals should meet with a number of difficulties which, in general, become insuperable if the object structure is unknown. Dynamical n-beam calculations of diffraction patterns testify that the amplitude distribution to be expected at the back-focal plane of the objective lens depends strongly on crystal thickness, on the orientation, on the structure and composition of the specimen, on the unit cell dimensions, etc.

Since the crystallites under study may have a certain distribution of thicknesses, a set of different diffraction characteristics may correspond to the same crystal structure. For the same reason, an electron-microscopic image would change with sample thickness. Interpretation of images should be based on mathematical simulation of all processes and parameters associated with the image formation. Precise knowledge of the object crystal structure is the necessary condition for such simulation. In general, there is no direct relationship between the crystal structure and the image observed since, even for relatively thin crystals, $q(xy)$ is an exponential function of the crystal potential. Then for an unknown structure the structural information obtained by simulation would be hardly too reliable since the number of variables is large.

A complex dependence of the diffracted amplitude distribution on numerous factors has made the impression that only the n-beam theory could explain the contrast in images even for very thin crystals. It is of interest that experimental studies rather than theoretical predictions have revealed certain conditions that allow direct observation of a crystal structure (Allpress and Sanders 1973; Buseck and Iijima 1974, 1975; Iijima 1971).

Iijima (1971) was the first to demonstrate the possibility of direct visualization of a structure motif in studying the crystals of $Ti_2Nb_{10}O_{29}$. This has marked a new stage in the development of high-resolution electron microscopy, and one of the implications was that electron-microscopic images could be interpreted using simpler approximations than the n-beam theory. Although the latter gives the most comprehensive description of imaging, it is nevertheless desirable to use some other theories requiring no a priori knowledge of the structure and providing criteria for a reliable and unambiguous interpretation of experimental data.

The phase-object approximation (POA) which is apparently valid for thicknesses of about $50-100$ Å (Cowley and Iijima 1972; Iijima 1971) proved to be most effective in this respect. If we consider only very thin crystals and neglect the Fresnel diffraction, then $q(xy) = \exp i\sigma\varphi(xy)$.

This is possible if the propagation function is close or equal to unity, i.e., if

$$p(xy) = \exp\left[-2\pi ik\frac{x^2+y^2}{2t}\right] = \exp\left[-2\pi i\frac{r^2}{2t\lambda}\right] \simeq \exp - \pi i.$$

Hence the maximum possible thickness is defined by the expression $r^2/t\lambda = 1$, or $t = r^2/\lambda$ where r is the limit of interpretable resolution. For resolution of 2 Å and $V = 100$ kV, $t = 100$ Å.

Here it is appropriate to discuss the so-called column approximation. In the small-angle case, the lateral divergence of scattered waves is limited to a very narrow column. For the given λ, the width of the column is defined, first of all, by the crystal thickness (Hirsh et al. 1977). In the above example the width of the column is about 2 Å for $\lambda = 0.04$ Å and $t = 100$ Å. This implies that the amplitude at each point at the exit face of the crystal depends, in fact, only on the $\varphi(xy)$ averaged along z for the given x, y. This determines extremely high locality in the transmission of information on the structure of the object. Therefore, both periodic and aperiodic thin phase-objects may be equally well described by a function of the type $\exp i\sigma\varphi(xy)$. In out-of-focus images it is possible to satisfy the condition $q = \exp i\sigma\varphi'(xy)t$ [see Eq. (7.21)] if $n = r^2/\lambda\Delta f$, i.e., for the column width $r = (n\Delta f\lambda)^{1/2}$. The amplitude of the wave at any point in the exit face of the crystal is defined by averaging over the column volume, while the column width depending on $n\Delta f$ may be as large as a unit cell parameter. For example, for $n\Delta f = 2000$ Å and $\lambda = 0.04$ Å the column width is 9 Å. The smearing of details becomes considerable under such conditions. The difference between the images for periodic and aperiodic objects will be also substantial.

Weak-Phase-Object Approximation (WPOA). The amplitude distribution in the image plane for a thin phase-object is described, according to (7.26), by the function

$$q(xy) = \exp\{-i\sigma\varphi'(xy)t\} * \mathcal{F}\{A(x^*y^*)\} * \mathcal{F}\{\exp i\kappa\}, \qquad (7.38)$$

where $\varphi'(xy)$ is the projection of the electrostatic potential on the unit length along Z. In the expression (7.38) the convolution with the Fourier transform of the aperture function $A(x^*, y^*)$ includes, actually, not only the aperture limitation, but also the influence of chromatic aberration and the incident beam divergence. All these effects may be allowed for with the help of the effective absorption function $\mu'(x, y)$, so that (7.38) becomes

$$q(xy) = \exp\{-i\sigma\varphi'(xy)t - \mu'(xy)t\} * \mathcal{F}\{\exp i\kappa\}. \qquad (7.39)$$

To visualize in the image the projection of the crystal structure, or, to be more precise, of the electrostatic potential, it is necessary that there should be a linear relationship between $q(xy)$ and this projection. This is possible if we assume

$$\exp\{-i\sigma\varphi'(xy)t - \mu'(xy)t\} \simeq 1 - i\sigma\varphi'(xy)t - \mu'(xy)t, \qquad (7.40)$$

i.e., $\sigma\varphi'(xy)t < 1$ and $\sigma\varphi'(xy) > \mu'(xy)$. The second condition suggests that variations of $\varphi'(xy)$ with respect to the average inner potential are relatively weak. Then $\varphi'(xy)$ may be replaced by φ_0 and

$$t \leqslant 1/\sigma\varphi_0 = \lambda V/\pi\varphi_0.$$

For 100 kV electrons, $\lambda V = 4060$ V, and if $\varphi_0 = 10$ V, $t = 120$ Å. Thus the approximation (7.40) seems to be satisfied exactly enough only for very thin samples of light atom materials leading to minor phase changes for $q(xy)$. This is the so-called weak phase-object approximation (WPOA) (Cowley 1975b). Although it does not apply for most of the crystal structures, it is nevertheless useful to discuss it, since it provides a qualitative explanation for many effects observed.

The expression for the diffracted amplitude distribution in the back-focal plane of the objective lens may be written, with account taken of (7.40), as

$$Q(x^*y^*) = [\delta(x^*y^*) - i\sigma\,\mathcal{F}\{\varphi(xy)\} - \mathcal{F}\{\mu(xy)\}]\exp i\kappa$$
$$= [\delta(x^*y^*) + \sigma\sin\kappa\,\mathcal{F}\{\varphi(xy)\} - \cos\kappa\,\mathcal{F}\{\mu(xy)\}]$$
$$- i[\sin\kappa\,\mathcal{F}\{\mu(xy)\} + \sigma\cos\kappa\,\mathcal{F}\{\varphi(xy)\}].$$

The image intensity distribution is given by

$$J(xy) = |\mathcal{F}\{Q(x^*y^*)\}|^2 = [1 + \sigma\varphi(xy) * \sin\kappa - \mu(xy) * \cos\kappa - i\varphi(xy)$$
$$* \cos\kappa - i\mu(xy) * \sin\kappa]^2 + \text{second order terms}.$$

Suppose that $\sin\kappa \to 1$, $\cos\kappa \to 0$ for all reflections contributing to the image. Then

$$Q(x^*y^*) = [\delta(x^*y^*) + \sigma\,\mathcal{F}\{\varphi(xy)\}] - i\,\mathcal{F}\{\mu(xy)\}$$
$$= \delta(x^*y^*) + \sigma\,\Phi(x^*y^*) - iM(x^*y^*).$$

Taking the reverse Fourier transform of the function $Q(x^*y^*)$ with $\sigma\varphi(xy) > \mu(xy)$ we obtain

$$J(xy) = |1 + \sigma\varphi(xy)|^2 + |\mu(xy)|^2 \simeq 1 + 2\sigma\varphi(xy). \qquad (7.41)$$

Thus, to obtain in the image a linear relationship of the intensity J and the projection of the electrostatic potential, it is necessary to adjust the experimental conditions so as to satisfy the condition $\sin\kappa \to 1$ within a sufficiently wide diffraction angular range. Scherzer, who had determined the optimum defocus conditions, was the first to solve this problem (Scherzer 1949).

Putting $\rho = [(x^*)^2 + (y^*)^2]^{1/2}$ we have, according to (7.23)

$$\kappa = \pi\lambda\Delta f\rho^2 - \frac{\pi}{2}C_s\lambda^3\rho^4. \qquad (7.42)$$

This expression has been derived for negative Δf. Otherwise, there should be a minus sign before the first term in (7.42). The problem is to find a value for Δf such as to ensure that $\sin\kappa$ is close to unity in a sufficiently wide range of ρ values. The relationship between the optimum defocus and the aperture dimensions may be determined using the condition $d\kappa/d\rho = 0$. According to (7.42) we have

$$\Delta f_0 = C_s\lambda^2\rho_0^2 = C_s\beta_0^2. \qquad (7.43)$$

If it is assumed that $\kappa_0 = 2\pi/3$ and (7.43) is satisfied, we obtain from (7.42)

$$\Delta f_0 = -1.15C_s^{1/2}\lambda^{1/2}, \quad \rho_0 = 1.08C_s^{-1/4}\lambda^{-3/4}. \qquad (7.44)$$

In modern microscopes $C_s = 0.7$ mm. Therefore, for $V = 100$ kV ($\lambda = 0.037$ Å), the optimum defocus is -550 Å. Figure 31 shows $\sin\kappa$ plotted against ρ for $\Delta f_0 = -550$ Å. It is seen that within the interval 0.12 Å^{-1} $\leqslant \rho \leqslant 0.31$ Å^{-1} ($8.3 \div 3.2$ Å in terms of interatomic distances or d-spacings) $\sin\kappa$ is sufficiently close to unity. To find the optimum aperture we make use of (7.44) which gives $\rho_0 = 0.25$ Å^{-1}. It is seen in Fig. 31 that this value corresponds to $\sin\kappa = 0.9$. However, the figure shows that the same value of $\sin\kappa$ is obtained with $\rho = 0.31$ Å^{-1}. Then $\rho_0 = 1.34C_s^{-1/4}\lambda^{-3/4} = 0.31$ Å^{-1}.

Since the increase in the aperture dimensions leads to the increase in the number of reflections contributing to the image, as well as to the improvement in resolution, the optimum aperture should be the largest for the given parameters of the electron microscope.

Numerous high-resolution images have been obtained in microscopes having $C_s = 1.8 - 2.0$ mm. A calculation similar to the one given above shows that the optimum defocus is then -1000 Å for $\rho_0 = 0.27$ Å^{-1}. Buseck and Iijima (1974, 1975, 1976) have found that a direct relationship between an image and a structure projection is obtained, on a qualitative level, with $\Delta f = -900$ Å and $\rho = 0.33$ Å^{-1}. The choice of a somewhat larger aperture is justified by the fact that incident beam divergence and chromatic aberration decrease the actual aperture diameter. The actual resolution in the images observed corresponds, in fact, to the aperture diameter of 0.27 Å^{-1} instead of 0.33 Å^{-1}.

Thus a simple and rather rough treatment of $q(xy)$ in the form (7.40) has allowed sufficiently precise prediction of the optimum experimental Δf_0 and ρ_0,

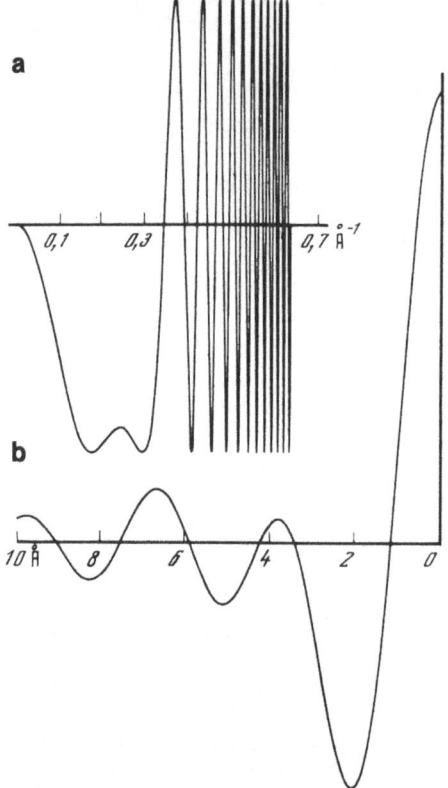

so that the intensity distribution may be directly interpreted in terms of the potential projection. It should be stressed that periodic contrast is observed with other Δf values, and sometimes with a better resolution. However, for values of Δf ranging from -900 to -1000 Å ($C_s = 1.8$ mm) images are in the best agreement with the two-dimensional electrostatic potential distribution of the object under study.

In the general case, where the condition for the optimum defocus is not met but the WPOA is valid, we have

$$q(xy) = 1 - i\sigma\varphi(xy) * \mathscr{F}\{A(x^*y^*)\exp i\kappa(x^*y^*)\}$$
$$J(xy) = |q(xy)|^2 = 1 + 2\sigma\varphi(xy) * \mathscr{F}\{A(x^*y^*)\sin\kappa\}. \tag{7.45}$$

The Fourier transform of the function describing the mage intensity leads to the following expression

$$j(x^*y^*) = \delta(x^*y^*) + 2\sigma\Phi(x^*y^*)A(x^*y^*)\sin\kappa. \tag{7.46}$$

The expressions (7.45) and (7.46) relate the intensity distribution in the image of a WPO and the relevant structure amplitudes. This may serve to solve the phase problem (Klug 1979). The delta function can be eliminated by omitting the term of the transform with $x^* = y^* = 0$ and

$$\Phi(x^*y^*) = -j(x^*y^*)/[2\sigma\sin\kappa(x^*y^*)] \,. \tag{7.47}$$

$A(x^*y^*) = 1$ for all the beams transmitted through the aperture. Therefore, the expression (7.47) may be used to determine the phases only for diffracted beams contributing to the image. The maximum errors in the determination of $\Phi(x^*y^*)$ appear for small values of $\sin\kappa(x^*y^*)$, and the minimum errors are possible with the optimum defocus where $\sin\kappa = 1$. One of the ways to improve considerably the image quality and resolution is the compensation for aberrations by a posteriori Fourier image processing (Erickson and Klug 1971). The Fourier transform of the intensity distribution in the image leads to the expression (7.46), which shows how the phase contrast transfer function modulates the amplitude of $\Phi(x^*y^*)$ and its phases. It is obvious that each diffraction maximum $\Phi(x^*y^*)$ at the point x^*y^* obtained by the Fourier transform of the image will be of the maximum amplitude for a definite defocus Δf implying $\sin\kappa = 1$ for the given reflection. Therefore, if a set of images is obtained experimentally with different Δf values, computer generated Fourier transforms of each image allow determination of the maximum amplitudes and phases for all diffraction maxima. The conventional Fourier synthesis then gives readily an idealized image of $\varphi(xy)$. Thus, limitations resulting from both the spherical and chromatic aberrations can be compensated, which improves substantially the resolution of structural details in the image. Moreover, this procedure may eliminate possible ambiguity in intuitive interpretation of a single image obtained under the Scherzer optimum defocus in terms of the crystal structure. Klug (1979) has described an application of a posteriori image processing to the interpretation of $GeNb_9O_{25}$ images obtained under different through-focus values. It is of interest that his approach has been purely kinematical, i.e., all dynamical effects have been ignored, although the structure under study contained very heavy atoms, and even thin $GeNb_9O_{25}$ crystals can hardly be treated as phase-objects. It should be noted that this procedure allows a substantial increase in the number of diffracted beams contributing to the image, and it is this increase that improves resolution.

Despite the fact that the above approximations provide an adequate explanation of the effects observed, there is a serious drawback, namely, the assumption of a small phase change, i.e., $\sigma\varphi(xy) \ll 1$. Cowley and Iijima (1976) have calculated that for $Ti_2Nb_{10}O_{29}$ there may be a phase change varying by 10π around the heavy atom positions for a thickness of 100 Å. A reasonably adequate interpretation of the images based on an invalid assumption seems therefore surprising, as noted by these authors. Fejes (1977) attempted to eliminate the contradiction involved. He has allowed for the smearing effect of the aperture in the image formation, which is one of the basic reasons for strong phase-objects acting as weak ones under experimental conditions.

The wave function in the image plane is equal, according to (7.10), to

$$\Psi(r) = \exp[-i\sigma\varphi(r)] * \frac{J_1(2\pi pr)}{(2\pi pr)}$$
$$= \cos\sigma\varphi(r) * \frac{J_1(x)}{(x)} - i\sin\sigma\varphi(r) * \frac{J_1(x)}{(x)} \tag{7.48}$$

where r is the coordinate vector lying in the plane of observation and $x = 2\pi pr$.

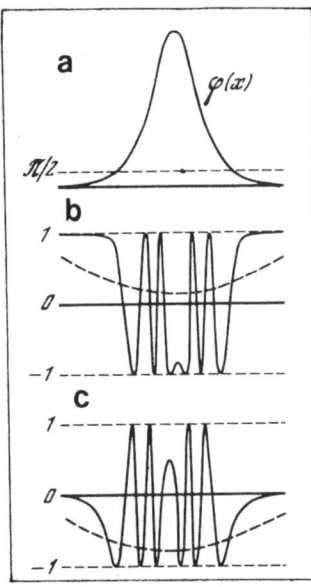

Fig. 32a–c. Approximation of a strong phase object by a weak one: **a** the phase change in $\sigma\varphi(x)$ resulting from a row of atoms parallel to the incident beam; **b** *solid curve:* $\cos\sigma\varphi(x)$; *dashed curve:* the convolution of $\cos\sigma\varphi(x)$ with $J_1(x)/x$; **c** *solid curve:* $\sin\sigma\varphi(x)$, *dashed curve:* the convolution of $\sin\sigma\varphi(x)$ with $J_1(x)/x$ (Fejes 1977)

Figure 32 illustrates the influence of the aperture limitation. Fig. 32a shows the potential $\sigma\varphi(r)$ distribution curve defined by a linear row of atoms which is normal to r. It is clearly seen that the maximum value of $\sigma\varphi(r)$ is much greater than $\pi/2$. In Fig. 32b and c, solid lines represent the curves describing the variation of the real and the imaginary components of $\exp-i\sigma\varphi(r)$ with r for the function $\sigma\varphi(r)$ given in Fig. 32a. Within the limits of r where $\varphi(r) \neq 0$ the curves $\cos\sigma\varphi(r)$ and $\sin\sigma\varphi(r)$ vary repeatedly from -1 to 1. The dashed lines in Fig. 32b and c, representing the curves $\cos\sigma\varphi(r) * J(x)/x$ and $\sin\sigma\varphi(r) * J(x)/x$ respectively, show the smoothing influence of the aperture. It was assumed in the calculations that $\rho = 0.33$ Å^{-1} and the half-width of the curve $\varphi(r)$ is 0.25 Å.

The smooth variation of these curves with r with extreme values at the point corresponding to the maximum in $\varphi(r)$ permits us to write the wave function in the image plane as

$$\Psi(r) = \cos\sigma\varphi(r) * J_1(x)/x - i\sin\sigma\varphi(r) * J_1(x)/x = \exp-i\sigma\varphi_{\mathrm{ef}}(r), \quad (7.49)$$

where $\varphi_{\mathrm{ef}}(r)$ is the effective potential distribution having maxima at the same positions as $\varphi(r)$. However, in contrast to $\varphi(r)$, the maxima of $\varphi_{\mathrm{ef}}(r)$ are always less than $\pi/2$.

The expression (7.49) implies $|\Psi(r)|^2 = 1$. However, $|\Psi(r)|^2 < 1$, since the averaging of $q(r)$ over a large number of oscillations decreases substantially the real and imaginary components of $\Psi(r)$. Therefore, it is more appropriate to describe the amplitude distribution in the image plane by the function

$$\Psi(r) = \exp\{-i\sigma\varphi_{\mathrm{ef}}(r) - \mu(r)\}$$

where $\mu(r)$ allows for effective absorption. The maxima of $\mu(r)$ coincide with those of $\varphi(r)$ and increase in magnitude with crystal thickness.

Thus, thin crystals having transmission functions described by (6.5) act as weak phase-objects with an absorption effect regardless of the value of $\sigma\varphi(r)$. It should be once more stressed that this approximation is a qualitative one, so that the intensity distribution in the image cannot be used for direct evaluation of $\sigma\varphi(xy)$.

Li and Tang (1985) have recently proposed a more rigorous and explicit explanation of the fact that the contrast distribution may correspond to the projected potential distribution in the case where the WPOA does not hold. They have made use of the *n*-beam dynamical theory to devise the so-called.

Pseudo-Weak-Phase-Object Approximation (PWPOA). If a crystal is divided into $n + 1$ equal slices of a minor thickness Δz, each slice will act as a weak phase-object. According to (6.12), the wave function transmitted through one slice may be written

$$q(xy) = 1 - i\sigma\varphi(xy).$$

Then, for Fresnel diffraction the expression (6.9) for a wave at the exit surface of the second slice becomes

$$q(xy) * p(xy) = 1 - i\sigma\varphi(xy) * p(xy)$$

where $p(xy)$ is defined by (6.13).

The wave function for electrons having left the second slice becomes, according to (6.10) and (6.12)

$$q_2(xy) = q_1(xy)[q_1(xy) * p(xy)]$$

$$= 1 - i\sigma\varphi(xy) - i\sigma\varphi(xy) * p(xy) - \sigma^2\varphi(xy)[\varphi(xy) * p(xy)] . \quad (7.50)$$

The basic PWPOA assumption is that the Fresnel diffraction between the successive slices is taken into account while multiple scattering is not. Therefore the term in (7.50) allowing for multiple scattering is neglected, so that

$$q_2(xy) = 1 - i\sigma\varphi(xy) - i\sigma\varphi(xy) * p(xy) .$$

Consequently, the expression (6.14) for the transmission function at the exit face of the crystal becomes

$$q_{n+1}(xy) = 1 - i\sigma\varphi(xy) - i\sigma\varphi(xy) * p(xy) - i\sigma\varphi(xy) * p(xy)$$

$$* p(xy)\ldots - i\sigma\varphi(xy) * \underbrace{p(xy) * p(xy)\ldots * p(xy)}_{n \text{ times}} . \quad (7.51)$$

The Fourier transform of the above expression is the diffracted wave function

$$Q(x^*y^*) = \delta(x^*y^*) - i\sigma\Phi(x^*y^*) - i\sigma\Phi(x^*y^*) \sum_{m=1}^{n} \mathscr{P}^m(x^*y^*) \quad (7.52)$$

where $\Phi(x^*y^*)$ and $\mathscr{P}(x^*y^*)$ are defined by (6.15) and (6.20), respectively.

Taking account of (6.20) and (2.16), the expression (7.52) may be modified

$$Q(x^*y^*) = \delta(x^*y^*) - i\sigma\Phi(x^*y^*)\left[\sum_{m=1}^{n}\exp(2\pi imu)\right]$$

$$= \delta(x^*y^*) - i\sigma\Phi(x^*y^*)\left[\frac{\sin\pi(n+1)u}{\sin\pi u}\exp(-\pi imu)\right] \quad (7.53a)$$

where $u = \Delta z\lambda[(x^*)^2 + (y^*)^2]/2$. For WPOA, the kinematical approximation holds and the reflection intensity $I(hk)$ is proportional to $|\Phi(hk)|^2$ where $h = x^*/a^*$, $k = y^*/b^*$.

As implied by (7.53), the reflection intensity in terms of PWPOA depends not only on $|\Phi(hk)|^2$, but also on the crystal thickness and the reflection coordinates x^*, y^*, so that

$$I(hk) \simeq |\Phi(hk)|^2\frac{\sin^2\pi(n+1)u}{\sin^2\pi u}. \tag{7.53b}$$

For the optimum defocus $\exp i\kappa = -i$ and

$$Q(x^*y^*)\exp i\kappa = \delta(x^*y^*) - \sigma\Phi(x^*y^*) - \sigma\Phi(x^*y^*)\sum_{m=1}^{n}\mathscr{P}^m(x^*y^*). \tag{7.54}$$

The Fourier transform of (7.54) is

$$\Psi_{n+1}(xy) = 1 - \sigma\varphi(xy) - \sigma\varphi(xy)\cdot\sum_{m=1}^{n}\mathscr{F}^{-1}[\mathscr{P}^m(x^*y^*)]$$

$$= 1 - \sigma\varphi(xy) - \sigma\varphi(xy)*\sum_{m=1}^{n}S_m(xy) - \sigma\varphi(xy)*\sum_{m=1}^{n\cdot}C_m(xy)$$

where

$$S_m(xy) = (1/m\Delta z)\sin[(\pi/m\Delta z)(x^2 + y^2)].$$

$$C_m(xy) = (1/m\Delta z)\cos[(\pi/m\Delta z)(x^2 + y^2)].$$

Li and Tang (1985) have analyzed the results of the convolution of $S_m(x, y)$ and $C_m(x, y)$ with $\varphi(x, y)$. For convenience, the function $\varphi(x, y)$ has been written as a Gaussian function

$$G(r) = A\exp(-r^2/R^2) \quad \text{where} \quad r^2 = x^2 + y^2,$$

representing, e.g., the potential distribution resulting from a column of atoms projected on the plane x, y. The values $G(r)*C_m(r)$ and $G(r)*S_m(r)$ are proportional to $\exp[-r^2/(R^2 + m\Delta z^2/\pi^2 R^2)]$, i.e., they are also Gaussian functions. The widths of these, however, increase with m.

The maximum values of these functions for $r = 0$ are (Li and Tang 1985)

$$G(r)*S_m(r)\,|_{r=0} = A\pi^2R^4/(m^2\Delta z^2 + \pi^2R^4)$$

$$G(r)*C_m(r)\,|_{r=0} = A\pi R^2m\Delta z/(m^2\Delta z^2 + \pi^2R^4).$$

These expressions imply that the first function decreases monotonically with m, while the second one increases from zero (with $m = 0$) to a certain maximum and then falls off. It is also seen that

$$G(r) * S_m(r) |_{r=0} > G(r) * C_m(r) |_{r=0} \quad \text{for} \quad m < R^2/\Delta z \qquad (7.55a)$$

$$G(r) * S_m(r) |_{r=0} < G(r) * C_m(r) |_{r=0} \quad \text{for} \quad m > R^2/\Delta z . \qquad (7.55b)$$

Thus, with the optimum defocus, the convolution of $\varphi(x, y)$ with $S_m(x, y)$ and $C_m(x, y)$ should lead to a smearing of the $\varphi(r)$ maxima. The condition (7.55a) will hold near the positions of these maxima if the crystal thickness does not exceed a certain critical value.

Making PWPOA we omit the terms $\sigma^2\varphi^2$, and the intensity distribution in the image is described by (Li and Tang 1985)

$$J_{n+1}(xy) = | \Psi_{n+1}(xy) |^2 = 1 - 2\sigma\varphi_{n+1}(xy) \qquad (7.56)$$

where

$$\varphi_{n+1}(xy) \quad = \varphi(xy) + \Delta\varphi_{n+1}(xy) \quad \text{and}$$

$$\Delta\varphi_{n+1}(xy) = \varphi(xy) * \sum_{m=1}^{n} S_m(xy) - \sigma\varphi(xy) \left[\varphi(xy) * \sum_{m=1}^{n} S_m(xy) \right]$$

$$- \frac{\sigma}{2} \left[\varphi(xy) * \sum_{m=1}^{n} S_m(xy) \right] - \frac{\sigma}{2} \left[\varphi(xy) * \sum_{m=1}^{n} C_m(xy) \right] .(7.57)$$

For relatively small n, the height of the $\varphi(x, y) * S(x, y)$ maxima is much greater than that for $\varphi(x, y) * C(x, y)$, the first term in (7.57) is much greater than the other ones, i.e.,

$$\Delta\varphi_{n+1}(xy) = \varphi(xy) * \sum_{m=1}^{n} S_m(xy) . \qquad (7.57a)$$

Therefore, under such conditions $\Delta\varphi_{n+1}(x, y)$ should be approximately proportional to $\varphi(x, y)$. That is, $\varphi_{n+1}(xy)$ should have maxima at the same positions as $\varphi(x, y)$, while the corresponding heights should be related by an approximate linear dependence. With the increase in n, these relationships no longer apply. For a certain critical value n_c the values $\varphi_{n+1}(x, y)$ at the positions of the heaviest atoms should be zero, since the contribution of the terms having the negative sign in (7.56) becomes sufficiently large. It should be stressed that the increase in n leads, on the one hand, to a decrease in the heights of the heavy atom maxima and, on the other hand, to an increase in those of the light atoms. Hence, there should be an optimum crystal thickness allowing the visualization of both light and heavy atoms in the image.

Thus, in terms of PWPOA, a real crystal may be replaced by an imaginary one, and real atoms by imaginary atoms. Atoms of both types occupy the same positions, but the maxima for imaginary atoms are more smeared. In addition, their scattering powers are redistributed in such a way that light atoms become "heavier" and vice versa. Note that the above formulae may be used to evaluate the crystal thicknesses for which the PWPOA would apply.

The results obtained show that it is not always that one should endeavor to use the minimum possible crystal thickness, since, in that case, the contribution of light atoms to the image will be very small. Therefore, simultaneous visualization of both heavy and light atoms is not feasible. For certain thicknesses, dynamical effects favor more adequate imaging of crystal structures.

Charge Density Imaging. Cowley and Moodie (1960) and later, in greater detail, Lynch et al. (1975) have shown that, under certain conditions the intensity distribution in a high-resolution image represents the projection of the charge density (PCD) distribution in the crystal along the incident beam direction. This approximation contains no physically serious limitations, and the results based on it provide direct information on the structure of the crystal under study. Furthermore, in terms of the PCD approximation, images of various crystal structures may be readily simulated, since the formulae for the intensity distribution in images are sufficiently simple as regards calculations. The perturbing effect of defocus on the transmission function $q(r)$ is described in terms of the Fresnel theory by the convolution

$$\Psi(r) = q(r) * \exp[-i\pi r^2/\lambda \Delta f] . \tag{7.58a}$$

Taking account of the spherical aberration, the total defocus for each ray scattered at the angle β is $\Delta f_{tot} = \Delta f - \frac{1}{2}C_s\beta^2 = \Delta f - \frac{1}{2}C_s\lambda^2(r^*_{hk})^2$. Then the diffracted amplitude distribution may be obtained by taking the Fourier transform of (7.58a) where Δf should be replaced by Δf_{tot}

$$\mathscr{F}\{\Psi(r)\} = Q(r^*)\exp i\pi \Delta f_{tot}\lambda(r^*)^2 . \tag{7.58b}$$

Assume that the crystal is thin enough to be regarded as a phase-object. Then

$$q(r) = \exp - i\sigma\varphi(r) = \int Q(r^*)\exp[-2\pi i r r^*]dr^* .$$

Hereafter r and r^* are defined by the coordinates of the points x, y and x^*, y^* in the plane of observation and the back-focal plane, respectively, so that $(r)^2 = x^2 + y^2$ and $(r^*)^2 = (x^*)^2 + (y^*)^2$.

Taking the reverse Fourier transform we obtain, applying (7.58b)

$$\Psi(r) = \int Q(r^*)\exp[i\pi \Delta f_{tot}\lambda(r^*)^2]\exp[-2\pi i r r^*]dr^* . \tag{7.59a}$$

The main assumption in the approximation involved is that

$$\exp i\pi \Delta f_{tot}\lambda(r^*)^2 \simeq 1 + i\pi \Delta f\lambda(r^*)^2 - \frac{i}{2}\pi C_s\lambda^3(r^*)^4 . \tag{7.59b}$$

This implies that the argument of the function in (7.59b) should be always less than $\pi/2$.

Consider first the particular case that $C_s = 0$. If the condition (7.59b) is satisfied, (7.59a) becomes

$$\Psi(r) = q(r) + i\pi \Delta f\lambda \int (r^*)^2 Q(r^*)\exp[-2\pi i r r^*]dr^* .$$

It is obvious that

$$\frac{d^2q(r)}{dr^2} = \frac{d^2}{dr^2} \int Q(r^*) \exp[-2\pi i r r^*] dr^*$$

$$= -4\pi^2 \int (r^*)^2 Q(r^*) \exp[-2\pi i r r^*] dr^* .$$

Therefore

$$\Psi(r) = q(r) - \frac{i\Delta f\lambda}{4\pi} \frac{d^2q(r)}{dr^2}$$

$$= \exp - i\sigma\varphi(r) \left\{ 1 + \frac{\Delta f\lambda\sigma}{4\pi} [\varphi''(r) + i\sigma\varphi'(r)] \right\}, \qquad (7.60)$$

where $\varphi'(r) = \partial\varphi/\partial x + \partial\varphi/\partial y$, $\varphi''(r) = \partial^2\varphi/\partial x^2 + \partial^2\varphi/\partial y^2$.

Neglecting the minor terms σ^n and $(\Delta f\lambda)^n$ with $n > 1$, we obtain, using (7.60), the expression for the intensity distribution in the image

$$J(r) = |q(r)|^2 \{1 + (\Delta f\lambda\sigma/2\pi)\varphi''(r)\} .$$

According to the Poisson equation we have

$$\frac{\partial^2\varphi_t}{\partial x^2} + \frac{\partial^2\varphi_t}{\partial y^2} = -4\pi[\rho_+ - \rho_-] = -4\pi\rho(r), \qquad (7.61)$$

where φ_t is the potential per unit length in the direction z, i.e., $\varphi(r) = t\varphi_t(r)$; $\rho(r)$ is the two-dimensional charge density distribution per unit length in the same direction.

The charge distribution in crystals is described by a periodic function. Therefore

$$\rho(r) = \sum_{h,k} V_{hk} \exp - 2\pi i \left(h\frac{x}{a} + k\frac{y}{b} \right) = \sum_{h,k} V_{hk} \exp[-2\pi i r r_{hk}^*] . \qquad (7.62)$$

Finally, with account taken of (7.61), we have

$$J(r) = |q(r)|^2 \{1 - 2\Delta f\lambda\sigma t\rho(r)\} . \qquad (7.63)$$

Thus, if the condition (7.59b) applies, the contrast in the image is directly interpretable in terms of the charge density in the crystal under study. The formula (7.63) has been derived for $C_s = 0$ and the absence of aperture limitation. For $|q(r)|^2 = 1$, the contrast in the image is proportional to the defocus of the objective lens and the crystal thickness. The latter is also valid for the kinematical approximation. In precisely focused images there should be no contrast ($\Delta f = 0$).

Since the aperture smears the oscillations in the transmission function for a strong phase object, the concept of *the effective transmission function* with effective potential and effective absorption may be introduced, i.e.,

$$q_{ef}(r) = \exp i\sigma\varphi(r) * \mathscr{F}\{A(r^*)\} = \exp\{i\sigma\varphi_{ef}(r) - \mu_{ef}(r)\} . \qquad (7.64)$$

Therefore the intensity distribution in the image for a sufficiently large aperture is given by (Lynch et al. 1975)

$$J(r) = |q_{ef}(r)|^2 \{1 - 2\Delta f\lambda\sigma t\rho_{ef}(r)\} , \qquad (7.65)$$

where $\rho_{ef} = \rho * \mathscr{F}\{A(r^*)\}$ is the charge density distribution modulated owing to the aperture limitation.

The expression (7.65) is similar to (7.63) and may be readily calculated analytically for any aperture function. The properties of the function $|q_{ef}|^2 = \exp - 2\mu_{ef}(r)$ have been discussed above.

To allow for the spherical aberration, we insert (7.59b) into (7.59a), that is

$$\Psi(r) = q(r) - \frac{i\Delta f\lambda}{4\pi}\frac{d^2q(r)}{dr^2} + \frac{i\pi}{2}C_s\lambda^3\int(r^*)^4 Q(r^*)\exp[-2\pi irr^*]\,dr^*$$

$$= q(r) - \frac{i\Delta f\lambda}{4\pi}\frac{d^2q(r)}{dr^2} - \frac{iC_s\lambda^3}{32\pi^3}\frac{d^4q(r)}{dr^4}. \tag{7.66}$$

Using the Poisson equation (7.61) and omitting terms with σ^n for $n > 1$ we obtain, on the basis of (7.66), the formula for intensity

$$J(r) = 1 - 2\Delta f\lambda\sigma t\rho(r) + \frac{C_s\lambda^3\sigma t}{4\pi^2}\frac{d^2\rho(r)}{dr^2}. \tag{7.67}$$

Taking account of (7.62) we rewrite (7.67) as

$$J(r) = 1 - 2\lambda\sigma t\rho_{ef}(r)$$

$$= 1 - 2\lambda\sigma t\sum_{h,k}\Delta f_{tot}(hk)V_{hk}\exp2\pi i\left(\frac{hx}{a} + \frac{ky}{b}\right) \tag{7.68}$$

where $\Delta f_{tot}(hk)$ is defined by the expression (7.20) and $r^*_{hk} = ha^* + kb^*$.

The expression (7.68) is similar in appearance to (7.63). In this case, contrast results from a certain effective charge density distribution, since the coefficients V_{hk} in (7.68) are modulated by the value Δf_{tot} depending on the indices hk.

It was assumed for the derivation of the Eqs. (7.63), (7.65) and (7.68) that $k = \Delta f_{tot}\lambda(r^*)^2 < \pi/2$. This implies that Δf, C_s and the magnitude of the aperture limitation are interrelated in terms of the present approximation. Lynch et al. (1975) have calculated the range of variation for these parameters consistent with the charge density approximation.

Figure 33 shows k plotted against $r^* = 2\sin\vartheta/\lambda$ for different C_s and Δf. Dashed lines show the boundary k values of $-\pi/2$ and $\pi/2$ marking the limits within which the condition (7.59b) is satisfied. It is seen that, for $C_s = 0$, a larger Δf should correspond to a smaller aperture. The use of aperture for $C_s = 0$ is necessary since k should be $< \pi/2$. This is the case where the PCD approximation is satisfied most exactly. With the increase in C_s the acceptable aperture size decreases, the absolute value of defocus increases, etc. For example, for an aperture with $r^*_{max} = 0.23\,\text{Å}^{-1}$ and $C_s = 1.8$ mm the range of the possible defocus values is limited by the interval $-800\,\text{Å} < \Delta f < -600\,\text{Å}$. It follows from Fig. 33 that, for $C_s = 1.8$ mm, the aperture radius should not exceed $0.25\,\text{Å}^{-1}$. Otherwise the main assumption in the present approximation might be violated. For the same reason, the aperture radius $r^*_{max} = 0.23\,\text{Å}^{-1}$ appears unacceptable for $C_s = 2.00$ mm.

The expression (7.63) implies that there is a linear relationship between the contrast and both Δf and the crystal thickness t. Therefore it seems natural that

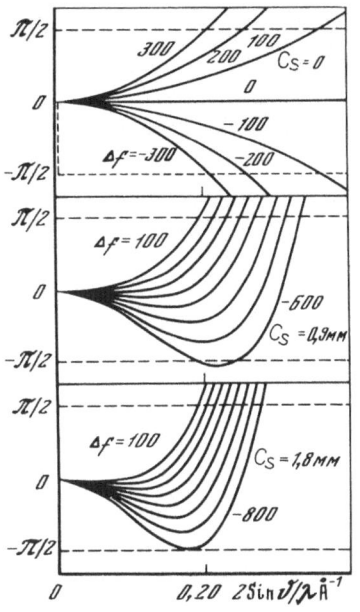

Fig. 33. κ plotted versus r for different C_s and Δf (Lynch et al. 1975)

Δf should decrease with t. Calculations for $Nb_{12}O_{29}$ crystals of thicknesses 30, 50 and 70 Å (Lynch et al. 1975) have led to the optimum Δf values of -700, -600 and -450 Å, respectively, whereas the Sherzer optimum defocus condition implies $\Delta f = -900$ Å irrespective of crystal thickness. Thus the given approximation seems to suggest more adequate Δf values than those proposed the WPOA. The PCD approximation is exactly satisfied with $C_s \to 0$. However, in modern microscopes, such as JEM-100CX, $C_s = 0.7$ mm. Therefore, for a fixed C_s the choice of the aperture should be a compromise between the requirement of the highest possible resolution and the need to minimize the spherical aberration. The chromatic aberration and the incident beam divergence should be also taken into account since these effects impose certain limitations on the aperture size. In all cases, the aperture should be large enough to transmit a sufficient number of diffracted beams that would ensure high contrast and high resolution. Another limitation in realizing the approximation involved is the use of only very thin crystals of thicknesses less than 100 Å containing heavy atoms.

The Transfer Function and the Smearing Function: The Role in the Image Formation. Consider a phase-object with the transmission function $q(r) = \exp[-i\sigma\varphi(r)]$ and the diffracted amplitude distribution $Q(r^*) = \mathscr{F}\{\exp[-i\sigma\varphi(r)]\}$.

Under experimental conditions, the diffracted amplitude distribution at the back-focal plane may be written analytically as

$$Q(r^*)A_{ef}(r)\exp i\kappa, \tag{7.69}$$

where

$$\kappa = \pi\lambda(r^*)^2[\Delta f - \tfrac{1}{2}C_s\lambda^2(r^*)^2], \quad (r^*)^2 = (x^*)^2 + (y^*)^2,$$

$$A_{ef}(r^*) = A(r^*) \exp[-\pi(\delta f)^2 \lambda^2(r^*)^4] \frac{J_1(2\pi\alpha\Delta f_{tot}r^*)}{2\pi\alpha\Delta f_{tot}r^*},$$

and δf is the standard deviation from Δf_0 for chromatic aberration.

The factor $\exp i\kappa$ allows for spherical aberration and defocus, and $A_{ef}(r^*)$ is the effective aperture function that includes the effects of aperture limitation, chromatic aberration, incident beam divergence, etc. [see Eqs. (7.36) and (7.37)]. The function taking account of the resultant effect of all aberration factors on $Q(r^*)$ is called the *transfer function*.

The Fourier transform of (7.69) leads to a modulation of the transmission function by convolution with the *spread function* $t(r)$, which is therefore the Fourier transform of the transfer function

$$\Psi(r) = \exp - i\sigma\varphi(r) * t(r), \tag{7.69a}$$

where

$$t(r) = \mathscr{F}\{A_{ef}(r^*)\exp i\kappa\} = \frac{J_1(2\pi\rho_{ef}r)}{2\pi r\rho_{ef}} * \mathscr{F}\{\exp i\kappa\},$$

and ρ_{ef} is the effective aperture radius.

The intensity distribution in the image plane is given by

$$J(r) = |\exp - i\sigma\varphi(r) * t(r)|^2. \tag{7.70}$$

The function $t(r)$ smears each value $q(r)$ at the point r within a certain region. The radius of this region is defined by the distance between the maximum in $t(r)$ for $r = 0$ and the first minimum. This function allows for the perturbing influence of all factors that distort the transmission function as it is transferred from the object plane to the image plane. The function $t(r)$ is complex and may be written as $t(r) = c(r) + is(r)$. Then (7.70) becomes (Cowley 1975b)

$$J(r) = [\sin\sigma\varphi(r) * s(r) + \cos\sigma\varphi(r) * c(r)]^2$$

$$+ [\sin\sigma\varphi(r) * c(r) + \cos\sigma\varphi(r) * s(r)]^2$$

$$= 1 + 2\sigma\varphi(r) * s(r) + [\sigma\varphi(r) * s(r)]^2 + [\sigma\varphi(r) * c(r)]^2 - \sigma^2\varphi^2(r) * c(r). \tag{7.71}$$

To derive the expression (7.71), $\sin\sigma\varphi(r)$ and $\cos\sigma\varphi(r)$ have been expanded into power series omitting terms of orders higher than 2. Figure 31 shows the function $s(r)$ under the Scherzer optimum defocus conditions for $\Delta f = -550$ Å, $C_s = 0.7$ mm, $\rho_0 = 0.33$ Å^{-1} and $\lambda = 0.037$ Å. Whereas $c(r)$ fluctuates only slightly about the zero value, the function $s(r)$ has a narrow maximum at the origin and then, beginning with about 1.0 Å, it decreases rapidly, fluctuating about the zero value. It is obvious that all the image points will be smeared over the area of a radius of about 2 Å. For electron microscopes having $C_s = 1.8 - 2.0$ mm with the optimum defocus ($\Delta f = -1000$ Å, $\rho_0 = 0.27$ Å^{-1}), the half-width of $s(r)$ is approximately 3 Å, while $c(r)$ fluctuates again around the zero value. Hence for the given C_s, ρ_0, and λ, the optimum defocus is most favorable for obtaining high-resolution images, since the basic maximum in $s(r)$ is localized within a narrow interval commensurable with individual interatomic or interplanar distances.

If the resolution limit is $3-4$ Å, we may assume $s(r) \simeq 1$, $c(r) \simeq 0$. Then $J(r) = 1 + 2\sigma\varphi(r) + \sigma^2\varphi^2(r)$. For $\sigma\varphi(r) \ll 1$ we obtain the WPOA, since the contribution of $\sigma^2\varphi^2(r)$ to the image intensity may be then neglected. Since this is a very rough approximation, in analyzing contrast in images all terms in (7.71) should be taken into account.

The above examples show that the half-width of the function $t(r)$ is essential for revealing structural features of the object, since it determines the limit of interpretable resolution that can be achieved in structure analysis (Cowley 1978).

For fixed half-widths of the functions $s(r)$ and $c(r)$, i.e., for the given experimental conditions, the image depends largely on the crystal structure. For example, the relation between the half-width of $t(r)$ and the unit cell parameters is important. If the unit cell parameters in a periodic structure exceed the half-width of $t(r)$, the image of the atomic pattern within the unit cell is independent of periodicity. Consequently, a group of atoms will give the same contrast in the image regardless of whether the object structure is periodic.

Practically, this is essential for simulation of high-resolution images of defect structures. It is possible to define an arbitrary atomic arrangement including any type of imperfections and then repeat the pattern periodically in two directions with periods of about $20-30$ Å. This would provide a model ready for calculation of images using a standard computer program for crystals.

This approach has been applied by a number of authors to the imaging of crystal defects of various types (Cowley 1978; Iijima 1978). On the other hand, the requirement that the unit cell parameters should exceed the half-width of the spread function is a necessary condition for the use of high-resolution images for structure analysis. One may hope to succeed in determining the atomic pattern and the nature of defects on the scale of interatomic distances only if this condition is satisfied. If unit cell parameters are of the same order of magnitude as the $t(r)$ half-width, $t(r)$ substantially perturbs the image, compared with the one expected for $t(r)$ being the δ-function. The contrast in the image is then extremely sensitive to any change in defocus and in aberration coefficients, as these determine directly the $t(r)$ half-width. It is obvious that structure analysis becomes in this case practically impossible, so that one has to analyze crystal defects by lattice imaging.

Aberration Free Focus Conditions (AFFC). Assume that a specimen is illuminated by a parallel beam from a practically coherent source. The effects of chromatic aberration, astigmatism, mechanical and electric vibrations minimized, the main factors perturbing the transmission function are defocus and spherical aberration. The aperture size may be chosen however large depending on the resolution required, and C_s can be varied by changing the lens current.

The expression (7.26) implies that the perturbing effect of defocus and spherical aberration will be minimized if $\exp i\kappa = 1$. Then the intensity distribution for an electron wave at the exit face will be exactly reproduced in the image. It is seen that without aperture $J = |q(r)|^2$. The condition $\exp i\kappa = 1$ is satisfied for

$$\kappa = \pi\Delta f\lambda\left(\frac{h^2}{a^2} + \frac{k^2}{b^2}\right) - \frac{\pi}{2}C_s\lambda^3\left(\frac{h^2}{a^2} + \frac{k^2}{b^2}\right)^2 = 2\pi(N+M), \qquad (7.72)$$

where N and M are integers, which is valid if

$$a^2 = pb^2, \quad \Delta f \lambda / a^2 = n, \quad C_s \lambda^3 / 2a^4 = m, \tag{7.73}$$

where m, n and p are integers without a common factor.

Kuwabara (see Cowley 1978) has noted that since $\exp i\kappa = 1$ the satisfaction of the conditions (7.73) for $\Delta f = 0$, $C_s = 0$ implies that there is a set of values $\Delta f = na^2/\lambda$ and $C_s = 2ma^4/\lambda^3$ such that the corresponding images will be identical to that for $\Delta f = C_s = 0$. The highest possible resolution may then be achieved if a and λ are strictly constant.

Cowley (1978) has suggested the following estimation of acceptable variations in a and λ for fixed Δf and C_s. For reflections of the maximum orders h_{max} contributing to the image, the condition (7.73) may be written $(C_s \lambda^3 / 2a^4) h_{max}^4 = M$ (where M is an integer). Suppose that, for variations in λ of $\pm \Delta \lambda$ and a of $\pm \Delta a$, the possible deviation from the integer M value is $\pm 1/6$. Then

$$\frac{3 C_s \lambda^2 h_{max}^4}{2a^4} \Delta \lambda = \frac{1}{6} \quad \text{and} \quad \frac{4 C_s \lambda^3 h_{max}^4}{2a^5} \Delta a = \frac{1}{6}.$$

Taking account of the condition $m = C_s \lambda^3 / 2a^4$ we obtain relationships for the limits for possible variations in λ and a depending on the number of reflections contributing to the image, i.e.,

$$\frac{\Delta \lambda}{\lambda} = \frac{1}{18 m h_{max}^4}, \quad \frac{\Delta a}{a} = \frac{1}{24 m h_{max}^4}. \tag{7.74}$$

The satisfaction of the conditions (7.74) is necessary for m and n in (7.73) to be integer, i.e., for the validity of the equation $\exp i\kappa(hk) = 1$ for all reflections contributing to the image. Assuming $m = 1$ and $h_{max} \leqslant 3$ we have $\Delta a/a = 0.05\%$, which implies that the crystal analyzed should be almost perfectly periodic.

One of the main reasons for deviations from perfect periodicity is the limited size of coherent scattering domains. As it has been shown in Chapter 2, the smaller the coherent scattering domain size Na, the greater the half-width of the basic maxima in the interference function. On the other hand, the smearing of the reciprocal lattice nodes may be formally treated as resulting from variations of reciprocal lattice parameters by $\Delta a^* = 1/Na$. Hence, for a perfectly periodic region consisting of N unit cells, there is an uncertainty in the values for parameters which is expressed by the relationship $\Delta a^* / a^* = \Delta a/a = 1/N$. Thus, to ensure that m and n are integers, the number of unit cells in a region having perfect periodicity which is coherently illuminated by an electron beam should be $N \geqslant 24 m h_{max}^4$ (Cowley 1978). For experimental determination of the AFFC, the equation (7.72) may be written

$$\Delta f = 2md^2(hkl)/\lambda + C_s \lambda^2 / 2d^2(hkl) \tag{7.75}$$

putting $\kappa = 2\pi m$ and $1/d^2(hkl) = h^2/a^2 + k^2/b^2$. The problem is then to adjust Δf and C_s so that the condition (7.75) should hold for all the diffraction beams contributing to the image. If, for example, the incident beam coincides with the direction [110] of a gold crystal for $\Delta f = 1573$ Å and $C_s = 0.545$ mm, the condition (7.75) will be satisfied for 37 waves having the indices {111}, {200}, {220}, {113}, {222},...up to {044} (Hashimoto et al. 1979). To ensure satisfaction of the AFFC for crystals having different unit cell parameters or for different orientations of the crystal, the values Δf and C_s should be varied.

Hashimoto et al. (1977), in studying thin gold crystals, with the conditions (7.75) satisfied, have obtained a resolution of detail in the image of atoms of about 0.5 Å, i.e., less than the diameter of an atom. The condition $\exp i\kappa(hk) = 1$ applied almost exactly for all reflections corresponding to d-spacings greater than $d(620)$. The experimental conditions were the following: $\lambda = 0.037$ Å, $\Delta f = 553$ Å, $C_s = 0.75$ mm. In the electron micrograph obtained, each individual row of gold atoms parallel to the direction [001] coincident with the incident beam is represented by a ring of higher intensity with a fine structure of contrast distribution.

Cowley (1978) has noted that, for a perfectly periodic crystal, details could be observed on a much finer scale than for an aperiodic object, but then there will no longer be a direct relationship between each point in the image and each point in the structure. Thus, by averaging over a comparatively large periodic domain in gold crystals, it became possible to resolve the finest details in the contrast variations in the image of projected individual rows of gold atoms.

In the light of the above treatment of relationships between unit cell dimensions and parameters characterizing experimental conditions, it becomes necessary to refine the concept of resolution (Cowley 1978). Distinction should be drawn between resolution of fine detail in the image (instrumental resolution), on the one hand, and resolution of two closely spaced details in the object structure (structural resolution), on the other hand. In the first case, fine detail in the image may result from interference effects or from additional maxima and minima accompanying the basic intensity maxima. These contrast details cannot be directly interpreted in terms of the object structure. The finest details in the contrast distribution in images obtained under the conditions (7.75) apparently may reflect some real structural features, although modified by the transmission function, only in the case where the structure is uniform and periodic within sufficiently large three-dimensional domains.

7.5 Direct Crystal Structure Determination Methods Under the WPOA

If the WPOA applies, there is a theoretical possibility for a direct crystal structure determination with the resolution of detail down to 1 Å (Klug 1979). No a priori information on the object structure is needed in this case. SAED patterns and structure images obtained under the optimum defocus and the resolution of 2 – 3 Å are used as the initial data. The significance of this approach is determined by the fact that, with the resolution of 2 – 3 Å, it is impossible to localize positions corresponding to the projections of the individual atom columns and it is difficult to reveal light atoms in the presence of heavy ones, etc. The decrease in the resolution limit down to 1 Å is almost equivalent to the determination of the projected electrostatic potential distribution by conventional methods of structure analysis using the formulae of the type (4.29).

The intensities in SAED patterns are independent of defocus and aberrations. Therefore, they may be accurately measured down to $d(hkl) \simeq 1$ Å and then used to calculate the structure amplitudes $|\Phi_0(hkl)|$ using the formulae given in Chapter 2.

To determine the structure, it is necessary to solve the phase problem, i.e., to find the phases for each value $|\Phi_0(hkl)|$. Analysis of high-resolution micrographs permits evaluating, at least approximately, the phases for those reflections that satisfy the optimum defocus condition. If, using some iterative procedure, one then succeeds in determining all the phases, formulae of the type (4.29) can be used to reconstruct the projected potential distribution.

Until recently, the works on this problem have been basically of methodological character. Their main aim has been to elaborate the optimum methods for the solution of the phase problem. X-ray direct methods make use of a three-dimensional data set, whereas with electrons we deal with a two-dimensional one. Under such conditions, the authors used as initial data those calculated in the kinematical approximation for test objects with well-known structures.

Ishizuka et al. (1982) have taken as the basis the structure of the copper perchlorophthalocyanine projected along the c axis. They have discussed the following phase correlation method.

The initial values $|\Phi_0|$ and the signs of the phases α_0 are obtained from ED and EM, respectively, using formulae of the type (2.55), (7.45) and (7.46). The Φ_0 values are included only for reflections whose phases have been obtained from EM. These data are used in the first-cycle inverse Fourier transform that gives the two-dimensional potential distribution function $\mathscr{P}_c$. This function is modified to exclude the negative potential regions as well as to change the peak heights. To do this, the *modification function in the direct space* MFD of the type $\mathscr{P}_c^* = \mathscr{P}_c - 0.1$ is used. The Fourier transform of the function $\mathscr{P}_c^*$ permits to determine Φ_c^* having more reasonable amplitudes, which have been changed by the *modification function in the reciprocal space* MFR of the type

$$\Phi_c = \begin{cases} \Phi_0 \exp(i\alpha_c^*) & \text{for} \quad \Phi_c^*/\Phi_0 \geqslant 0.5 \\ \Phi_c \exp(i\alpha_c^*) & \text{otherwise.} \end{cases}$$

The new Φ_c set is used to obtain the new function $\mathscr{P}_c$ which is again modified by the MFD, etc. To minimize the factors

$$R = \frac{\sum\{|\Phi_0| - |\Phi_c|\}}{\sum |\Phi_0|} \quad \text{and} \quad R' = \frac{\sum\{|\Phi_0 - \Phi_c|\}}{\sum \Phi_0}$$

10 – 20 cycles are needed.

Fan et al. (1985) used the same substance as a test object and the same initial data. The image intensity was calculated according to (7.45) with the resolution of 2 Å and 2.5 Å, while the initial phases of the structure amplitudes have been determined using the expression (7.46). The authors have made use of the direct phasing method developed in X-ray crystallography and employed the modified computer program MULTAN (Ishizuka et al. 1982).

One of the main difficulties in the practical use of direct methods is the absence of reliable criteria for the validity of the WPOA based on the experimental data only.

In the case of PWPOA, the intensity distribution at the back-focal plane of the objective lens, as implied by (7.53b), depends not only on the modulus of the structure amplitude but also on the sample thickness. The latter value is also required for determining the phases by the analysis of electron-microscopic images. For PWPOA and $\exp i\kappa = -i$, (7.56) and (7.57) imply

$$J(xy) = 1 - 2\sigma\varphi(xy) - 2\sigma\varphi(xy) * \sum_{m=1}^{n} S_m(xy) .$$

Taking the Fourier transform we have

$$Q'(x^*y^*) = \delta(x^*y^*) - 2\sigma\Phi(x^*y^*)\left\{1 + \sum_{m=1}^{n} \cos[\pi m\lambda\Delta z(x^{*2}+y^{*2})]\right\}$$

$$= \delta(x^*y^*) - 2\sigma\Phi(x^*y^*) \sum_{m=0}^{n} \cos\{\pi m\lambda\Delta z[(x^*)^2+(y^*)^2]\} . \quad (7.76)$$

Therefore, to determine the phases from the image intensity distribution, account should be taken of the cosine sum in (7.76).

Thus, one of the basic problems in the direct phasing methods is the precise determination of the crystal thickness.

7.6 High-Resolution Electron Microscopy (HREM) of Crystals

The first successful high-resolution studies were those of compounds synthesized from the systems $Nb_2O_3-WO_3$ and $Nb_2O_3-TiO_2$. The basic structural units in these compounds are MO_6-octahedra connected with one another through shared apices (Fig. 34). There are two types of the structures in question. In structures of the first type, the basic structural fragments are two-dimensional blocks containing $m \cdot n$ octahedra, where m and n are the numbers of octahedra linked together through shared apices in two orthogonal directions (Fig. 34). These blocks are connected with one another through shared octahedral edges. Blocks lying in the same plane are also connected by cations in tetrahedral coordination.

Each given plane is represented by blocks and vacant rectangular cavities alternating regularly in two directions. All the edges in each block are adjacent to empty cavities. Variations in block dimensions as well as various possibilities to link the blocks together lead to a wide diversity in structural motifs. Figure 34 shows several examples of structural models of this type. In structures of the second type there are channel-like cavities, while octahedra are linked together in the same way as in the first-type structures (Fig. 34). Channel-like cavities may be occupied by cations or vacant. The diversity of the structures is governed by the concentration of vacant cavities and by their arrangement. In all the structures of the second type the b parameter is 3.9 Å, while the other two parameters are quite large, reaching 36 Å.

Naturally, in all electron-microscopic studies the crystals were oriented so that the incident beam was parallel to the axis b. With resolution of $3-4$ Å, one could hope to visualize details in atomic structure within one unit cell. Iijima

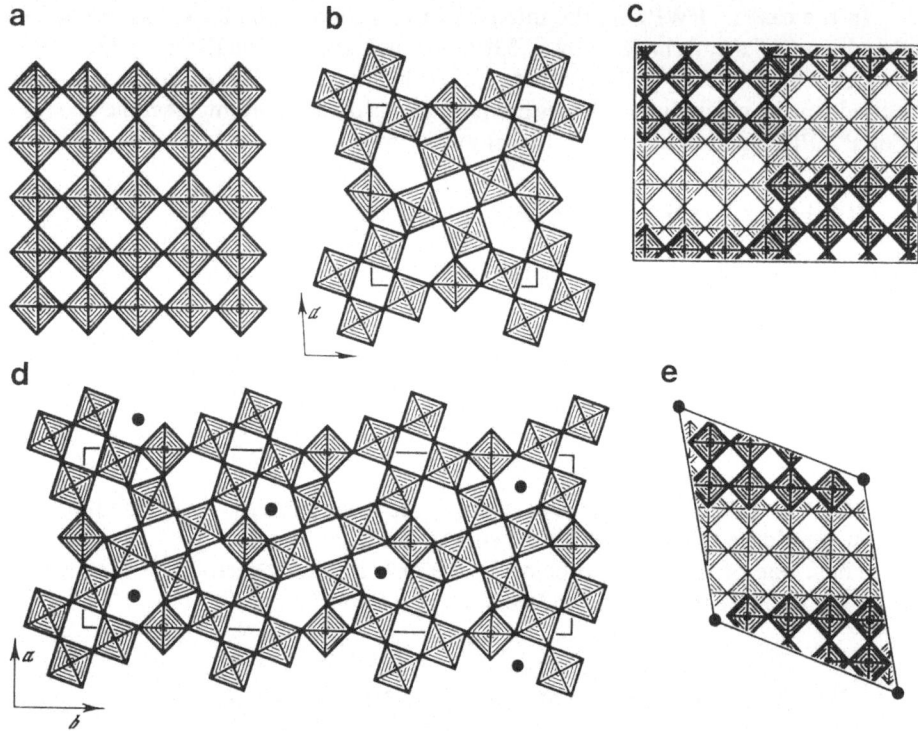

Fig. 34a – e. Two types of structural patterns in compounds obtained from the systems $Nb_2O_3 – WO_3$ and $Nb_2O_3 – TiO_2$: **a** a (5×5) block of octahedra linked through shared apices; **b** and **c** two possible channel structures differing in the distribution of pentagon channels; **d** and **e** two possible structures illustrating different packings of (3×4) octahedral blocks (Allpress and Sanders 1973)

(1971) was the first to succeed in this for crystals of $Ti_2Nb_{10}O_{29}$. In the image obtained, dark spots are clearly seen, representing the lines connecting the blocks. Metal atoms superposed on one another are arranged along these lines at distances that are less than the resolution. Grey points represent single octahedral positions. White spots correspond to electrostatic potential minima (Fig. 35). Iijima and Allpress (1974) have studied Nb and W oxides with channel structures. A compound with the known distribution of channel cations has been studied first. For this substance, an image corresponding to the structure projection was obtained with a resolution of 3.8 Å. Then the same experimental conditions have been employed to study the structure of $2 Nb_2O_7 \cdot 7 WO_3$ with an unknown distribution of vacant channels and those occupied by cations. The analysis of the image has allowed the determination of the crystal structure involved (see Fig. 36). This was the first crystal structure determined by HREM.

A similar approach has been applied by Li and Hashimoto (1984) to the determination of a complex structure of cebaite $Ba_3Ce_2(CO_3)_5F_2$ having the unit cell parameters $a = 21.2$ Å, $b = 5.06$ Å, $c = 13.1$ Å and $\beta = 95°$. The optimum experimental conditions have been chosen by analyzing electron micrographs of

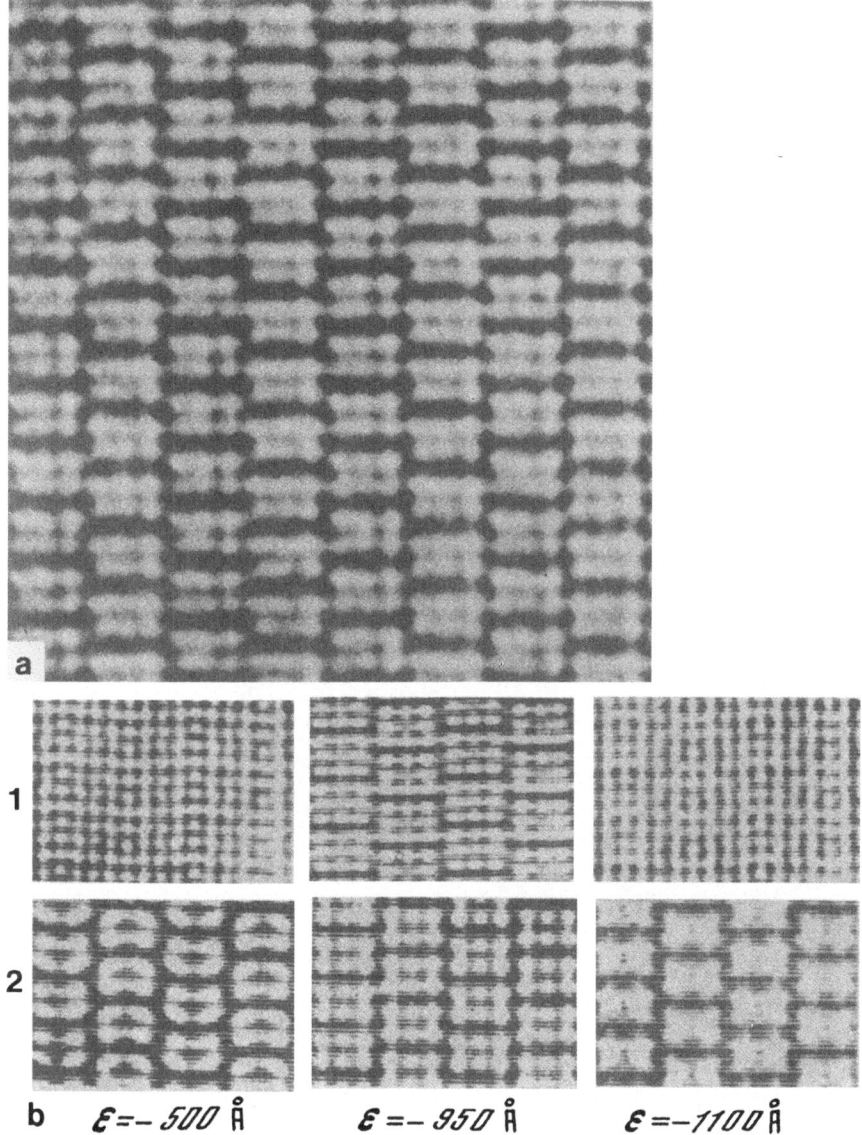

Fig. 35. a Electron-microscopic image of the structure projection of $Ti_2Nb_{10}O_{27}$ (Iijima 1971); **b** Calculated images for $Ti_2Nb_{10}O_{27}$ structure projection for different Δf (Fejes 1977): *1* with account of chromatic aberration; *2* without account of chromatic aberration ($\Delta f = \varepsilon$)

huanghoite $BaCe(CO_3)_2$ whose atomic structure was known. The authors have deliberately used the crystals whose thickness was greater than the maximum thickness consistent with WPOA. Therefore, dynamical extinction effects were present but the image contrast of light atoms was enhanced. The distribution of heavy and light atoms in the cebaite structure, determined by analyzing electron micrographs, has been confirmed by calculations of the image intensity.

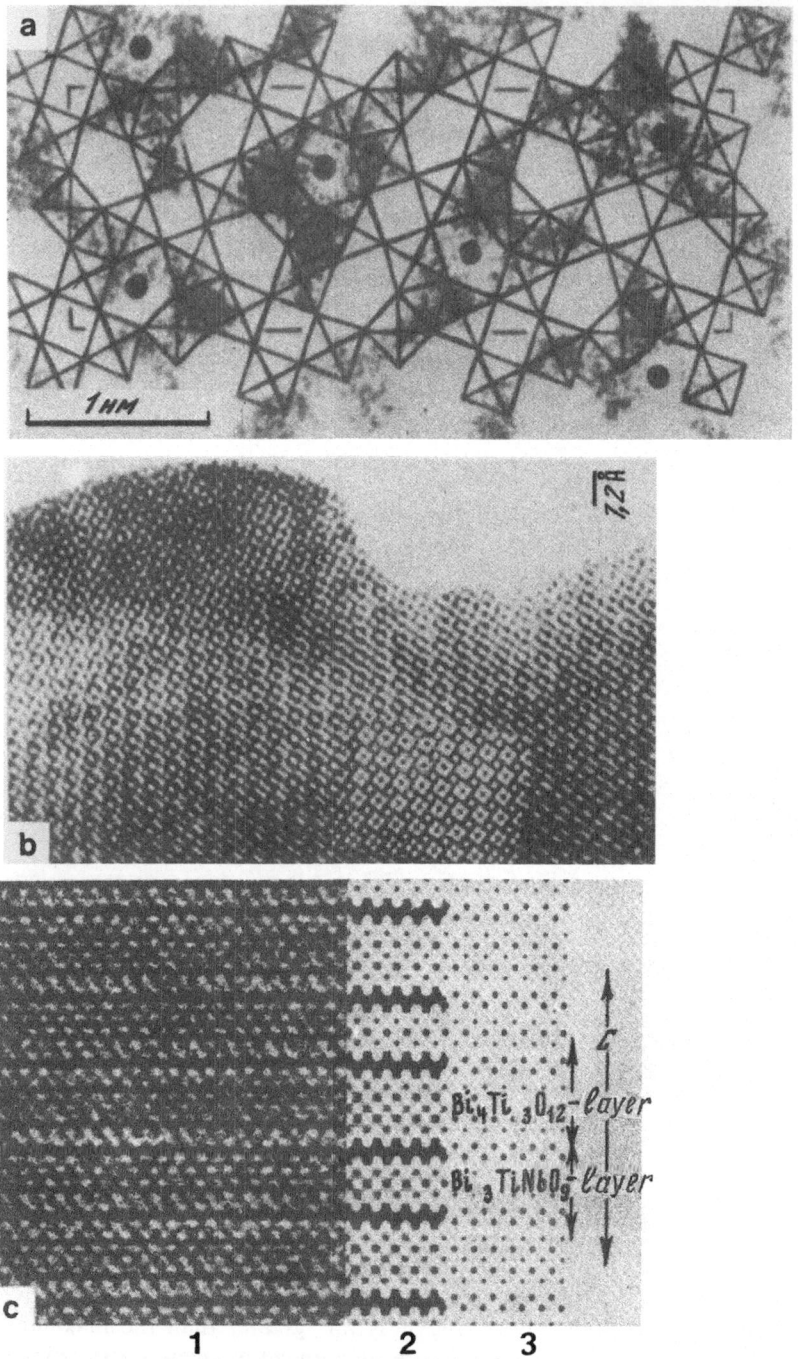

Fig. 36. a Direct structure determination of $2Nb_2O_7 \cdot 7WO_3$ using a HREM image (Iijima and All-press 1974); **b** a HREM image of hollandite structure projection (resolution is about 2.5 Å, a fragment of a calculated image is shown (Bursill and Wilson 1977); **c** *1* HREM image; *2* interpretation for the contrast; *3* idealized structure model for mixed-layer $Bi_7Ti_4NbO_{21}$ (Horiuchi et al. 1977)

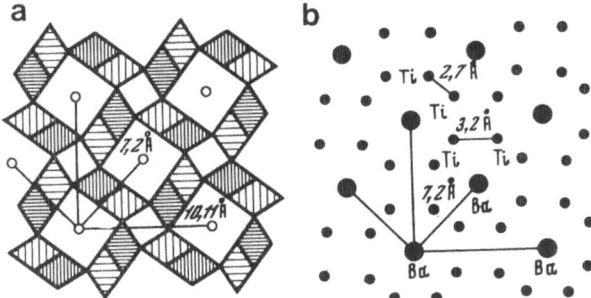

Fig. 37a, b. Crystal structure of hollandite: **a** projection down [001]; **b** distribution of Ba and Ti in the projection down [001] (Bursill and Wilson 1977)

Smith and Parise (1985) have proved, using HREM and basing on the knowledge of the $SnSb_2Se_4$ structure, that $SnSb_2S_4$ is isostructural to $SnSb_2Se_4$.

Bursill and Wilson (1977) achieved substantial improvement in resolution of detail by decreasing the incident beam divergence down to $2 \cdot 10^{-4}$ rad and minimizing the thermal vibrations in the electron gun (the deviation from the mean defocus was about 50 Å). Note that the aberration coefficients for the microscope involved were rather high ($C_s = 2.5$ mm, $C_c = 2.0$ mm), which did not favor very high resolution. Thin crystals of tetragonal hollandite $BaMgTi_7O_{16}$ were used as the object. The structure in question is channel-like with an anionic framework and unit cell parameters $a = 10.110$ Å and $c = 2.986$ Å. The square channels are occupied by Ba cations (Fig. 37). The figure shows the cation distribution projected along c with the characteristic distances $Ba - Ba = 7.2$ Å, $Ti - Ti = 3.2$ Å (Type I) and $Ti - Ti = 2.7$ Å (Type II) marked in.

To obtain an image of this cation distribution, the resolution should have been higher than or equal to 2.7 Å, which is much better than the resolution of 3.8 Å achieved by Iijima and Allpress (1974). With the optimum defocus of -1150 Å and the aperture with $r^*_{max} = 0.5$ Å^{-1}, the authors obtained the resolution required, which may be clearly demonstrated by comparing the image contrast and the scheme of the cation distribution (Fig. 36).

Bursill (1979) has shown that with $V = 100$ kV it is possible to obtain an instrumental point-to-point resolution of 1.7 Å for $C_s = 0.7$ mm, $\sigma \leqslant 30$ Å and an almost parallel beam with the semicone angle $\leqslant 0.2$ mrad. The structural resolution is then 2.9 Å, although such a resolution could be obtained using a focused beam ($\alpha = 0.8 - 1.0$ mrad).

One of the most essential points in the work of Bursill and Wilson (1977) is that they have once again demonstrated the necessity to study crystals with known structures so that these could be used as tests for the resolution achieved. The observation of images of the same crystal structure under different experimental conditions allows evaluation of the contribution of each separate factor affecting the resolution. This favors the choice of optimum ways to improve resolution. The paper discussed deals with the influence of incident beam divergence, chromatic aberration, defocus etc.

Tanaka and Jouffrey (1984) have analyzed the problem of interpretation of lattice images of relatively small-unit-cell crystals. To obtain a structure image of such a crystal, it is necessary to visualize the atoms themselves instead of atomic groups. The authors have noted that, for such crystals, the optimum Scherzer conditions do not always lead to images with as much directly interpretable information as possible. Thus, the problem arises how to determine the conditions ensuring one-to-one correspondence between the image and the structure. The solution becomes much simpler if the crystal thickness and the defocus are known. These values can be obtained with the help of an optical diffractometer and by analyzing thickness fringes at the edge of a crystal under study. Therefore, the authors consider that it will be necessary to equip the microscope with an optical system on-line with the computer in order to Fourier transform the image intensity.

At present, every year sees rapid progress in experimental HREM crystal structure studies of various substances including minerals. (See, e.g. Wenk 1976; Marinder and Sundberg 1984; Proc Int Congr Electron Microsc 1986.)

7.7 HREM and Real Structure of Crystals

The main contribution of HREM to structural mineralogy and crystal chemistry as well as to solid state physics consists in allowing the study of the real crystal structure including miscellaneous types of imperfections. Previously, views on localized crystal defects were based on indirect methods of crystal structure analysis and therefore were often of a hypothetical nature. Various defects provide complex diffraction effects, including diffuse scattering and selective modifications of diffraction maxima with certain indices. Since the influence of defects is usually "spread" all over the reciprocal space, the intensity of the corresponding diffraction effects often appears small.

Special methods have been elaborated for interpretation of diffraction patterns from defect crystals. However, they can provide only a general outline of the nature of defects, their concentration and distribution. Moreover, all these parameters are averaged over the diffracting crystal volume.

For EM methods, the situation is substantially different. The role played by the analysis of diffraction contrast in defining the nature of dislocations, dislocation nets, stacking defects, interstitial groups of atoms, etc., is well known (Hirsh et al. 1977).

A new level in the study of defects has been achieved with the use of HREM. It has been found that an electron wave at the exit face of the crystal contains local information on the structure of a column of the width equal to the resolution limit. Under specific conditions, electrons having left different parts of the exit face do not overlap and do not interact in the image plane. Thus visualization is ensured both of the periodic part of the crystal and of isolated atomic groups. It is only the defects oriented along the incident beam that cannot be revealed.

Thus, it is feasible to analyze deviations from an idealized structure having the dimensions of a vacancy, an interatomic distance, a part of a unit cell or several unit cells. Therefore it is not surprising that the use of HREM has ensured determination of a great variety of order-disorder patterns in crystals and provided fundamentally new data on the mechanism of structure transformations, etc.

It should be noted, however, that any micrograph reflects the specific structure of only one very small part of the object. Therefore, to make sure that the defect type observed is really characteristic of the substance under study, repeated experiments are needed.

The number of works on the nature of defects in crystals increases constantly (see, e.g. Buseck and Iijima 1974, 1975, 1976; Buseck and Cowley 1983; Iijima 1973, 1975a, b, 1978; Iijima and Allpress 1974; Iijima and Buseck 1978; Iijima et al. 1973, 1974; Ijima and Zhu 1982; Pierce and Buseck 1976; Veblen 1980, 1981, 1983a, b; Veblen and Buseck 1979a, b, 1980, 1981; Veblen and Ferry 1983; Veblen et al. 1977). Therefore we confine our attention to just a few examples of the advantages of HREM.

Nakazawa et al. (1974), and later Pierce and Buseck (1976), have observed directly the rows of vacancies projected along c in pyrrotine structures $Fe_{1-x}S$ using bright- and dark-field images. Iijima (1975a, b) has found an ordered distribution of point defects in the images of nonstoichiometric crystals $Nb_{12}O_{26}$.

Hashimoto et al. (1977) have managed to reveal individual Al for Au substitutions, thus demonstrating once again the unique possibilities of HREM.

Analysis of electron micrographs obtained from thin AuAl and GaAs crystals under AFFC conditions has shown that intuitive interpretation of the contrast is impossible (Hashimoto et al. 1979). For example, Cu and Au atoms in the Al crystal were displayed in the image as light atoms or vacancies. Difficulties also arise in revealing the positions of different atoms in $G - P$ zone in $Cu - Al$ alloys.

Buseck and Iijima (1974, 1975, 1976) have established an abundance of stacking defects in enstatites and of out-of-phase domains in some amphiboles, in skapolite, etc. They have found intrusions of clinoenstatite of the width of several unit cells in an orthoenstatite matrix. Analyzing variations of contrast in the images of cordierite crystals, they have inferred ordered distribution of Si and Al over tetrahedral sites.

A new model for the distribution of Ca and Na in the labradorite superstructure was suggested by Kumao et al. (1981) on the basis of high resolution structure images and the comparison of experimental point ED patterns with those obtained by Fourier transforming the images with the help of an optical diffractometer.

A number of authors (Amouric et al. 1978, 1981; Wenk 1976, 1978; Iijima and Buseck 1978; Iijima and Zhu 1982; Spinner et al. 1984; Veblen 1983a, b) used HREM to study the stacking sequences in various phyllosilicates, such as talcs, micas, chlorites, etc. The basic structural unit in these minerals is a 2:1 layer made up by condensation of a central octahedral and two tetrahedral sheets forming sequences $T - O - T$. The tetrahedral sheets belonging to one layer are shifted by the vector $a_i/3$ with respect to one another in normal projection on the plane (001) (where a_i is one of the six vectors connecting the centers of two neighboring hexagons in a tetrahedral sheet).

As shown by Iijima and Buseck (1978), and Amouric et al. (1978), EM images of talc and mica obtained with the incident beam parallel to a, contain linear rows of bright spots with the spacing 4.5 Å. Each row corresponds to an interlayer, and the bright spots are associated with interlayer channels having the minimum electron density. Amouric et al. (1981) have shown that under certain optimum conditions, the arrangement of bright spots in the image can be directly interpreted in terms of the stacking sequence in the crystal analyzed.

Spinner et al. (1984) have used HREM to study stacking defects in clinochlore chlorite. Chlorites consist of 2:1 talc-like layers (T-layers) alternating regularly with brucite octahedral layers (B-layers). Various layers stacking sequences are possible for the same chemical composition. On the one hand, this leads to poly-typism, and, on the other hand, to stacking faults in chlorite structures. The sample studied by Spinner et al. (1984) corresponds to a one-layer IIb-2 polytype (Bailey and Brown 1962). Therefore, the octahedra in T- and B-layers are of op-posite orientations. There is no superposition of nearest cations belonging to ad-jacent layers. The lower apices of the B-layers project into the centers of T-layer octahedra. The enantiomorphic polytypes IIb-4 and IIb-6 differ from IIb-2 (in the case of the same orientation for T-layers) in $\pm b/3$ shifts of neighboring T-layers along the axis b chosen for IIb-2.

EM images of chlorite crystals permitted to observe the projected shift vec-tors across the T- and B-layers in the chlorite structure and thus to obtain infor-mation on the layer stacking on the unit-cell level. For example, analysis of one of the images has shown the following pattern in the sequence of T- and B-layers: IIb-4, -2, -2, -4, -2, -4, -4, -4, -6, -4, -4, -2, -6, -2, -4, -2, -4, -6. In other words, the crystal involved had a semirandom stacking sequence with stacking faults result-ing from shifts of chlorite packets by $\pm b/3$ for the same orientation of T- and B-layers. The authors have emphasized that unambiguous interpretation of HR images requires either preliminary X-ray investigation or imaging of the crystal for its various orientations.

Veblen (1983a, b) used HREM to study micro-intergrowths of chlorite, wone-site (Na-trioctahedral mica) and biotite. He found a semirandom stacking se-quence in chlorite. Wonesite was shown to contain fragments corresponding to 1-layer, 2-layer, 3-layer and disordered polytypes. The biotite structure consisted of alternating 1-layer and 2-layer polytypes.

The geological and mineralogical significance of the study of the nature of de-fects in minerals and relationship with thermal history, structure and composi-tion of minerals are discussed in detail in (Wenk 1976).

HREM proved to be an effective means for the study of minerals with modu-lated and intergrowth structures (Buseck 1983, 1984; Buseck and Cowley 1983; Cowley et al. 1979).

Specific examples of the analysis of defects in mineral structures by HREM are given in the concluding chapters of the present book.

7.8 Simulation of HREM Images

Crystals with unknown structure and thickness may produce periodic images having no direct relationship with the structure motif even under the optimum defocus. For this reason, simulation of images for structural models is especially significant. Such simulation is based on numerical computing methods employ-ing one of the formulations of the n-beam dynamical theory.

The computing method of Goodman and Moodie (1974) modified and im-proved by a number of authors (Ishizuka and Uyeda 1977; O'Keefe 1973; O'Keefe and Buseck 1979; O'Keefe et al. 1978; Self et al. 1983) is the most widely used. For a known crystal structure, this method permits to calculate the transmission function $q(xy)$ and the diffraction function $Q(x^*y^*)$. The expressions (7.26), (7.69) and (7.69a) that allow for spherical and chromatical aberrations, aperture limitation, defocus, and incident beam divergence are used to calculate the function $\Psi(xy)$. The values $|\Psi(x, y)|^2$ represent the intensity distribution in the simulated image, which can be directly compared with the experimental one. This method was used to calculate diffraction patterns and to simulate images for a number of objects (Allpress and Sanders 1973; Bursill and Wilson 1977; Iijima 1978; Klug 1979; Lynch and O'Keefe 1972; Skarunlis and Cowley 1976). It should be stressed that the complete practical solution of this problem is laborious.

Therefore, it is often more appropriate to use approximations, such as the phase-grating and the charge-density approximations. The work of Fejes (1977) is an example of a successful application of the phase-grating approximation. The main assumption in this approximation is that the scattered amplitude distribution at the exit face of a crystal of the thickness t may be described by the function $\exp i\sigma\varphi t$, where t is usually assumed $10-15$ Å. This approach treats the dynamical interactions of diffracted beams somewhat more adequately than the kinematical approximation. However, there is a serious drawback in the phase-grating approximation, as the relative contribution of external diffracted beams to the image increases due to the Ewald sphere being approximated by a plane.

Fejes (1977) used this approach for the analysis of images of thin crystals of $Ti_2Nb_{10}O_{29}$ and $Nb_{10}W_{18}O_{94}$ obtained under varied experimental conditions. He has shown that, if proper account is taken of spherical aberration and aperture limitation, as well as of the incident beam divergence and chromatic aberration, then for definite defocus values close agreement is achieved between the simulated image and the calculated one.

Figure 35 shows examples of such image calculations for a thin parallel-sided $Ti_2Nb_{10}O_{29}$ crystal for different defocus values. The agreement between the calculated image and the observed one is achieved with $\Delta f = -950$ Å which is in excellent conformity with predictions based on rougher approximations. Moreover, a direct relationship between the structure projection (Fig. 34) and the image was found for $\Delta f = -950$ Å.

Note that the agreement between calculated and observed images is reasonably good for all Δf. It seems remarkable that although high contrast is preserved with considerable variations of Δf, the images obtained differ substantially from the structure projection even for deviations of $150-200$ Å from the optimum Δf.

However, even for very thin crystals under the optimum defocus, fine details of the contrast in calculated images do not always correspond to the actual electrostatic potential distribution. Amouric et al. (1981) have calculated images for perfect mica polytypes, 1M-biotite and $2M_1$- and 3T-muscovite-phengite micas. For crystals $50-100$ Å thick under the optimum defocus, images of projections viewed down the close-packing directions of the mica structure have been calculated in terms of the multi-slice method. Agreement between the image contrast and the projected potential distribution has been obtained only for 1M-biotite having all octahedral sites (M1 and M2) occupied by cations. For $2M_1$- and 3T-micas the contrast distribution in the calculated images did not correspond to the real cation distribution in these structures Moreover, it was found that, for comparatively small deviations from the optimum defocus, bright spots no longer coincided with the positions of interlayer channels. Therefore, an intuitive determination of the stacking sequence including stacking defects was not feasible. The authors emphasize that n-beam dynamical calculations, however complex, expensive and time-consuming, are necessary for revealing the optimum conditions for the given microscope that permit to obtain structure images and to test intuitive inferences. Image simulation in terms of the charge density approximation (CDA) requires less computer time and leads to a reasonable agreement between the observed and the calculated data if all instrumental factors are taken into account (Lynch et al. 1975).

However, the most detailed information on the contrast distribution to be expected in an image is obtained where the transmission function is calculated in terms of the n-beam dynamical theory. Image calculations for hollandite (Bursill and Wilson 1977) and Cu-phthalocyanine (Ishizuka and Uyeda 1977) may serve as examples. The authors managed to evaluate separately the effects of defocus, incident beam divergence, crystal thickness, chromatic and spherical aberrations, etc. on the quality of image simulated. As a result, some general criteria for the optimum conditions for HR imaging have been formulated. Li and Tang (1985) have shown on the basis of image calculations for chlorinated Cu phtalocyanine that, within a certain interval of thicknesses, calculations in terms of the PWPOA and those based on a more complex multi-slice method of Goodman and Moodie lead to very close results.

Unique possibilities of EM for revealing various point defects have been already mentioned. Image simulation also plays an important part in interpreting the nature of these defects. Image pattern calculations for various hypothetical defect models and their comparison with the observed contrast distribution permit evaluating the validity of each model. Dynamical effects that may arise for waves scattered by the defects are thus simultaneously taken into account.

The structure refinement of $GeNb_9O_{25}$ has been carried out by Skarnulis and Cowley (1976) who employed a similar method. On the whole, the comparison of the calculated and the experimental data implies that, under the optimum experimental conditions and for the crystal thickness of about 100 Å (i.e., for the phase-object case) the images obtained may be treated, to a certain degree of approximation, as crystal structure projections.

Bursill (1979) has described criteria for distinction between instrumental and structural resolution. To determine the instrumental resolution, it is necessary to calculate, in terms of the n-beam approximation, a set of images with different apertures defining the resolution for a crystal with the known thickness and structure under fixed experimental conditions, i.e., for the given C_s, C_c, Δf, ripple defocus σ, beam divergence α. Agreement between the experimental image and the calculated one defines the instrumental point-to-point resolution.

However, this value does not always coincide with the structural resolution, since a number of images may correspond to the same structure. For example, heavy atoms may be represented by dark or bright spots, the image may be shifted, two dark or bright spots may appear instead of one atom position. The structural resolution may be determined in the case of a one-to-one correspondence between the contrast distribution in experimental and calculated images as well as between the experimental image and the structure.

The main drawback in the image simulation method involved is that it may be used only for the known atomic pattern, while the actual objective is to determine the structural motif by means of HREM. Therefore, problems of unambiguous interpretation of electron-microscopic images of crystalline objects are still expecting complete solution.

On the other hand, if the crystal structure is known and experimental conditions are controllable, it is not at all necessary to obtain a one-to-one correspondence between the image and the structure. This is the case with thicker crystals when the POA does not hold. In such a situation, the contrast distribution in the image becomes more sensitive to certain fine peculiarities in the electrostatic potential distribution, e.g., the degree of ionization for atoms, the nature of interatomic bonding, order disorder in cation distribution, real crystal symmetry, etc. (Cowley 1978).

The present level of the experimental equipment does not allow complete realization of the potential possibilities of HREM.

7.9 High-Resolution High-Voltage Electron Microscopy (HRHVEM)

Experience testifies that the use of electron microscopes with voltage greater than 100 kV often provides new possibilities for the study of inorganic substances by HR imaging. Using the Eqs. (7.27) and (7.37a) one may evaluate the first approximation for the resolution limit defined by the spherical and the chromatic aberrations with accelerating voltages of 100 kV and 1 MV. The relevant data are given in Table 1. It is seen that in 1 MV microscopes the resolution in micrographs can be increased considerably especially if chromatic aberration effects and irradiation damage are minimized. Thus, conditions are created for HREM study of thicker crystals with relatively small unit-cell parameters (≤ 5 Å). While with $V = 100$ kV, groups of atoms are often observed, higher voltages facilitate visualization for individual cations and anions (Uyeda et al. 1979).

Horiuchi (1979) has reported a number of examples illustrating the advantages of a 1 MV electron microscope for structure determinations as well as for revealing the nature of defects associated with, e.g., irradiation damages, structure modulations, clusters etc.

Table 1. Values for the resolution limit (d_s and d_c) determined by the spherical and the chromatical aberrations, respectively, for different voltages

	100 kV		1 MV	
C_s, mm	0.7	1.8	6	10
d_s, Å	2.8	3.6	1.5	1.9
σ, Å	30	200	200	400
d_c, Å	1.26	3.45	1.5	2.2

Tanaka and Jouffrey (1980) have compared lattice images calculated for 100 kV and 1 MV. The V_2O_3 structure having relatively small unit-cell parameters ($a = 4.95$ Å, $c = 14.0$ Å) has been used as an example. For 100 kV, the number of diffracted beams contributing to the image is insufficient for a structure image to form. For $V = 1$ MV, $C_s = 4.2$ mm and the optimum defocus, rays corresponding to $d \leq 1.6$ Å can pass through the aperture, so that an image is formed containing cations and anions that do not overlap in the projection along $[2\bar{2}01]$. Calculations have also shown that, in this case, the maximum possible crystal thickness allowing a structure image is 20 Å for 100 kV with the ideal phase-contrast lens, and 35 Å for 1 MV and $C_s = 4.2$ mm.

CHAPTER 8

Oblique-Texture Electron Diffraction

8.1 General

Oblique-texture electron diffraction (OTED) is widely used in the USSR for structure studies of finely dispersed crystalline substances including minerals (Pinsker 1949; Vainshtein 1956; Zvyagin 1967; Zvyagin and Vrublevskaya 1974; Imamov 1977; Imamov and Pinsher 1965; Imamov et al. 1982; Tsipursky and Drits 1977a, b, 1984; Tsipursky et al. 1978). In contrast to SAED, a great number of small particles contribute ot the formation of an OTED pattern. A significant contribution to the elaboration of OTED has been made by Pinsker (1949). Vainshtein (1956) has developed the theoretical grounds for the method and proved its high efficiency as an independent means of structure analyis for various objects. Zvyagin (1967) has revealed vast opportunities of the method in application to clays and other finely dispersed minerals with platy particles. He has shown that OTED can be used for identification of clay minerals, determination of unit-cell parameters, estimation of the degree of ordering, etc.

OTED has a number of advantages over X-ray powder diffraction.

- An OTED preparation requires quite a small amount of material due to strong interaction between electrons and substance;
- An OTED pattern contains a two-dimensional distribution of reflections in contrast to a one-dimensional one in the case of an X-ray powder pattern;
- A rapid decrease in the scattering factor with the increase in the scattering vector minimalizes the usual problems in Fourier transformations;
- The possibility to obtain almost perfectly oriented samples leads to diffraction patterns with the minimum overlapping of reflections which improves the resolution of neighboring reflections.

The factors listed above favor the application of OTED to structure studies of polycrystalline materials including determination and refinement of unknown crystal structures. However, there is some ambiguity concerning the role of dynamical effects for polycrystalline materials consisting of very thin crystallites (50 – 200 Å). On the one hand, dynamical effects should be rather small since the objects generally have large unit-cell parameters (see Chaps. 5 and 6). Moreover, averaging of the intensity over a great number of crystals differing in thickness and orientation should also decrease dynamical effects. On the other hand, large unit-cell parameters ensure simultaneous excitation of a large number of waves in the crystals, which should increase dynamical effects.

Unfortunately, there is still no analytical solution with all these aspects taken into account. Therefore, one has to employ another approach that consists in testing the possibility to use experimental intensities for structure studies.

There are, at present, numerous works confirming the inference of Vainshtein that the kinematical approximation is generally sufficient to describe reflection intensities. These papers concern, e.g., structure determinations for thin films used in microelectronics, as well as numerous structure refinements of layer silicates (Drits et al. 1984; Drits 1982; Imamov 1977; Imamov and Pinsker 1965; Imamov et al. 1982; Plançon et al. 1985; Zvyagin 1967; Zvyagin et al. 1979).

There are, however, two problems limiting the reliability of the experimental foundation for the applicability of the kinematical approximation for OTED. The first problem is associated with the photographic registration of diffraction patterns leading to limitations for the accuracy in intensity measurements and, consequently, in atomic coordinates and the corresponding scattering powers. Thus, variations in reflection intensity resulting from dynamical effects might be within the experimental error in the intensity measurements. Naturally, this hinders the revealing of dynamical effects and may lead to considerable errors in structure details. The answer to this has been the use of scintillator counters and filters which eliminate incoherent diffusion background and improve the accuracy of the intensity measurements (Avilov et al. 1976; Imamov et al. 1982). This ensured the refinement of several structures with a $4\% - 5\%$ reliability factor (Drits et al. 1984; Imamov et al. 1982; Plançon et al. 1985; Tsipursky and Drits 1977a, b) and confirmed that the kinematical theory can be applied using the two-beam dynamical approximation for only a small number of the strongest reflections.

The second problem which arises mainly with crystals having rather large unit cells is the overlapping of reflections. This prevents accurate measurement of local integrated intensities for a large number of reflections. A solution to this problem can be the simulation of diffraction patterns. The diffracted intensities depend not only on the structure of the unit cell but also on the degree of particle disorientation and the size of coherent domains. It is therefore necessary to elaborate a mathematical formalism that would describe diffraction taking these parameters into account.

It will be shown in this chapter that the solution of these two problems has again confirmed the validity of the kinematical approximation for structure studies of finely dispersed minerals.

8.2 OTED Patterns: Peculiarities of Geometrical Arrangement of Reflections and Integrated Intensities

Aggregations of thin plate-like particles are used as objects in OTED studies. If a drop of a suspension containing thin crystallites is put on a thin colloid film, a textured preparation is obtained after drying. The crystallites are arranged with their best-developed faces parallel to the support surface and are rotated at random about the texture axis which is normal to the support plane. There are also some other ways to prepare a texture mosaic film.

Reciprocal lattices of separate crystallites forming a textured preparation should be brought into coincidence within a single reciprocal lattice with the texture axis passing through the origin. In other words, the reciprocal lattice of an oriented polycrystalline sample may be treated as the result of rotation of a single-crystal reciprocal lattice about the texture axis. We shall now confine our attention to layer mineral structures. This is the case where the ab plane is the best-developed one, so that the texture axis is parallel to $c*$. It has been shown in Chapter 3 that the reciprocal lattice for such crystals is represented by a two-dimensional set of parallel rods containing hk lattice points with a repeat distance $c*$, where hk are constant for each rod. The reciprocal lattice being rotated about $c*$, the lattice points located along the lattice row with the given hk indices circumscribe circles located at the surface of a cylinder.

Therefore, the resultant reciprocal lattice of the texture may be imagined as a set of coaxial cylinders inserted into one another (Fig. 38). The surface of such a cylinder is formed by a set of parallel rings with intensity spread over these. The position of a point at the cylinder surface may be described then by the cylinder radius s_0 equal to the projection of s on the plane ab and by the distance Z between the point involved and the plane ab, which is equal to the projection of s on $c*$. If the point in question coincides with the center of a reciprocal lattice node, we shall denote $s_0 = b_{hk}$ and $Z = D_{hkl}$. Then, if $\gamma = 0$,

$$b_{hk} = \left(\frac{h^2}{a^2} + \frac{k^2}{b^2} \right)^{1/2} \quad \text{and} \quad D_{hkl} = ha* \cos \beta* + kb* \cos \alpha* + lc* . \quad (8.1)$$

In phyllosilicates, $b = a\sqrt{3}$ and $b_{hk} = (3h^2 + k^2)^{1/2}/b$. This means that lattice rows having equal values of $(3h^2 + k^2)^{1/2}$ are situated at equal distances from the texture axis (e.g., $b_{02} = b_{11} = b_{\bar{1}1} = b_{\bar{1}\bar{1}}$ or $b_{20} = b_{13}$ etc.). Therefore, the surface of one cylinder may contain nodes having different hk but equal values of $(3h^2 + k^2)^{1/2}$.

Assume that a texture is tilted by the angle ϑ_s with respect to the incident beam. The diffraction pattern then can be pictured by replacing the Ewald sphere

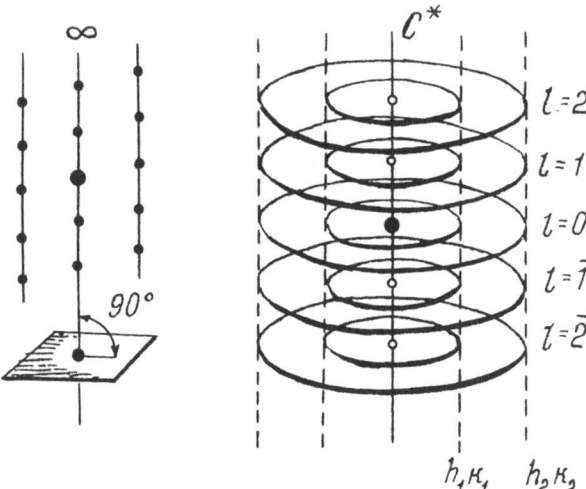

Fig. 38. Reciprocal lattice of a textured sample (Vainshtein 1956)

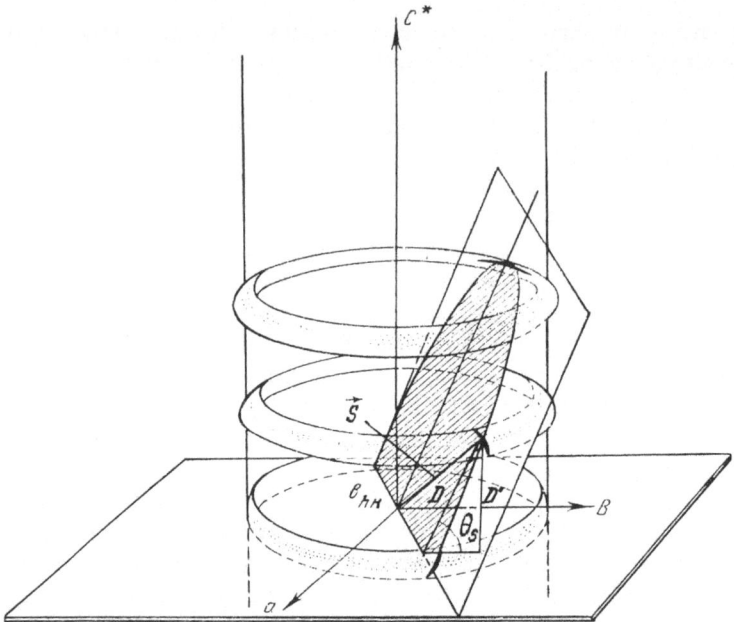

Fig. 39. An oblique section of the texture reciprocal lattice by the Ewald sphere (Vainshtein 1956)

by a plane passing through the origin and making an angle $(90 - \vartheta_s)$ with the axis c^* (Fig. 39). This plane should contain point reflections located on planar oblique sections of the cylinder, i.e., on ellipses (Pinsker 1949; Vainshtein 1956; Zvyagin 1967).

In reality, texture samples are never ideal. Therefore, texture disorientation can be described by a microcrystal orientation function $N(\alpha)$, which determines the portion of crystallites whose normals make an angle α with the texture axis. Each reciprocal lattice point is transformed into a spherical belt. As a result, the coaxial cylinders mentioned become "average" texture cylinders. If this is the case, diffraction patterns consist of arcs centered at the reciprocal lattice origin, and their angular spread increases with texture disorientation.

Analysis of the arrangement of reflections provides information on the dimensions and the shape of the unit cell, and the indices of reflections much more readily than in the case of X-ray powder diffraction. It is evident that the cylinder radius b_{hk} in the reciprocal lattice corresponds to the minor axis of the ellipse in the diffraction pattern on the scale $L\lambda$ (see Chapt. 3), while the distance D_{hkl} corresponds to the distance between the hkl reflection and the minor axis which is given by (Zvyagin 1967)

$$D_{hkl} = hp + ks + lq \qquad\qquad\qquad (8.2)$$

where

$$p = \frac{L\lambda}{\sin\vartheta_s}\, a^* \cos\beta^* \,; \quad s = \frac{L\lambda}{\sin\vartheta_s}\, b^* \cos\alpha^* \,; \quad q = \frac{L\lambda c^*}{\sin\vartheta_s}\,.$$

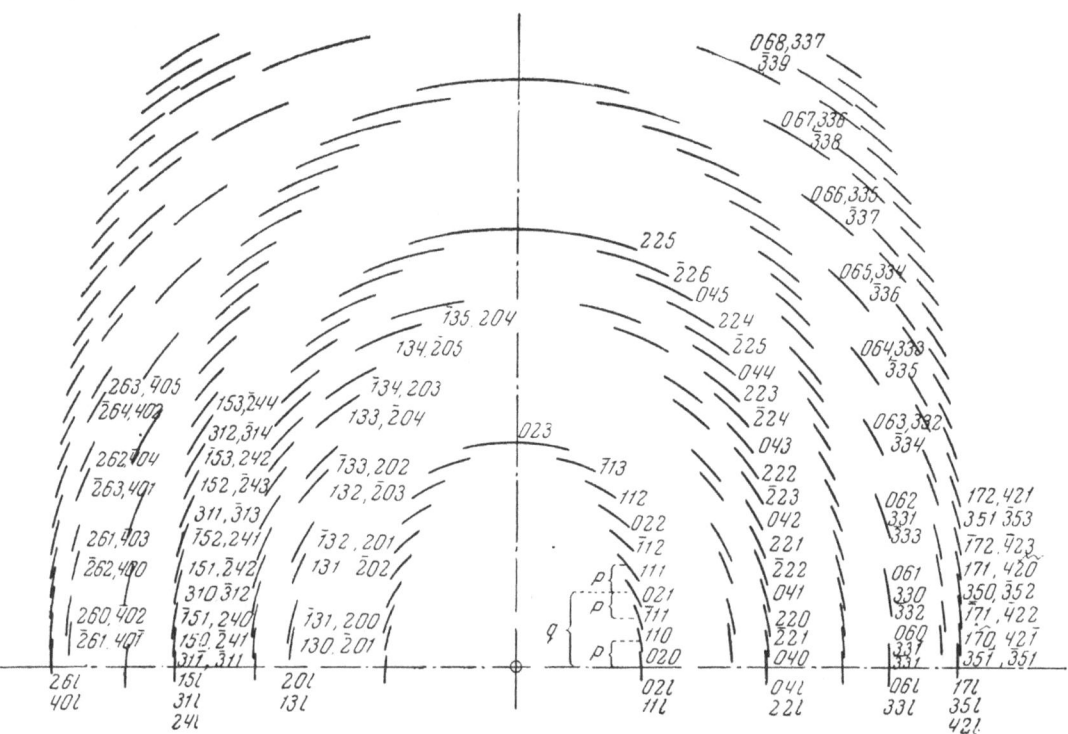

Fig. 40. Arrangement of hkl reflections in an OTED pattern from 1 M-mica (Zvyagin 1967)

Thus the problem is to determine p, s, and q that are directly related to the arrangement of reflections on each ellipse.

For phyllosilicates, the problem of determination of unit-cell parameters and indexing is facilitated by the fact that each ellipse contains reflections with a known set of hk indices, and $h + k$ is always even, as the corresponding crystal lattices are C-centered. For orthorhombic lattices $p = s = o$, $D_{hkl} = lq$, and reflections on each ellipse are equally spaced along the normal to the minor axis. For monoclinic lattices, $s = 0$, $D_{hkl} = hp + lq$. In phyllosilicates the first ellipse contains reflections $02l$ and $\overset{\pm}{1}\overset{\pm}{1}l$, so that reflections $02l$, $11l$ and $\bar{1}1l$ repeat with the period q with the increase of l. Between each two reflections $02l$ and $02(l + 1)$ with a fixed l there is a pair of reflections $11l$ and $\bar{1}1(l + 1)$ at the distances $ql + p$ and $q(l + 1) - p$ from the minor axis, respectively. The reflections $11l$ and $\bar{1}1l$ are located at distances $ql + p$ and $ql - p$, respectively, from the $02l$ reflection, where l is fixed (Fig. 40).

These considerations simplify determination of the characteristics of the mineral under study.

Rational rules for geometrical analysis of OTED patterns for a general case of triclinic lattices have been worked out by Zvyagin (1967).

Vainshtein (1956) has derived a formula describing integrated intensity of arc reflections in OTED patterns

$$I_{hkl} = I_0\lambda^2 \left|\frac{\Phi(hkl)}{V_0}\right|^2 V \frac{L\lambda p}{2\pi R' \sin\vartheta_s} \tag{8.3}$$

where V is the diffracting volume, p is the multiplicity factor, ϑ_s is the angle between the texture axis and the incident beam, R' is the distance from the arc center to the major axis. Other notations are equivalent to those already used in the preceding chapters.

In some cases it is convenient to evaluate the so-called local intensity of a reflection (Vainshtein 1956). The angular length of the arc is defined by the angular spread α_m of the normals n to basal planes of individual particles with respect to the normal n_s to the support. The actual intensity distribution along the arc can be replaced by a distribution of constant intensity along the effective arc length. The latter is equal to $r\alpha'$ where $\alpha' \sim \alpha_m/2$, and $r = L\lambda/d(hkl)$ is the distance between the arc center and the pattern. The local intensity is equal to the integrated intensity corresponding to a minor length $\varDelta$ of the slit of the detector or the microphotometer, i.e.,

$$I_{\text{loc}}(hkl) = I_I(hkl)\frac{\varDelta d(hkl)}{\alpha' L\lambda} = I_0\lambda^2 \left|\frac{\Phi(hkl)}{V_0}\right|^2 V\frac{\varDelta d(hkl)d(hko)p}{r\pi\alpha'}. \tag{8.4}$$

The formulae (8.3) and (8.4) are used to determine the moduli of structure amplitudes $|\Phi(hkl)|$ on the basis of measured integrated or local intensities. To allow for dynamical effects in terms of the two-beam approximation, corrections may be introduced by employing either the expression (5.34) or Bethe potentials.

However, for minerals with large unit cell parameters, reflections in OTED patterns are often located so close to one another that measurement of local intensity becomes difficult. Besides, it is sometimes of interest to compare profiles for reflections with different hkl. For photometric registration of intensities, this meets serious difficulties that can be eliminated by using scintillator counters. Therefore, in the kinematical approximation, it is advantageous to take account of intensity variation within a reflection which depends on the coherent domain dimensions in the ab plane and the degree of particle disorientation in the sample.

8.3 Two-Dimensional Intensity Distribution in OTED Patterns

Plançon (1980) has proposed a method for calculating intensities diffracted by partially oriented powders of tabular crystals with layer structure. He has considered the so-called asymmetric X-ray transmission diffraction where the sample is fixed and intensities are recorded by a scintillator counter moving around the goniometer axis. For such transmission, the expression $\vartheta_c = 2\vartheta_s$ for the rotation angles ϑ_c of the counter and ϑ_s of the sample does not hold.

Diffracted intensities in asymmetrical transmission are described by the formula (Plançon 1980)

$$I_{hk}(s) = \frac{1}{s\,\Omega_c\,\sigma}\int\bar{N}_e(s,\varphi)\,i_{hk}(Z)\,T(X)\,d\varphi \tag{8.5}$$

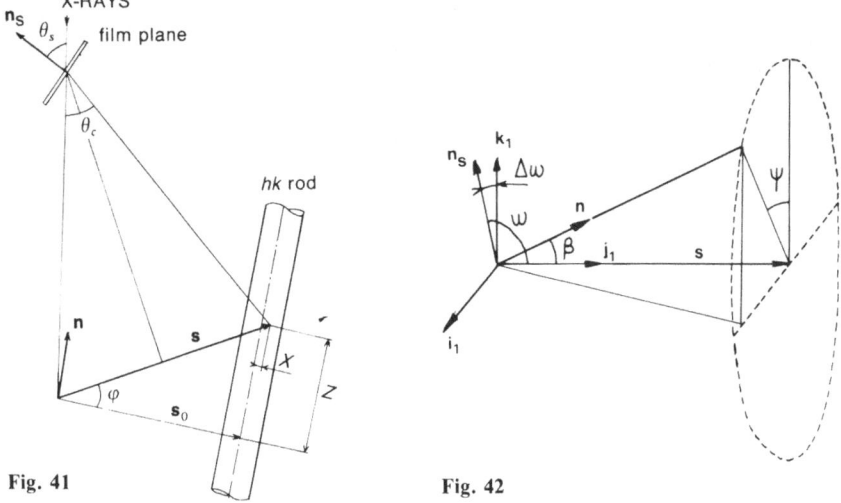

Fig. 41. Description of parameters used in intensity calculation (Plançon et al. 1985)

Fig. 42. Description of parameters for calculation of the particle orientation effect (Plançon et al. 1985)

where the integration is over all the possible microcrystal orientations with respect to the scattering vector s. The parameters involved are shown in Fig. 41. Ω_c is the area of the layer unit cell, σ is the scattering domain area, φ is the angle between s and its projection on the basal plane of the crystallite, $Z = s \sin \varphi$ is the coordinate of the tip of s along the rod of the crystallite and $X = s \cos \varphi - s_0$ is that with respect to the rod center (s_0 is the distance between the rod and the reciprocal space origin); $i_{hk}(Z)$ is the intensity distribution along the hk rod axis in the kinematical approximation, $T(X)$ is the function characterizing the dimensions and the shape of scattering domains, $\bar{N}_e(s, \varphi) = (1/2\pi) \int_0^{2\pi} N(\alpha) d\psi$ is the disorientation function where ψ specifies the orientation of a microcrystal normal with respect to s (see Fig. 42).

Intensity recording in OTED, in contrast to asymmetrical X-ray diffraction, is two-dimensional, i.e., the counter moves not around an axis, but in two directions, so that intensities are measured in a plane (Fig. 43). Thus the intensity distribution depends not only on s but also on ϑ, i.e., the angle between the scattering vector and the plane containing n_s and the incident beam.

For two-dimensional scattering, we may also apply the approach of Plançon (1980) where (8.5) is replaced by

$$I_{hk}(s, \vartheta) = (1/s\,\Omega_c\,\sigma) \int \bar{N}_e(s, \vartheta, \varphi)\, i_{hk}(Z)\, T(X)\, d\varphi, \tag{8.6}$$

where

$$\bar{N}_e(s, \vartheta, \varphi) = (1/2\pi) \int_0^{2\pi} N(\alpha)\, d\psi = (1/2\pi) \int_0^{2\pi} N(s, \vartheta, \psi, \varphi)\, d\psi . \tag{8.7}$$

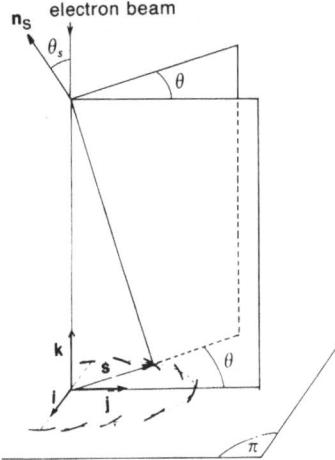

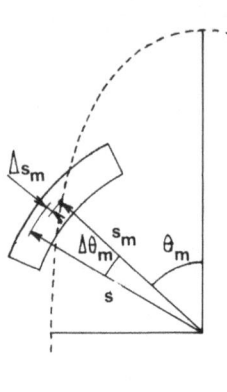

Fig. 43   **Fig. 44**

Fig. 43. Geometrical conditions in the case of OTED (Plançon et al. 1985)

Fig. 44. Parameters characterizing a point in a reflection relative to reflection center (Plançon et al. 1985)

The orientation term $\bar{N}_e(s, \vartheta, \varphi)$ is determined as follows. Vainshtein has proved that in direct texture patterns (texture axis parallel to the incident beam) intensity distribution along each arc characterizes the $N(\alpha)$ distribution. In OTED patterns, intensity distribution along arcs located on the minor axis of an ellipse characterizes the function $N(\alpha/\sin \vartheta_s)$. Thus it is possible to obtain the precise experimental function $N(\alpha)$ using different ϑ_s values.

The plane passing through n_s and the incident beam contains the major axis of the ellipse and makes an angle ϑ with the plane containing the incident beam and s (Fig. 43). The angle between n_s and s can be determined using a coordinate system (i, j, k) where i and j are in the plane π and are parallel to the minor and the major axis of the ellipse, respectively, and k is anti-parallel to the incident beam. Then

$$n_s = -\sin \vartheta_s j + \cos \vartheta_s k$$

$$s = s(-\sin \vartheta i + \cos \vartheta j) .$$

The angle between n_s and s is then given by

$$\omega = \arccos(n_s \cdot s/s) = \arccos(-\cos \vartheta \sin \vartheta_s) .$$

The formula (8.7) can be calculated using another coordinate system (i_1, j_1, k_1) (see Fig. 42), where j_1 and k_1 belong to the (n_s, s) plane, $j_1 \| s$, $k_1 \perp s$ and $i_1 \perp (\cdot j_1, k_1)$. We have

$$n = \sin \beta \sin \psi i_1 + \cos \beta_1 j_1 + \sin \beta \cos \psi k_1$$

$$n_s = -\sin(\Delta \omega) j_1 + \cos(\Delta \omega) k_1 ,$$

where $\beta = n_s = \pi/2 - \varphi$, ψ describes the deviation of n from the plane (n_s, s), and $\Delta \omega$ is the deviation of ω from $\pi/2$, i.e.,

$$\Delta \omega = \arccos(n_s \cdot s/s) - \pi/2$$
$$= \arccos(-\cos \vartheta \sin \vartheta_s) - \pi/2$$
$$= \vartheta_s - \vartheta_c/2 = \vartheta_s - \arcsin \lambda s/2 \; .$$

This leads to

$$\alpha = \arccos(n \cdot n_s) = \arccos(-\cos \beta \sin \Delta \omega \cos \vartheta_s + \sin \beta \cos \psi \cos \Delta \omega)$$

$$\alpha = \arccos(-\sin \varphi \sin \Delta \omega \cos \vartheta_s + \cos \varphi \cos \psi \cos \Delta \omega) \; . \tag{8.8}$$

One can calculate $N(s, \vartheta, \varphi)$ from (8.8) and then $I_{hk}(s, \vartheta)$ using (8.6).

It is appropriate to characterize intensity distribution with respect to arc centers, which are defined by s_m and ϑ_m so that

$$s_m = (s_0^2 + Z^2)^{1/2}$$
$$\vartheta_m = \arcsin[(s_0/s_m \sin \vartheta_s)^2 - \cot^2 \vartheta_s]^{1/2} \; .$$

Any other point on an arc defined by s and ϑ will be specified by Δs_m and $\Delta \vartheta_m$ with respect to the center, so that

$$\Delta s_m = s - s_m$$
$$\Delta \vartheta_m = \vartheta - \vartheta_m$$

as shown in Fig. 44.

8.4 Factors Affecting Diffracted Intensities

The influence of the spread and the shape of the particle orientation function and the scattering domain size on the intensity distribution in OTED patterns has been discussed by Plançon et al. (1985).

For simplicity it has been assumed that the structure factor is the same for all reflections, i.e., $\Phi^2(Z) = 1$. The scattering domains have been approximated by discs 200 Å thick and of the radius $R = 200$ or 1000 Å. This is in agreement with previous observations (e.g., in clay minerals). Two shapes have been considered for the function $N(\alpha)$. In the first case, (triangular shape), $N(\alpha)$ decreases linearly with α until $N(\alpha) = 0$ for $\alpha = \alpha_{max}$. In the second case (rectangular shape), $N(\alpha)$ is constant in the range $0 < \alpha < \alpha_{max}$ and zero for $\alpha > \alpha_{max}$. The values of the half-width at half maximum (HWHM) have been chosen as 5° and 10°, which is in agreement with the range of values observed in most of the samples prepared for oblique texture electron diffraction. Values for s_0 have been chosen as 0.22 Å and 0.67 Å^{-1}, respectively, which corresponds to the first and the fifth ellipses in clay minerals. The tilt angle ϑ_s has been fixed at 70°, which is the maximum value that can be achieved under normal experimental conditions.

Figure 45 compares intensity distribution $I(\Delta s_m)$ for reflections (a) belonging to the same ellipse (the same h, k indices) but having different Z values, and (b) belonging to different ellipses but having the same Z. Calculations are made for two values for R. It is seen that
- as expected, reflections belonging to the same ellipse are broadened with Z. Thus the ratio of HWHM for reflections at $Z = 0.1$ and 0.2 Å^{-1} for the first ellipse ($s_0 = 0.22$ Å^{-1}) is 0.81 for $R = 200$ Å and 0.64 for $R = 1000$ Å;

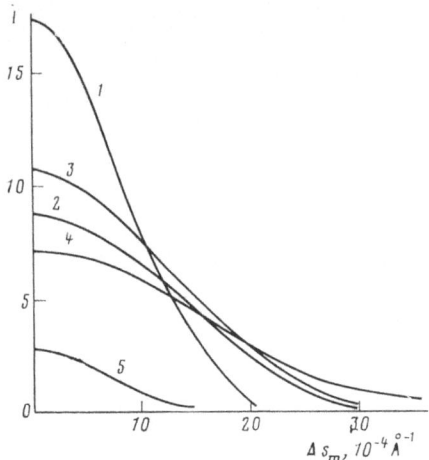

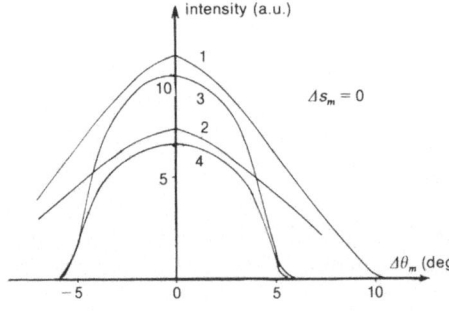

Fig. 45. Intensity versus Δs_m for different reflections in the case of OTED ($\Delta \vartheta_m = 0$): 1 $s_m = 0.22$ Å^{-1}, $Z = 0.1$ Å^{-1}, $R = 1000$ Å; 2 $s_0 = 0.22$ Å^{-1}, $Z = 0.2$ Å^{-1}, $R = 1000$ Å; 3 $s_0 = 0.22$ Å^{-1}, $Z = 0.1$ Å^{-1}, $R = 200$ Å; 4 $s_0 = 0.22$ Å^{-1}, $Z = 0.2$ Å^{-1}, $R = 200$ Å; 5 $s_0 = 0.67$ Å^{-1}, $Z = 0.2$ Å^{-1}, $R = 200$ Å

Fig. 46. Intensity versus $\Delta \vartheta_m$ for different reflections and two different shapes of the orientation function $N(\alpha)$($\Delta s_m = 0$, $R = 200$ Å): 1 $s_0 = 0.22$ Å^{-1}; $Z = 0.1$ Å^{-1}, $N(\alpha)$ triangular, HWHM = 5°; 2 $s_0 = 0.22$ Å^{-1}, $Z = 0.2$ Å^{-1}, $N(\alpha)$ triangular, HWHM = 5°; 3 $s_0 = 0.22$ Å^{-1}, $Z = 0.1$ Å^{-1}, $N(\alpha)$ rectangular, HWHM = 4.5°; 4 $s_0 = 0.22$ Å^{-1}, $Z = 0.2$ Å^{-1}, $N(\alpha)$ rectangular, HWHM = 4.5°

- HWHM for reflections having small Z depend more strongly on R than for those having high Z values. For example, the HWHM for a reflection at $Z = 0.1$ Å^{-1} (the first ellipse) decreases by the factor 1.46 with the increase in R from 200 Å to 1000 Å, while the factor is only 1.15 for a reflection at $Z = 0.2$ Å^{-1}. This effect may be used for qualitative evaluation of the scattering domain size;
- there is no direct relationship between the coherent domain radius and the HWHM of $I(\Delta s_m)$ for reflections at $Z \neq 0$;
- $I(\Delta s_m)$ is practically independent of the shape of $N(\alpha)$.
  Figure 46 shows $I(\Delta \vartheta_m)$ for the two $N(\alpha)$ functions mentioned.

The main conclusions following from this figure are:
- Intensity distribution $I(\Delta \vartheta_m)$ depends on the shape of the orientation function but is almost independent of reflection indices.
- The HWHM values for reflections divided by $\sin \vartheta_s$ practically coincide with the HWHM of $N(\alpha)$.

The conclusions can be connected with the use of local intensity for determining structure factors, i.e.

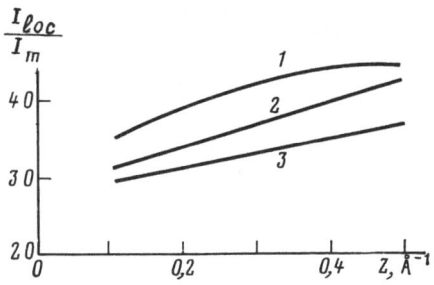

$$I_{loc}(s_m, \vartheta_m) = \int I(\Delta s_m)\, d(\Delta s_m) = K\, \Phi^2(s_m, \vartheta_m)/s_0 s_m. \tag{8.9}$$

According to (8.9), $I(\Delta s)$ curves for different reflections have been used to determine $\Phi(s_m, \vartheta_m)$. Relative deviations of values thus calculated from the real ones are less than 5% even under the most unfavorable conditions (small s_0 and large Z). This justifies application of (8.9) to reflections having intensities estimated with an error greater than 10%.

It should be stressed, however, that it is difficult to use local integrated intensity for determining structure factors in practice. The usual experimental conditions are: wavelength $\lambda = 0.037$ Å, distance between sample and recording equipment $L = 0.60$ m. Distances of 0.2 or 0.3 mm between reflections are difficult to obtain since the samples most suited to structural analysis generally have large unit cells and small scattering domains (a few hundred Å). This leads to overlapping of reflections, thus preventing accurate determination of local integrated intensities and the structure factor.

Therefore, a more complicated but a more accurate method has been proposed (Plançon et al. 1985) for evaluation of structure factors. Several reflections that do not overlap are chosen, and $I(\Delta s_m)$ curves are calculated for them assuming different dimensions for scattering domains. The width of $N(\alpha)$ used in these calculations is obtained from intensity distribution along the arcs located at the minor axis. Comparison of the calculated and the experimental curves allows evaluation of the coherent domain size. Assuming $|\Phi|^2 = 1$, one can calculate the values for $I(\Delta s_m)$ for all the values of s_m and ϑ_m required. Then $I_I(\Delta s_m)/I_{max}$ is plotted versus Z for all reflections belonging to the given ellipse ($s_0 = $ const). $I_I(\Delta s_m)$ is equal to the area under the $I(\Delta s_m)$ curve for fixed s_m, ϑ_m. Thus a calibration graph is obtained (Fig. 47) which is used to introduce corrections by multiplying experimental maximum intensities at the points s_m, ϑ_m by the corresponding I_I/I_{max} values.

8.5 A Technique for OTED Intensity Measurements

Electron diffractometers for OTED are produced only in the USSR. Electrons emitted by the cathode are accelerated by the anticathode voltage, pass through a system of diaphragms and are focused on the screen by an electromagnetic lens. The incident beam is transmitted through a sample that can be rotated about it. Diffracted beams are focused on a luminescent screen or registered on a photo-

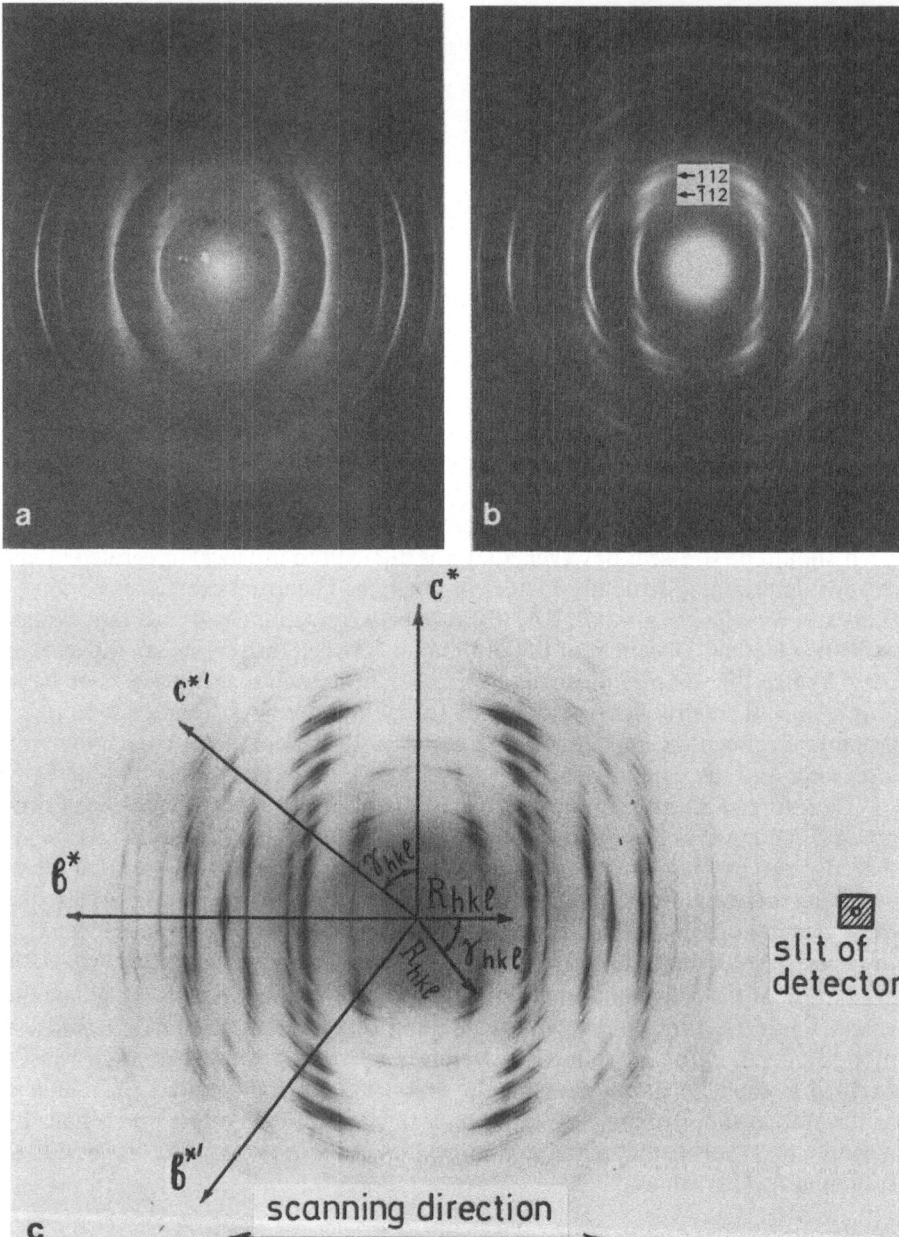

Fig. 48a–c. OTED patterns of **a** natural smectite with turbostratic structure; **b** Fe^{3+}-montmorillonite (sample 11, Table 6); **c** 1 M-phengite: rotation of the pattern to bring the hkl reflection to the scanning direction is shown

plate. In the diffractometer ER-100 intensities can be measured with the help of a scintillator counter which is placed into the center of the diffraction pattern instead of a photo plate or a screen. ER-100 has a deflecting system to which voltage is applied changing with a certain velocity. The diffraction pattern is simultaneously shifted in a fixed direction with a related velocity. After a number of amplifying cycles, the signal registered by the detector is recorded on a chart strip that moves in step with the diffraction pattern. Initially, the device was intended for measuring intensity in ring patterns obtained from completely disoriented polycrystalline samples.

Avilov et al. (1976) have shown that the system for measurement of intensities in ER-100 can be used for OTED if the object holder is modified so as to allow rotation of the sample about the incident beam with constant angle ϑ_s between the incident beam and the texture axis. Rotation of the sample with ϑ_s fixed, will be accompanied by rotation of the diffraction pattern in the screen plane.

In the initial state the sample tilt axis coincides with a fixed scanning direction. The minor axes of the ellipses also coincide with these directions. If the unit cell parameters, the tilt angle ϑ_s and $L\lambda$ are known, the cylindrical coordinates $s_m = R_{hkl}$ and $\vartheta_m = \gamma_{hkl}$ with respect to the pattern center and the scanning direction can be readily found for any reflection (Fig. 48).

To measure the local intensity of a reflection, the sample tilt axis should be rotated about the incident beam so as to bring the reflection analyzed to the scanning direction. The reflection then should be drawn near the detector slit and scanned along it. A vernier is used to rotate the sample through angles to an accuracy of 0.1°.

8.6 Crystal Structure Refinements of Mica Polytypes on the Basis of Electrometric Intensity Measurement

Tsipursky and Drits (1977a) employed this method, having improved the intensity registration system. Initially, a natural finely dispersed $2M_1$-muscovite sample was chosen as a standard object. As muscovite structures had been previously refined precisely by X-ray and neutron diffractions (Güven 1971; Rothbauer 1971), it was possible to compare the results of the OTED study with reported data.

The structural formulae and unit-cell parameters for muscovites studied by OTED and neutron diffraction are, respectively, $K_{0.79}Na_{0.04}Ca_{0.03}(Al_{1.84}Fe^{3+}_{0.06}Fe^{2+}_{0.01}Mg_{0.10})(Si_{3.16}Al_{0.84})O_{10}(OH)_2$, $a = 5.190$, $b = 9.000$, $c = 20.048$ Å, $\beta = 95.73°$ and $K_{0.85}Na_{0.09}(Al_{1.90}Fe^{3+}_{0.12})(Si_3Al_1)O_{10}(OH)_2$, $a = 5.1918$, $b = 9.0153$, $c = 20.0457$ Å, $\beta = 95.735°$ (Rothbauer 1971).

Possible instability of the incident beam intensity was taken into account by normalizing the intensity of each reflection with respect to that of the reference reflection. The latter was measured immediately before and after measuring the intensity of the reflection analyzed.

The local intensity was recorded on the strip as $I = f(\Delta s_m)$.

Reflection intensity was evaluated from the maximum amplitude, since measurements of the area under the curve entailed large errors resulting from an un-

Table 2. Atomic coordinates for $2\,M_1$ muscovite

Atom	x/a	y/b	z/c
K	0.0	0.0976(5)	0.25
Al	0.2506(10)	0.0834(5)	0.0002(2)
T_1	0.4661(8)	0.9289(5)	0.1356(3)
T_2	0.4516(8)	0.2581(4)	0.1356(2)
O_1	0.4193(10)	0.0926(7)	0.1680(3)
O_2	0.2526(10)	0.8107(7)	0.1576(3)
O_3	0.2499(11)	0.3705(7)	0.1688(3)
O_4	0.4615(11)	0.9447(7)	0.0534(3)
O_5	0.3869(10)	0.2523(7)	0.0538(3)
O_6	0.4585(10)	0.5614(7)	0.0502(3)
H	0.366	0.662	0.061

certainty in the position of the background line. Intensities $I(hkl)$ were transformed to $\Phi(hkl)$ by the formula (8.4). Ninety-six reflections were recorded having $I_{max}:I_{min} = 125$ using the minimum scanning velocity and the slit diameter 0.05 mm. These data were used to obtain $\Phi_0(hkl)$ values that were compared with $\Phi_c(hkl)$ calculated using atomic coordinates determined by neutron diffraction (Rothbauer 1971). While the agreement between Φ_0 and Φ_c was, in general, satisfactory for each ellipse, there was a systematic increase in $\Delta\Phi = |\Phi_c| - |\Phi_0|$ with $\sin\vartheta/\lambda$. After introducing corrections R became 6.2%. Taking into account that the chemical composition of the sample under study differed from that in (Rothbauer 1971) we may conclude that the results obtained imply (a) high accuracy in intensity measurements (10% – 15% on average), and (b) the validity of the kinematical approximation.

Since the details of the structure under study could have been different from those of $2\,M_1$-muscovite in (Rothbauer 1971), it was refined on the basis of OTED data. Intensities for 60 weak reflections estimated visually have been added to the reflections measured precisely. For the 30 weakest reflections intensity was assumed, in accordance with (Vainshtein 1956), to be equal to half the intensity of the weakest reflection evaluated visually. The programm "X-ray 70" was used for refinement. R decreased to 0.05 and the standard deviations in interatomic distances were ≤ 0.005 Å. Table 2 contains the atomic coordinates obtained by OTED and Table 3 shows the corresponding interatomic distances. Minor differences in muscovite bond lengths obtained by X-ray, neutron and electron diffractions are readily explained by differences in chemical composition of the samples involved. A smaller amount of Al for Si substitution in tetrahedra and the presence of Mg in octahedra in the muscovite structure under study has led to a decrease in $(T-O)_{mean}$ and an increase in $(M-O)_{mean}$.

The results obtained by Plançon et al. (1985) have allowed the introduction of corrections into the maximum intensity according to a more rigorous procedure based on the relationship $I_{loc}/I_{max} = f(Z)$. This approach increased reliability in the estimation of structure factors although it did not lead to any substantial increase in accuracy for atomic positions. Below are given the results of structure

Table 3. Bond lengths for refined mica structures

2 M_1 muscovite

T_1-O_4 1.644(8)	T_2-O_5 1.635(8)	Al-O_4 1.945(7)	K-O_1 2.857(8)
T_1-O_1 1.637(7)	T_2-O_1 1.640(8)	Al-O_4 1.938(7)	K-O_2 2.878(8)
T_1-O_2 1.626(8)	T_2-O_2 1.655(8)	Al-O_5 1.963(7)	K-O_3 2.844(8)
T_1-O_3 1.645(8)	T_2-O_3 1.628(8)	Al-O_6 1.910(7)	
		Al-O_5' 1.961(7)	2.860 Å
1.638 Å	1.639 Å	Al-O_6' 1.914(7)	
		1.938 Å	

1 M phengite

T-O_2 1.607(6)	M 1.942(6)	K-O_4 2.903(5)	
T-O_3 1.632(6)	M 1.938(6)	K-O_3 2.947(5)	
T-O_4 1.628(7)	M 1.960(6)		
T-O_4' 1.639(7)		2.917(3)	
1.626(3)	1.947(3)		

1 M celadonite

T_1-O_1 1.612(7)	T_2-O_1 1.635(6)	M_1-O_4 2.012(6)	
T_1-O_2 1.623(9)	T_2-O_2 1.620(8)	M_1-O_5 2.058(7)	
T_1-O_3 1.617(7)	T_2-O_3 1.628(8)	M_1-O_6 2.010(6)	
T_1-O_5 1.610(7)	T_2-O_6 1.591(8)	2.027(4)	
1.616(4)	1.618(4)		
M_2-O_4 2.043(6)	K-O_1' 3.115(6)	K-O_1 3.089(6)	
M_2-O_5 2.098(7)	K-O_2' 3.128(7)	K-O_2 3.099(7)	
M_2-O_6 2.039(6)	K-O_3' 3.067(6)	K-O_3 3.126(6)	
2.060(4)			

refinements of 1 M-celadonite (Drits et al. 1984) and 1 M phengite (Plançon et al. 1985).

1 M-Celadonite. The mineral under study had the composition $K_{0.86}Ca_{0.10}(Al_{0.05}$ $Fe_{0.90}^{3+}Fe_{0.32}^{2+}Mg_{0.73})(Si_{3.96}Al_{0.04})O_{10}(OH)_2$ and the unit cell parameters $a = 5.223$, $b = 9.047$, $c = 10.197$ Å, $\beta = 100.43°$. The celadonite structure consists of 2:1 layers having identical azimuthal orientations. The 2:1 layers consisting of regular polyhedra have the symmetry C 2/m. They are stacked so that hexagonal rings in the tetrahedral sheet of one layer are above those of the other layer. Interlayer cavities are filled with K cations.

Intensities were measured for 112 strong and medium reflections according to the method described. A 0.05 mm slit was placed before the detector. Calculations have shown that such a slit is small enough not to distort reflection profiles. Analysis of $I(\Delta s_m)$ curves has refined the coherent scattering domain dimensions ($t = 150$ Å and $R = 200$ Å). The intensity distribution for the reflection at the

Table 4. Atomic coordinates for 1 M celadonite

Atom	x/a	y/b	z/c
K	0.5	0.0	0.5
M_1	0.5	0.16614(51)	0.0
M_2	0.5	0.83228(49)	0.0
Si_1	0.41432(69)	0.33287(54)	0.27331(50)
Si_2	0.41281(65)	0.66758(57)	0.27267(52)
O_1	0.18659(96)	0.25098(68)	0.33323(59)
O_2	0.44305(114)	0.50052(74)	0.33074(61)
O_3	0.68887(93)	0.25503(63)	0.33407(58)
O_4	0.39645(95)	0.00156(79)	0.11257(62)
O_5	0.37812(84)	0.33172(65)	0.11294(48)
O_6	0.85604(91)	0.17902(71)	0.11419(57)
H	0.247	0.002	0.093

Atom	u_{11}	u_{22}	u_{33}	u_{12}	u_{13}	u_{23}
K	0.0537	0.0509	0.0458	0.0	0.0054	0.0
M_1	0.0478	0.0116	0.0482	0.0	0.0087	0.0
M_2	0.0306	0.0103	0.0612	0.0	0.0231	0.0
Si_1	0.0563	0.0483	0.0485	−0.0002	0.0214	0.0121
Si_2	0.0628	0.0287	0.0488	0.0059	0.0193	0.0071
O_1	0.0838	0.0493	0.0298	0.0039	0.0193	0.0187
O_2	0.0530	0.0424	0.0608	−0.0029	0.0101	0.0047
O_3	0.0576	0.0451	0.0637	0.0004	0.0238	−0.0039
O_4	0.0463	0.0713	0.0516	0.0061	0.0186	0.0192
O_5	0.0710	0.0485	0.0488	0.0027	0.0235	0.0095
O_6	0.0699	0.0445	0.0490	0.0090	0.0130	0.0026

minor axis was used to find the width of the function $N(\alpha)$ at half maximum equal to 5°. Figure 45 shows correction curves that were used to evaluate local intensities from measured maximum intensities I^{max} (s_m, ϑ_m). Intensities of 94 weak and very weak reflections were estimated visually. The anisotropic refinement was performed using the program XRAY-72 (Stewart et al. 1972) for space groups C2 and C2/m. In the C2/m model, a disordered distribution of octahedral Fe and Mg was assumed, while in the C2 model bi- and tri-valent cations were supposed to be regularly distributed over the two independent cis-octahedral sites. Several refinement cycles have given preference to the C2 model. The refined atomic coordinates, the standard deviations and the thermal factors obtained with $R = 0.05$ are given in Table 4. Interatomic cation-anion distances are given in Table 3. Refinement of scattering powers for octahedral sites yielded the following compositions for the cis-octahedra M2 and M3

M2: $Mg_{0.13}(Fe^{2+}, Fe^{3+})_{0.87}$

M3: $Mg_{0.65}(Fe^{2+}, Fe^{3+})_{0.35}$.

Taking account of the mean cation-oxygen bond lengths for pure octahedral Mg, Fe^{2+}, Fe^{3+} and Al (Drits 1975), the octahedral compositions have been estimated as

M2: $Mg_{0.13}Fe^{2+}_{0.09}Fe^{3+}_{0.78}$

M3: $Mg_{0.60}Fe^{2+}_{0.23}Fe^{3+}_{0.12}Al_{0.05}$.

The calculated distances $M2 - O, OH = 2.02$ Å and $M3 - O, OH = 2.06$ Å are in agreement with the experimental bond lengths (Table 3). High contents of bivalent cations in octahedra lead to a decrease in cation repulsion. As a result, the octahedral bases rotation angles decrease substantially (3.9° and 1.3° for M2 and M3 respectively).

The mean $T - O$ bond lengths in T1 and T2 are equal within the precision of the determination. Interlayer K cations have, in fact, hexagonal prismatic coordination as the mean tetrahedral rotation is only 0.54°.

1M-Phengite. The mineral under study was finely dispersed. The chemical composition and the unit cell parameters were $K_{0.80}Na_{0.02}Ca_{0.01}(Al_{1.66}Fe^{3+}_{0.06}Fe^{2+}_{0.02}Mg_{0.28})[Si_{3.41}Al_{0.59}]O_{10}(OH)_2$; $a = 5.199$, $b = 9.005$, $c = 10.164$ Å, $\beta = 101.3°$. The reasons for choosing this particular structure were the following. One of the essential crystal-chemical problems is order/disorder in the distribution of Si and Al over tetrahedral sites. This problem is complicated not only by close scattering powers of Si and Al but also by a relatively low Al content in tetrahedra. On the other hand, order/disorder in the Si, Al distribution in a 1 M-mica is defined by the choice of the space group. If the refinement is carried out for C2/m, then only one symmetrically independent tetrahedral site is refined. This implies disordered distribution of Si and Al. The space group C2 allows some degree of ordering since there are two independent tetrahedral sites. Thus, the actual problem is to choose the right space group.

To reveal criteria for the choice of the true space group, the following methodological works has been carried out. On the basis of the 3T-muscovite refined structure (Güven and Burnham 1967) where the Si, Al distribution had been proved ordered, a 1M-muscovite structure with the space group C2 was simulated. Averaging the atomic coordinates by introducing symmetry planes and centers yields a 1 M muscovite structure belonging to the space group C2/m.

For each of these two models, the values $\Phi_c(hkl)$ have been calculated for 300 reflections. $\Phi_c(hkl)$ values calculated for the given hkl in different models differed, on average, by 5%. Hence the true space group can be determined only with precise intensity measurements. The atomic coordinates for the C2 model were then refined using the structure factors calculated for the C2/m model. Although the factor R was 0.01, the mean bond lengths in nonequivalent tetrahedra remained different, and the standard deviations for coordinates of atoms lying close to the symmetry plane were greater by a factor of 10 than those for other atoms. The Wilson criteria also permitted revealing the true space group. On the basis of these data, requirements have been formulated for the choice of the space group: (a) intensity should be measured with an accuracy of 10% – 15%; (b) the refined standard deviations should have their minimum values close for all atoms to the minimum R; (c) the Wilson statistical analysis of intensities should be employed. Application to the mineral under study has shown that the true space groups is C2/m.

Table 5. Structure refinement of a mica with muscovite-phengite composition

(a)

	x	y	z
K	0.5	0.0	0.5
T	0.41719(61)	0.67072(57)	0.26981(43)
M	0.0	0.66689(52)	0.0
O_1	0.41891(84)	0.0	0.10528(82)
O_2	0.34809(94)	0.69131(72)	0.10971(63)
O_3	0.48142(87)	0.5	0.31989(65)
O_4	0.17221(92)	0.72790(71)	0.33460(65)
H	0.212	0.0	0.141

(b)

	u_{11}	u_{22}	u_{33}	u_{12}	u_{13}	u_{23}
K	0.0541	0.0528	0.0514	0.0	0.0036	0.0
T	0.0516	0.0487	0.0495	0.0009	0.0154	0.0067
M	0.0544	0.0571	0.0492	0.0	0.0095	0.0
O_1	0.0846	0.0255	0.0368	0.0	0.0103	0.0
O_2	0.0644	0.0501	0.0464	0.0019	0.0098	−0.0083
O_3	0.0587	0.0391	0.0538	0.0	0.0124	0.0
O_4	0.0510	0.0600	0.0458	−0.0030	0.0077	0.0026

(a) Atomic coordinates with e.s.d's (in parentheses). T: tetrahedral cation; M: Octahedral cation; K: interlayer cation; O_1: oxygen of the hydroxyl group; O_2: inner oxygen; O_3, O_4: external oxygens. (b) Temperature factors (Å^2).

Intensities were measured with the help of the same method as in the case of celadonite. Two hundred and ten reflections were included into structure refinement using the program XRAY-72 (Stewart et al. 1972). Table 5 shows the refined atomic coordinates, the standard deviations, and the anisotropic thermal factors obtained with $R = 0.042$. The accuracy in the determination of interatomic distances is approximately the same as in single-crystal X-ray refinements of $2\,M_1$-muscovite (Güven 1971, Rothbauer 1971) and $2\,M_1$-phengite (Güven 1971).

8.7 Determination of Hydrogen Positions in Mica Structures by OTED

Electron diffraction has considerable advantages over X-ray diffraction so far as determination of hydrogen positions is concerned. However, photographic intensity measurement in phyllosilicate structure refinements did not allow the revealing of the hydrogen positions (Zvyagin et al. 1979). Considerable increase in precision in intensity measurements has ensured hydrogen localization in all the three mica structures involved (Tables 2, 4, 5). To do this, difference maps have been computed, by eliminating contributions of all the atoms except hydrogen. In a map thus obtained the region around the hydroxyl oxygen has been analyzed.

In the case of $2 M_1$-muscovite, the difference map shows only one relatively strong peak whose coordinates are close to those of the hydrogen position revealed by neutron diffraction (Rothbauer 1971). The O-H vector of the length 1.06 Å is tilted toward the vacant trans-octahedron, making an angle of 78° with the normal to the layers. For 1 M-celadonite, the region analyzed contains four peaks. One of these, which is eight times higher than the others, has been ascribed to the hydrogen position or, to be more precise, to the center of a potential distribution produced by the proton and the near-by electrons. The proton is located inside the vacant trans-octahedron, and the O-H vector of the length 0.8 Å makes an angle of 76° with the normal to the layer. Thus, there are two protons inside each celadonite trans-octahedron that may be regarded as an effective bi-valent cation favoring local charge compensation.

In the 1 M-phengite structure, the O-H vector orientation has been found to be close to that in $2 M_1$-muscovite although it makes an angle of 104° with the layer normal.

The O-H vector orientations in the micas under study obtained from OTED data have been compared with those yielded by electrostatic energy calculations (Bookin et al. 1982). The differences in the corresponding angular parameters do not exceed $1° - 2°$.

Thus, the use of a scintillator counter has ensured not only the determination of heavy atom positions with a accuracy unattainable before, but also the localization of hydrogen.

The main inference here is that the kinematical approximation treats quite adequately the interaction between electrons and matter at least for clay minerals having comparatively large unit-cell parameters, complex structures and relatively light atoms. In the structure refinements described, even the Blackman corrections for the strongest reflections were not required.

8.8 Study of Octahedral Cation Distribution in 2:1 Layers of Dioctahedral Smectites

To illustrate the possibilities of OTED for determining the structural features of finely dispersed and poorly crystallized minerals, we shall consider the OTED study of dioctahedral smectites (Tsipursky and Drits 1984). These minerals are abundant in nature and occur in various geological environments (Drits and Kossovskaya 1980). The smectite structure consists of 2:1 layers, and interlayer water molecules and exchangeable cations. In dioctahedral smectites, 2/3 of octahedral sites are occupied by cations, while in trioctahedral varieties all the octahedra are occupied. Dioctahedral smectites include montmorillonite M_x^+ $(Al_{2-x}Mg_x)Si_4O_{10}(OH)_2 \cdot n\,H_2O$, beidellite $M_x^+\,Al_2(Si_{4-x}Al_x)O_{10}(OH)_2 \cdot n\,H_2O$, nontronite $M_x^+\,Fe_2^{3+}(Si_{4-x}Al_x)O_{10}(OH)_2 \cdot n\,H_2O$, and varieites of intermediate compositions (Méring and Oberlin 1971).

Exceptionally high degree of dispersion and extremely low structural order are characteristic of smectites. Therefore, many crystal-chemical features of smectites are still uncertain. Below are given the results of an OTED study of octahedral cation distribution in dioctahedral smectite structures.

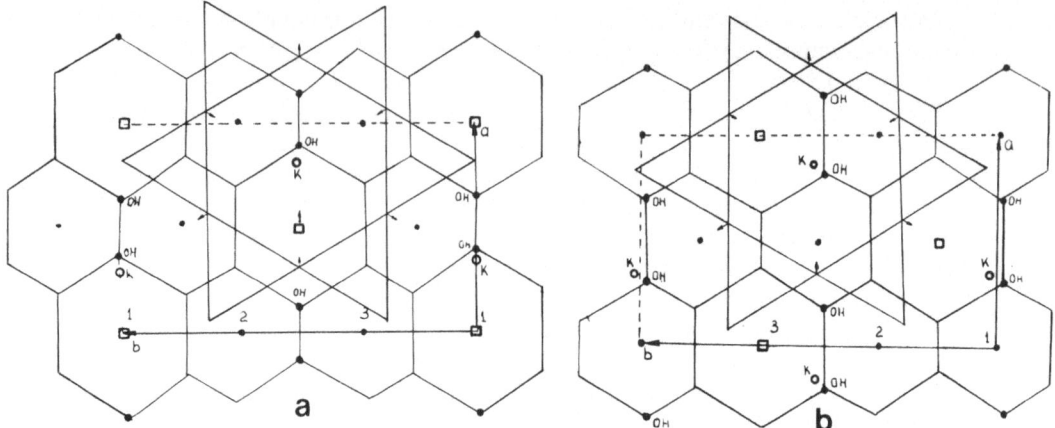

Fig. 49 a, b. (001) projections of octahedra and adjacent tetrahedra of a 2:1 layer. **a** *Cis*-octahedra are occupied (model 1); **b** *cis*-octahedral positions forming one of the two regular point systems are vacant; *trans*-octahedra are occupied (model 2)

The *trans*-octahedra in the octahedral sheet of a 2:1 layer are those having OH groups located at the apices connected by the space diagonal, as distinct from the *cis*-octahedra whose OH groups form edges shared by two *cis*-octahedra. A 2:1 layer is centrosymmetrical if the *trans*-octahedra are vacant (Model 1, Fig. 49), and acentric if the *trans*-octahedral sites, as well as the *cis*-octahedra forming one of the two symmetrically independent point systems, are occupied by cations (Model 2, Fig. 49).

The turbostratic structure of most dioctahedral smectites hinders direct X-ray study of the distribution of the octahedral cations. Méring and Oberlin (1971) approached this problem by analyzing point electron diffraction patterns from microcrystals with the thickness of a single 2:1 layer. They concluded that the structure of the 2:1 layer of Pfaffenreuth nontronite corresponded to Model 1 and that of Wyoming montmorillonite to Model 2. Further studies using SAED showed that there were beidellites with centrosymmetrical layers and the charge localized in the tetrahedral sheets, as well as with noncentrosymmetrical layers and the charge equally distributed over the tetrahedra and the octahedra (Besson 1980; Besson and Tchoubar 1972).

This experimental procedure is, however, difficult to carry out and it is also difficult to interpret the point diffraction patterns. First, one cannot be sure that the sample thickness represents one single layer. Second, distortions of the intensity distribution are inevitable because of the local deformations of single layers and their structure distortions resulting from the interaction with the beam. Finally, diffraction data obtained for separate single layers may fail to reflect the octahedral cation distribution in the 2:1 layers forming thicker particles.

One of the best ways of determining directly the distribution of the octahedral cations is the increase in structural ordering resulting from a rearrangement in layer stacking. Using OTED, Tsipursky et al. (1978) showed that K-saturation of nontronites led to an increase in structural ordering, which was accompanied by the appearance of distinct reflections on the diffraction pat-

Table 6. Crystallochemical formulae of the smectites studies on $O_{10}(OH)_2$ basis. Sample details are given below

Sample number	Nature and amount of cations									
	Interlayer				Tetrahedra		Octahedra			
	Ca	Na	K	Mg	Si	Al	Al	Fe^{3+}	Fe^{2+}	Mg
1	0.22	0.01	0.04		3.95	0.05	1.38	0.18		0.44
2	0.06	0.20			3.96	0.04	1.54	0.18		0.26
3	0.06	0.17	0.09		3.98	0.02	1.38	0.14	0.01	0.48
4	0.13	0.02	0.04		3.91	0.09	1.36	0.39	0.02	0.24
5	0.03	0.31	0.02		4.00		1.51	0.10		0.39
6	0.15	0.01	0.03		4.00		1.40	0.26		0.34
7	0.18	0.05	0.02		3.98	0.02	1.32	0.26		0.41
8	0.06	0.17	0.07		4.00		1.39	0.26	0.05	0.30
9	0.13	0.25			3.83	0.17	1.47	0.19		0.34
10	0.15	0.20	0.10		3.89	0.11	0.89	0.62	0.03	0.46
11	0.11	0.04	0.30		4.00		0.20	1.13	0.38	0.29
12	0.06	0.21	0.27		3.71	0.29	1.64	0.05	0.01	0.31
13	0.10	0.05	0.05	0.27	3.73	0.27	1.05	0.37		0.57
14		0.46[c]			3.86	0.14	1.68			0.32
15	0.07	0.01	0.11	0.15	3.41	0.59	1.57	0.37	0.01	0.05
16	0.11	0.03	0.16		3.53	0.47	0.96	0.88	0.02	0.26
17	0.12	0.05	0.01	0.17	3.45	0.55	0.33	1.59		0.08
18	0.23	0.03	0.01		3.49	0.51[a]		1.87	0.17	
19	0.16	0.03	0.01	0.05	3.65	0.35		1.92		0.08
20		0.02	0.40	0.04	3.46	0.54[b]	0.15	1.85		

[a] Fe^{3+} instead of Al in tetrahedral sites.
[b] Cation composition of tetrahedral sheet is $Si_{3.46}Al_{0.40}Fe^{3+}_{0.14}$.
[c] Sum of interlayer cations obtained by sodium-radioactivity method of G. Besson.

terns. Analysis of the intensities of reflections with $k \neq 3n$ confirmed the conclusion of Méring and Oberlin (1971) concerning the centrosymmetrical structure of the 2:1 nontronite layers. Mamy and Gaultier (1976) suggested a more effective way of improving the structural ordering of smectites which, in addition to K-saturation, involved a large number of wetting-drying cycles (WD). Besson et al. (1982) used this technique to study the Garfield nontronite by several diffraction methods. Comparison of the intensity distribution in the experimental diffraction patterns with those calculated showed a quantitative agreement, which led to the conclusion that the *trans*-octahedral sites of the 2:1 layers of nontronities were vacant. Besson (1980) also confirmed the observation of Méring and Glaeser that the 2:1 layers of the Wyoming montmorillonite were noncentrosymmetrical.

However, electron diffraction has not been used for a systematic study of structural features of smectites because most of them appear to have turbostratic structure (Fig. 48a).

We have used the technique of Mamy and Gaultier (1976) for structural rearrangement of more than 30 samples of dioctahedral smectites of different compositions (Table 6). Study of these by oblique-texture electron diffraction has shown that the cation distribution over available octahedral sites in 2:1 dioctahedral smectites varies more than expected.

Although a large number of smectites have been studied, only three natural Ca-nontronites and the Black Jack Mine beidellite proved to have a sufficiently high degree of ordering in layer stacking. Disordered structures of smectites in the dehydrated state may be attributed to the type of exchangeable cations. Many of the oceanic smectites studied proved to have K in interlayer positions, and their structural ordering in vacuum was higher than that of Ca- or Na-smectites. It may be assumed that, when the interlayer water is completely removed, relatively small Ca, Na and Mg ions are contained within the hexagonal rings of the 2:1 layers and therefore the process of dehydration is accompanied by random interlayer shifts and rotations of the layers at arbitrary angles. Such effects in the structures of dehydrated Ca- and Na-smectites may also be promoted by the fact that the mica-like stacking mode is not the most favorable one because of the electrostatic repulsion of the tetrahedral cations of the adjacent layers.

The replacement of Ca and Mg by K in interlayers leads to an increase in exchangeable cation content and favors the elimination of random defects due to the packing of adjacent layers according to the mica-like mode.

K-Smectites Without WD Cycles. The K_2CO_3 treatment proved to be most effective for nontronites. Initially, they had turbostratic or semi-random structures, and after the treatment distinct reflections appeared both on the first and the second ellipses. The data obtained indicated that all the 2:1 layers in the particles of natural montronites had the same azimuthal orientation. The replacement of Ca by K in vacuum decreased the number of translational defects. The ribbon-like form of the nontronite single crystals confirmed that rotational defects were absent.

It should be stressed that with ribbon-like crystals, the K_2CO_3 treatment is quite sufficient to produce in vacuum particles with a mica-like stacking mode of identically oriented layers (Zvyagin and Pinsker 1949). Fe^{3+}-smectites from Red Sea sediments having ribbon-like particles whose layer charge in contrast to nontronites, is localized in octahedra, may serve as an example (Fig. 48).

Electron diffraction study of K-saturated (Al, Mg)-smectites with isometric or poorly shaped particles revealed low structural ordering. On the first ellipse, only continuous diffuse scattering was present, and it was only on the second ellipse that discrete maxima with $k = 3n$ were usually discernable. Such patterns may correspond to structures where packing defects are caused either by random $n60°$ rotations of layers, or by their $\pm b/3$ shifts. It is impossible to give preference to any of the two models without precise analysis of the intensity distribution. However, SAED indicated that a relatively high concentration of faults caused by rotations of adjacent layers at arbitrary angles was another important reason for low structural perfection of the smectites in question.

K-Smectites After 70–100 WD Cycles. According to the data of Mamy and Gaultier (1976) the 2:1 layers in smectite particles after 70–100 WD cycles had either identical azimuthal orientations or were rotated through angles $n60°$.

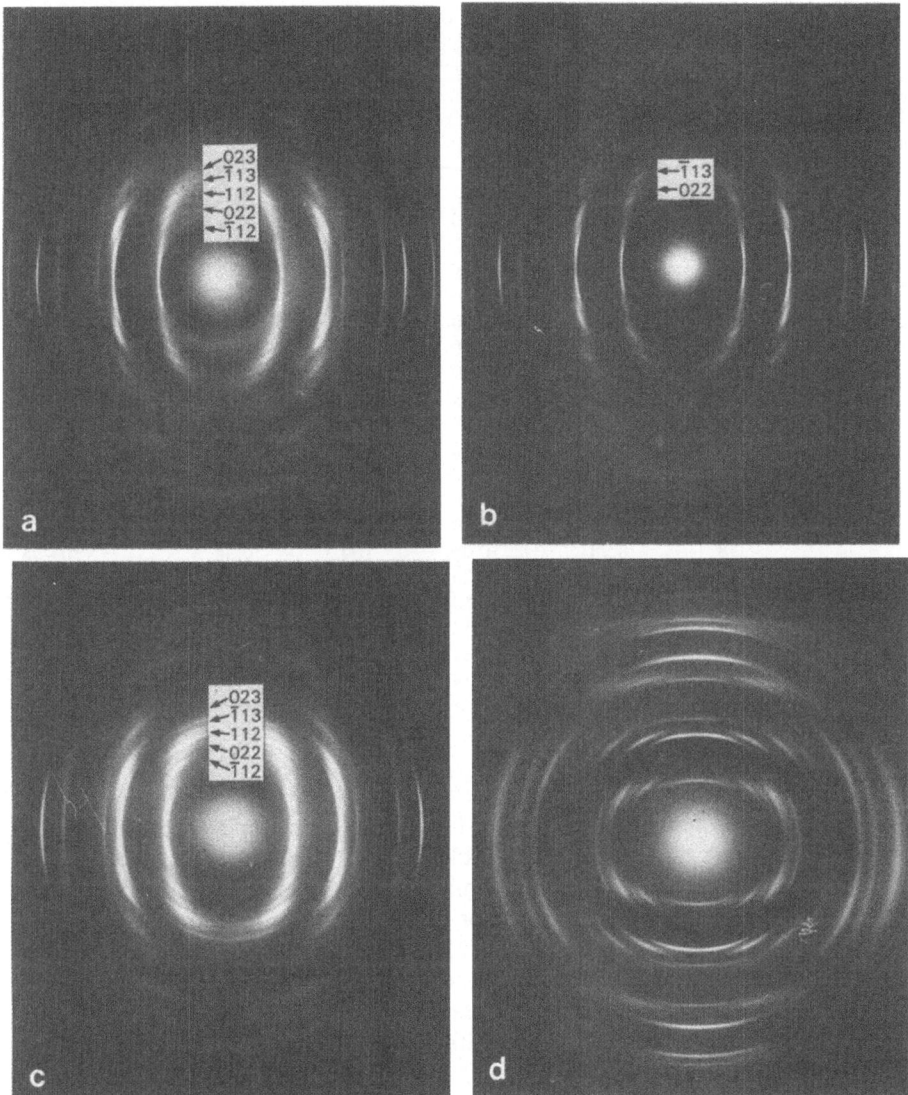

Fig. 50a–d. OTED patterns of **a** beidellite, Kamchatka (sample 15, Table 6); **b** montmorillonite, Wyoming (sample 2, Table 6); **c** clay from Panama Basin (sample 10, Table 6); **d** glauconite from Piltene

Increase in the number of WD cycles led to more distinct $11l$ and $02l$ reflections on the first ellipse with a small redistribution of intensities on the second one. After $70-100$ WD cycles the patterns of most smectites had distinct reflections on the first and the second ellipses (Fig. 50), the general quality of the diffraction pattern being similar, for example, to that of glauconite (Fig. 50).

Table 7. Unit cell parameters and occupancies of the *trans*-octahedreal sites (P_t%) of the smectites studied

Sample number	$a(Å)$	$b(Å)$	$c(Å)$	$\beta(°)$	$\dfrac{-c \cdot \cos\beta}{a}$	$P_t(\%)$
1	5.18	8.97	10.07	99.6	0.32	75 – 100
2	5.18	8.98	10.10	99.5	0.32	75 – 100
3	5.18	8.97	10.05	100.4	0.35	50 – 75
4	5.18	8.98	10.08	101.5	0.38	0 – 25
5	5.18	8.98	10.10	99.5	0.32	75 – 100
6	5.19	9.00	10.10	101.3	0.38	0
7	5.19	9.00	10.10	101.3	0.38	0
8	5.18	8.97	10.20	101.3	0.38	0
9	5.18	8.98	10.13	99.8	0.33	75 – 100
10	5.18	8.98	10.08	101.0	0.37	25 – 50
11	5.23	9.06	10.20	100.3	0.35	0
12	5.18	8.98	10.05	101.4	0.38	0
13	5.18	8.97	10.10	100.5	0.35	50 – 75
14	5.18	8.98	10.10	99.6	0.32	75 – 100
15	5.18	8.98	10.08	100.2	0.34	50 – 75
16	5.20	9.01	10.20	101.3	0.38	0
17	5.26	9.12	10.10	101.0	0.37	0
18	5.29	9.17	10.10	101.0	0.36	0
19	5.26	9.12	10.14	101.0	0.37	0
20	5.28	9.14	10.14	100.7	0.36	0

Analysis of the geometry of the reflections and their indices indicated that the smectites studied had monoclinic one-layer unit cells similar to those of 1 M micas, with the parameters listed in Table 7. It is evident from this table that all (Al, Mg)-smectites (beidellites and montmorillonites) may be subdivided into three groups according to the values of $c \cdot \cos\beta$. In the first group, $|c \cos\beta|$ is considerably greater than $a/3$ ($0.37a - 0.38a$), in the second group it is close to $a/3$ ($0.34a - 0.35a$), while in the third it is equal to or less than $a/3$. The intensity distribution on the first ellipse is similar for (Al, Mg)-smectites from the same group, but differs noticeably for smectites from different groups. A marked difference is observed for smectites with the maximum and the minimum $c \cdot \cos\beta$ values. For instance, the strongest reflections on the patterns of (Al, Mg)-smectites from the first group with the maximum $c \cdot \cos\beta = -0.38a$ are $\bar{1}12$ and 112, and between them there is a weak 022 reflection.

On the diffraction patterns of (Al, Mg)-smectites with the minimum $c \cdot \cos\beta$ values, the 022 reflection is, on the contrary, the strongest and the $\bar{1}12$ and 112 reflections are weak. The intensity of the $\bar{1}13$ reflection increases markedly. In the patterns of the smectites with $c \cdot \cos\beta$ of approximately $-(0.34a - 0.35a)$, the intensities of the reflections are practically equal.

Smectites with high octahedral Fe^{3+} contents ($Fe^{3+}/Al > 0.9$) have similar intensity distributions on the first ellipse, $\bar{1}12$ and 112 being the strongest reflections. The $c \cdot \cos\beta$ values range from $-0.35a$ for Fe-smectites (sample 11) to $-(0.36a - 0.37a)$ for nontronites. In sample 16, which shows a relatively high octahedral Al content ($Fe^{3+}/Al = 0.91$), $c \cdot \cos \beta = -0.38a$. The decrease of $c \cdot \cos \beta$ in the case of nontronites is expected, as the ionic radius of Fe^{3+} is greater than that of Al^{3+}, which reduces the difference between the sizes of *cis*- and *trans*-octahedra of 2:1 layers. As a consequence, the shift between the upper tetrahedral sheet and the lower one within the 2:1 layer is closer to $-a/3$ in nontronites than in montmorillonites and beidellites.

The OTED patterns of the Red Sea Fe-montomorillonite (sample 11, Tables 6 and 7), Wyoming montmorillonite (sample 2, Table 6), Kamchatka beidellite (sample 15, Tables 6 and 7) and Panama montmorillonite (sample 10, Tables 6 and 7) that are typical representatives of different groups are presented in Figs. 48, 50. The distribution of intensities in Fig. 50c is intermediate between those of Fig. 48b and Fig. 50a.

The data imply that a series of $c \cdot \cos \beta$ values from $-0.38a$ to $-0.32a$ is possible for (Al, Mg)-smectites with continuous transition in the intensity distribution between the extreme cases indicated. In the general case the intensity distribution of 11 l and 02 l reflections in the diffraction patterns of the above K-smectites should depend on their chemical composition, the charge localization, the distribution of octahedral cations over *cis*- and *trans*-octahedral sites, real distortions of the 2:1 layers, and their stacking sequence. In order to evaluate the role of the main factors affecting the diffraction characteristics of K-smectites, we should first consider in greater detail the structural features of the 2:1 layers when:

1) Only *cis*-octahedral positions are occupied by cations: layer symmetry C2/m (model 1);
2) *Cis*-octahedral sites belonging to one regular system of points are vacant: layer symmetry C2 (model 2);
3) All the octahedral sites are occupied with the probability of 2/3; the symmetry of a unit cell of the 2:1 layer is C2/m (model 3);
4) The cation distribution is intermediate between the three limiting cases indicated.
   Since direct determination of the atomic coordinates by diffraction methods for dispersed minerals with defect structures is impossible, we used the method of structure simulation (Drits 1975).

Coordinates for structures of the same composition as sample 12 with octahedral cations distributed as in models 1, 2, and 3 are given in Table 8. For these models, structural factors $\Phi^2(hkl)$ for the 02 l, 11 l, 20 l and 13 l reflections were calculated.

The intensities of 11 l and 02 l reflections appeared to be most sensitive to changes in cation distribution (Table 9). The gradual transition from model 1 to model 3 leads to a decrease in intensities of the strongest 020, $\bar{1}12$ and 112 reflections and an increase in intensities of the weaker $\bar{1}13$ and 022 reflections. With the equally probable cation distribution over octahedral sites, the calculated intensities for the whole group of these reflections are "equalized". With the transition from model 3 to model 2, the $\bar{1}12$, 112 and 023 reflections are weakened, and the 022 and $\bar{1}13$ reflections appear to be the strongest.

Table 8. Atomic coordinates for different structural models of smectites having the compositon of Ascan smectite (sample 12)

Atoms	Model 1			Model 2			Model 3		
	x/a	y/b	z/c	x/a	y/b	z/c	x/a	y/b	z/c
M_1	0.0	0.0	0.0	0.0	0.0	0.0	0.0	0.0	0.0
M_2	0.0	0.333	0.0	0.0	0.321	0.0	0.0	0.333	0.0
M_3	0.0	0.667	0.0	0.0	0.654	0.0	0.0	0.667	0.0
T_1	0.417	0.329	0.270	0.432	0.333	0.270	0.427	0.333	0.270
T_2	0.417	0.671	0.270	0.432	0.662	0.270	0.427	0.667	0.270
K	0.5	0.0	0.5	0.5	0.0	0.5	0.5	0.0	0.5
O_1	0.481	0.5	0.320	0.489	0.496	0.335	0.487	0.5	0.330
O_2	0.172	0.728	0.335	0.173	0.725	0.335	0.172	0.728	0.330
O_3	0.172	0.272	0.335	0.170	0.268	0.320	0.172	0.272	0.330
O_4	0.419	0.0	0.105	0.334	−0.024	0.105	0.369	0.0	0.108
O_5	0.348	0.691	0.110	0.417	0.656	0.109	0.369	0.667	0.108
O_6	0.348	0.309	0.110	0.343	0.347	0.109	0.369	0.333	0.108

Table 9. Φ^2 calc. (02l) and Φ^2 calc. (11l) for models of smectites with different octahedral cations distributions

hkl	1	2	3	4	5	6
020	840	690	240	140	160	140
110	20	50	100	140	270	400
$\bar{1}11$	350	250	60	10	60	100
021	60	30	20	20	60	90
111	–	10	60	90	160	260
$\bar{1}12$	580	480	470	430	430	320
022	170	230	420	530	590	730
112	630	540	530	480	470	360
$\bar{1}13$	130	170	230	370	380	480
023	210	170	180	200	210	160
113	10	20	40	100	90	120
$\bar{1}14$	10	–	10	10	10	10

Columns 1 – 6 correspond to the following cation occupancies of the *trans*, *cis*-1 and *cis*-2 octahedral sites. (1) 0.0; 1.0; 1.0 (model 1). (2) 0.25; 0.75; 1.0. (3) 0.5; 0.5; 1.0. (4) 0.67; 0.67; 0.67 (model 3). (5) 0.85; 0.25; 1.0. (6) 1.0; 0.0; 1.0 (model 2).

To evaluate the influence of structural distortions of the 2:1 layers on the 11l and 02l intensities for each of the patterns, idealized structural models were constructed using undistorted polyhedra. The chemical composition and the hexagonal symmetry of the tetrahedral sheets corresponded to that of sample 12. Φ^2(02l) and Φ^2(11l) were calculated for each model, and their relative values for the three extreme cases of the cation distribution are presented in Fig. 51.

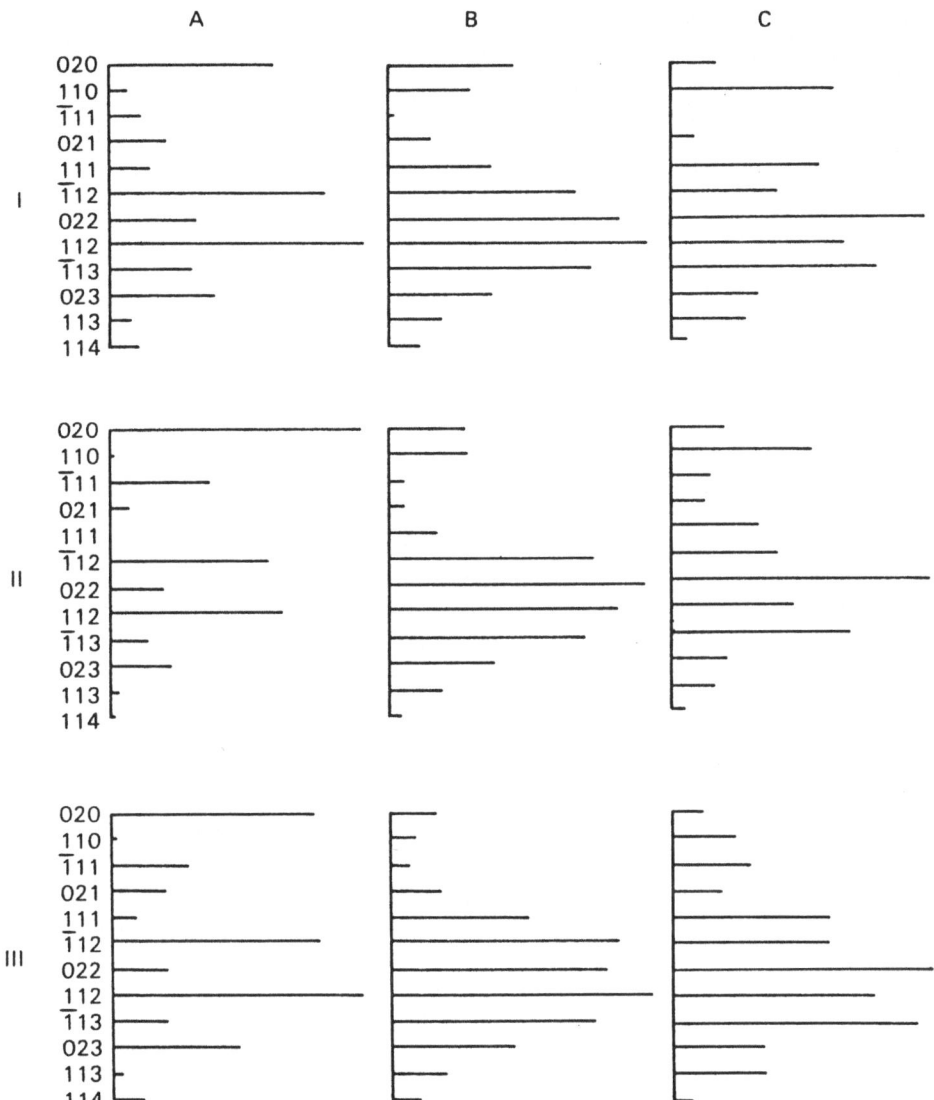

Fig. 51 A – C. Scheme for the distribution of relative $\Phi^2(02l)$ and $\Phi^2(11l)$. **A** Model 1; **B** All octahedral sites are occupied with equal probability (Model 3); **C** Model 2. *I* (Al, Mg) smectite without any distortions of structural polyhedra (idealized model); *II* (Al, Mg) smectite having distortions in 2:1 layers; *III* nontronite having distortions in 2:1 layers

In order to evaluate the influence of cation composition of octahedra, $\Phi^2(02l)$ and $\Phi^2(11l)$ were calculated for nontronite (sample 20) taking account of real distortions of 2:1 layers that may be expected for these minerals (Besson et al. 1982). Relative values of $\Phi^2(02l)$ and $\Phi^2(11l)$ obtained for nontronite with the same chemical composition as the Garfield nontroinite (sample 20, Tables 6 and 7) for the three extreme cases of cation distribution are presented in Fig. 51.

Table 10. Comparison of Φ^2 calc. ($02l$, $11l$) and Φ^2 obs. ($02l$, $11l$) for smectites with different occupancies of *trans*-octahedral sites (P_t)

hkl	$P_t = 0.0$ sample 12		$P_t = 0.67$ sample 15		$P_t = 1.0$ sample 2	
111	–	–	90	60	260	180
$\bar{1}12$	580	530	430	490	320	320
022	170	200	530	550	730	800
112	630	660	480	490	360	320
$\bar{1}13$	130	130	370	330	480	540
023	210	200	200	220	160	160
113	10	–	100	60	120	80

Comparison of the data obtained shows that the predominant factor determining $02l$ and $11l$ reflection intensity is the distribution of octahedral cations over *cis* and *trans* positions. The specific cation composition of the 2:1 layers and their structural distortions affect mainly the absolute intensity values and not their relative distribution. Analysis of the variations of $\Phi^2(02l)$ and $\Phi^2(11l)$, taking account of the above factors, shows that the cation distribution over the available octahedral sites may be established using the data presented in Table 9 and in Fig. 51 based on the comparison of calculated and experimental intensities.

As the intensities have been estimated visually the accuracy of the determination of occupancies does not exceed 20%.

From Tables 7, 10 it is clear that a wide variety of cation distribution modes over *cis*- and *trans*-octahedra is observed in dioctahedral smectites.

It may be noted that *trans*-octahedra tend to be occupied in montmorillonites, the charge being localized only in octahedra, whereas centrosymmetric layers are characteristic of smectites with the charge localized in tetrahedra.

CHAPTER 9

SAED and HREM Study of Mixed-Layer Minerals

9.1 Hybrid-Structure Minerals

In the last 10 – 15 years a number of works have been published on structural and crystal-chemical studies of minerals having extremely peculiar structure and composition. Evans and Allman (1968) have proposed the term of "hybrid structures" for all structures containing brucite-like layers of the composition $Me(OH)_2$ that alternate regularly along c with layers of an other type. There is a number of minerals having structures of this kind. The best-known and the most abundant one is *chlorite* whose structure consists of alternating brucite-like and talc-like layers. The *lithiophorite* structure consists of octahedral sheets of the idealized composition MnO_2 alternating regularly with $(Al_{2/3}Li_{2/3})(OH)_2$ layers (Wadsley 1952).

Some hybrid minerals have regular mixed-layer structures consisting of layers that differ not only in cationic but also in anionic composition. This leads to some serious crystal-chemical consequences that are displayed both in diffraction effects and in the description of their crystal lattices. On the one hand, in such minerals, unusual types of structural heterogeneity appear. On the other hand, the difference in ionic radii for anions in the alternating layers leads to the fact that different layer types are described by different sublattices.

The sublattices of the two components may differ not only in unit-cell parameters within layers but also in repeat distances along the axis c. In this case, in the reciprocal space there are two interpenetrating sublattices sharing ool lattice points. Consequently, the two reciprocal sublattices are displayed in diffraction patterns as two separate reflection systems.

Some years ago two minerals have been known whose structures consist of layers having different cationic and anionic composition, valleriite and koenenite. Koenenite has the composition 1.78 $[(Mg_{0.64}Al_{0.36})(OH)_2][(Na_{0.35}Mg_{0.65})Cl_2]$ (Allman and Lohse 1966; Allman et al. 1968). X-ray structure analysis has shown that the brucite component and the chloride one have different sublattices and different space groups. In the valleriite structure, brucite-like layers alternate with sulfide tetrahedral sheets filled by Fe and Cu cations (Evans and Allman 1968). Owing to a substantial difference in ionic radii for S and OH and since the anions are close-packed in layers of both types, the lateral dimensions of the unit cell in the brucite-like sublattice ($a = 3.07$ Å) differ from those in sulfide sublattice ($a = 3.792$ Å). Besides, the brucite-like layer have a one-layer repeat distance, whereas the sulfide layers are "close-packed", as if they were neglecting the presence of the second layer type. Therefore, the sulfide component is to be described in terms of a rhombohedral sublattice.

Further insight into the crystal chemistry of hybrid structure minerals, including the diversity in their forms and fine structure features required intense and comprehensive investigation.

Peculiarities of the structure and, apparently, formation conditions, lead to the fact that hybrid minerals generally occur in mineralogically heterogeneous samples and often in extremely small quantities. Therefore, X-ray powder diffraction is not effective for structure analysis. It is only after having found a single crystal that Evans and Allman (1968) managed to prove that valleriite, a mineral discovered one hundred years ago, is not a copper-layer sulfide.

Since hybrid minerals have layer structures, their particles are usually plate-shaped. As it has been shown, OTED is fruitful for analyzing structures of finely dispersed substances consisting of plate-shaped crystals. However, attempts to apply OTED to hybrid minerals failed due to poor crystallinity of the particles. Hence, SAED appeared to be the only means to gain information on hybride structures.

This chapter gives an account on the results of structure studies of a number of hybrid minerals of unusual structure and composition. The main contribution to the study of hybrid minerals having a sulfide component was made by N. I. Organova together with A. L. Dmitrik and the present author (Organova 1972; Organova et al. 1971a, b, 1972, 1973a, b, 1974). The results obtained are not only of crystal-chemical, but also of methodological interest. It is the hybrid minerals discussed below that have been the first example demonstrating the efficiency of SAED for structure determinations. Tochilinites have been lately receiving more attention owing to their presence not only in terrestrial environments but also in meteorites (Zolensky 1984; Tomeoka and Buseck 1985).

Asbolanes are another group of hybrid minerals whose unusual crystal-chemical nature has been revealed by SAED (Chuhrov et al. 1980a, b, 1982, 1983a, b). These mixed-layer structures consist of alternating incommensurate octahedral layers of different cationic composition. One of the layer types is usually composed by Mn-octahedra. Asbolanes, also known for about a hundred years, are of extremely poor crystallinity and high dispersion and occur mostly in poly-phase samples. Therefore, both X-ray diffraction and OTED failed to provide any structural information. HRTEM, too, can hardly be fruitful since the asbolane structure is to be described by two sublattices having different hexagonal unit cell parameters. SAED again proved to be the only means to obtain information on asbolane structure and revealed numerous asbolane varieties.

9.2 Structure Analysis of Hybrid Minerals

Application of SAED to the minerals involved requires some special conditions. First, the incident beam should coincide with one of the crystallographic axes. For plate-like crystals only one rational reciprocal lattice section can be readily obtained. Therefore, ultra-fine sections were studied and diffraction patterns from bent edges were obtained using the technique of Gorshkov (1970).

A goniometer was used to obtain the crystal orientation required. As a rule, a number of SAED patterns was obtained for each sample to make certain of reproducibility in the intensity distribution. The crystals under study were examined by dark-field imaging to select the parts that looked uniformly illuminated for all reflections used to form a dark-field image. Photographs having a symmetrical intensity distribution with respect to the central spot and therefore suitable for structure analysis were taken with different exposure times.

Strong and medium reflection intensities were measured by a microdensitometer. Other intensities were estimated visually. To reveal the nature of electron scattering, Σf_i and Φ^2 plotted against $\sin \vartheta / \lambda$ were analyzed (Vainshtein 1961) for each of the two possible relationships between intensity and structure factor, i.e., $I_{hkl} \sim \Phi^2_{hkl}$ and $I_{hkl} \sim \Phi^2_{hkl}/d_{hkl}$. The structure factors thus obtained were employed in two-dimensional Patterson maps. Analysis of these provided structure models that were refined by successive potential projections and the least-squares. All the calculations were performed with the help of the computer BESM-4, using standard programs.

It should be noted that previously the Patterson synthesis was not in general use for SAED studies of minerals. Therefore, special attention will be given below to the analysis of the interatomic function.

9.3 Crystal Structure of Tochilinite

Tochilinite is one of the first dispersed minerals whose structure has been determined by SAED (Organova et al. 1972). There are two natural tochilinite varieties that differ in morphology. One of these forms isometric aggregations consisting of needle-like or felt-like particles; the other is represented by non-isometric acicular crystallite aggregations. The terms *isometric* and *acicular* will be used hereafter only to indicate the corresponding morphological variety. The presence of magnesium and aluminum oxides and water, in addition to Fe and S, as indicated by the chemical analysis, has suggested alternation of sulfide layers and brucite-like ones, as in valleriite. This has been confirmed by the analysis of ool reflections in a X-ray rotation pattern from acicular tochilinite (Organova et al. 1971b).

Electron microscopy has shown that the shape of microcrystals is practically unaffected by macro-morphological differences. Strip-like and isometric particles are characteristic of both varieties, although the latter shape occurs more frequently in isometric tochilinite. Figure 52 shows point ED patterns from isometric tochilinite single crystallites. The reflections hko are arranged to form an orthorhombic centered pattern with $a = 5.37$ and $b = 15.60$ Å, space group Cmm and $h + l = 2n$. The elongation direction in microcrystals coincides with b^*. A single reflection network implies a single crystal lattice for the brucite-like and the sulfide components. Typical SAED patterns from acicular tochilinite crystals are shown in Fig. 52. It is of interest that the crystal elongation direction coincides here with a^*. In addition to an orthorhombic system of reflections ($a = 5.42$ Å, $b = 15.65$ Å), there is also a pseudo-square network of reflections with $a = 8.34$ Å, $b = 8.54$ Å and $\gamma = 86°$. Figure 53 shows idealized sketches illustrating relationships between the orthorhombic and the pseudo-square lattices in

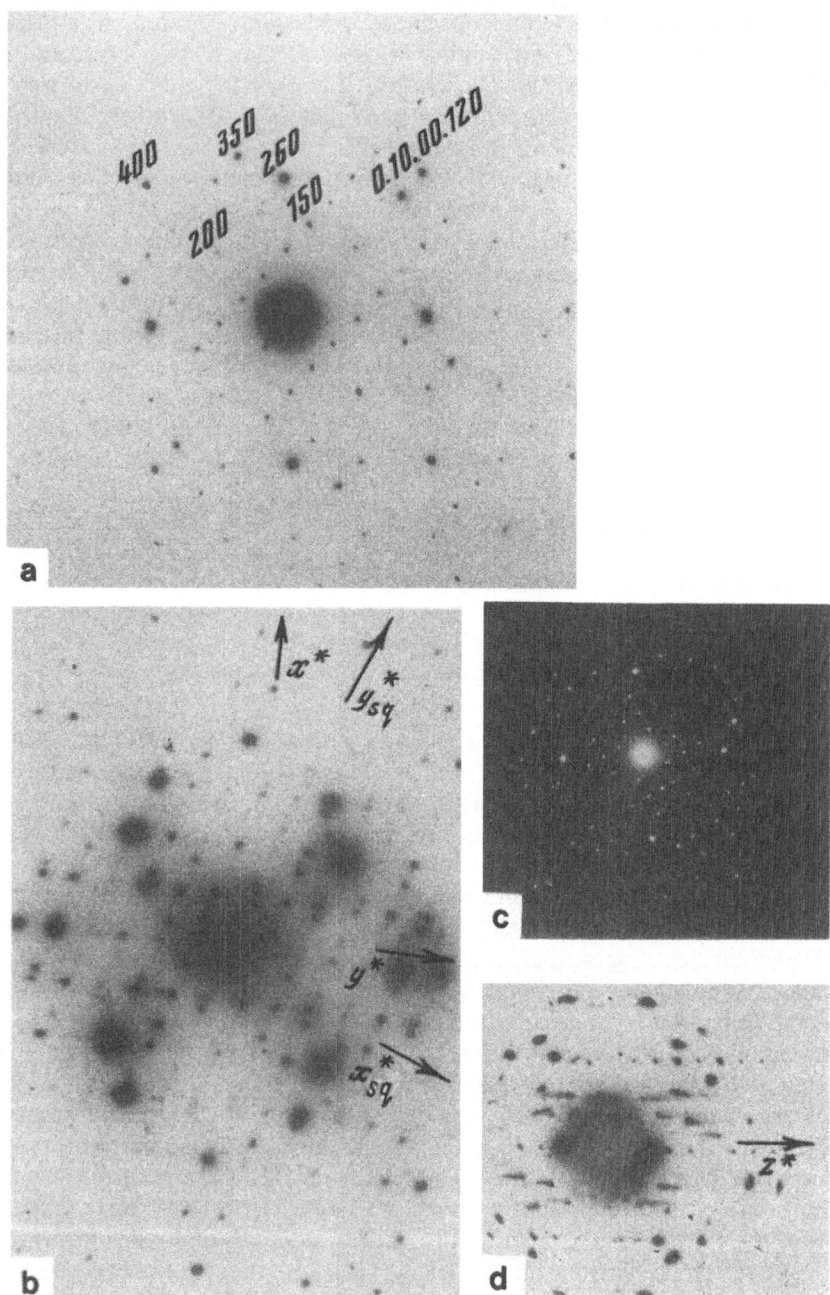

Fig. 52a – d. Point ED patterns from crystallites of **a** isometrical tochilinite (reflections hko); **b** acicular tochilinite (reflections hko); **c** the same as in **b**, but with a different ratio of intensities for hko reflections belonging to the pseudo-square and orthogonal networks; **d** tochilinite I (reflections hol and ool)

direct and reciprocal space. Idealization consists in assuming $3a = b$ for the orthorhombic lattice with result that the second lattice becomes tetragonal. Analysis of these schemes gives readily the transition matrices for unit cell parameters and reflection indices for the two lattices in question:

$$\begin{pmatrix} a_{sq} \\ b_{sq} \end{pmatrix} = \begin{pmatrix} -\frac{1}{2} & \frac{1}{2} \\ \frac{3}{2} & \frac{1}{6} \end{pmatrix} \begin{pmatrix} a_{or} \\ b_{or} \end{pmatrix}, \quad \begin{pmatrix} h_{sq} \\ k_{sq} \end{pmatrix} = \begin{pmatrix} -\frac{1}{2} & \frac{1}{2} \\ \frac{3}{2} & \frac{1}{6} \end{pmatrix} \begin{pmatrix} h_{or} \\ k_{or} \end{pmatrix},$$

$$\begin{pmatrix} a_{or} \\ b_{or} \end{pmatrix} = \begin{pmatrix} -\frac{1}{5} & \frac{3}{5} \\ \frac{9}{5} & \frac{3}{5} \end{pmatrix} \begin{pmatrix} a_{sq} \\ b_{sq} \end{pmatrix}, \quad \begin{pmatrix} h_{or} \\ k_{or} \end{pmatrix} = \begin{pmatrix} -\frac{1}{5} & \frac{3}{5} \\ \frac{9}{5} & \frac{3}{5} \end{pmatrix} \begin{pmatrix} h_{sq} \\ k_{sq} \end{pmatrix}.$$

As the orthorhombic and pseudo-square components have common reflections in the diffraction pattern (Fig. 53), the whole set of reflections can be described in terms of a single unit-cell having

$$\begin{pmatrix} A \\ B \end{pmatrix} = \begin{pmatrix} -3 & 3 \\ 9 & 1 \end{pmatrix} \begin{pmatrix} a_{or} \\ b_{or} \end{pmatrix} = \begin{pmatrix} 6 & 0 \\ 6 & 0 \end{pmatrix} \begin{pmatrix} a_{sq} \\ b_{sq} \end{pmatrix}.$$

Point patterns from a large number of acicular tochilinite single crystals have indicated that the relationships between intensities for reflections forming the two networks differ for different particles. As a rule, reflections from the orthorhombic and the pseudo-square networks have nearly the same intensity, but sometimes reflections corresponding to the pseudo-square pattern are stronger (cf. Fig. 52). Therefore, it has been concluded that acicular tochilinite crystals are formed by two epitaxially intergrown phases, one of which is practically identical to isometric tochilinite. This phase has been termed tochilinite I. The second phase, tochilinite II, will be shown below to have a tochilinite-like structure differing from tochilinite I in a number of features including the layer unit-cell parameters. Pure tochilinite II has been never found, although is has been always present in varied quantities in acicular samples.

Note that a rotation photograph obtained from a needle-like tochilinite aggregation contained, besides the basic system of layer lines with $a = 5.42$ Å, an additional system of lines corresponding to $a = 27.1$ Å (Organova et al. 1971b).

It is seen in Fig. 53 that, if the crystal is rotated around a_r, the spacing between the layer lines corresponding to the pseudo-square network will be five times less, which will result in a repeat distance of $5.42 \times 5 = 27.1$ Å. The basic reflections in the layer lines have been indexed by utilizing geometrical analysis of point patterns yielding the following unit-cell parameters: $a = 5.42$ Å, $b = 15.50$ Å, $c = 10.72$ Å, $\beta = 95°$ (Organova et al. 1971b). The β value implies that successive structural units are shifted with respect to one another along a by $-a/6$. (A structural unit consists of a sulfide layer and a brucite-like one). The cell is C-centered, since $h + k = 2n$ for all reflections.

Figure 52 shows an SAED pattern from a bent edge of a single crystal. Layer lines containing reflections ool and $20l$ are clearly seen with the hko reflections at the background. The absence of $10l$ reflections confirms that the cell is C-centered, while the arrangement of reflections indicates that $\beta^* = 85°$. The data ob-

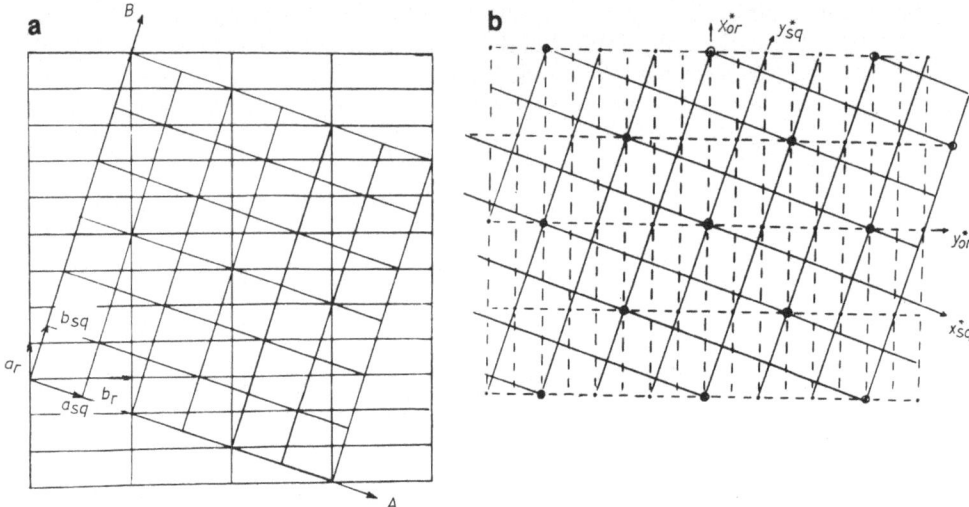

Fig. 53a, b. Relationships between the orthorhombic and the pseudosquare lattices in **a** the direct space, and **b** the reciprocal space

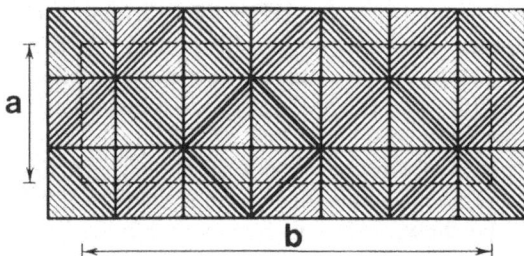

Fig. 54. Mackinawite layer structure

tained have been sufficient for indexing the reflections in powder patterns for both tochilinite varieties.

The unit cell parameters obtained have been used to devise a structure model for tochilinite I. The initial structure model for sulfide layers has been based on the structure of the tetragonal FeS (mackinawite) having $a = 3.679$ Å (Berner 1962). Figure 54 shows the normal projection of an idealized mackinawite tetrahedral sulfide layer. Each tetrahedron has two edges parallel to (001) and shares four edges with the adjacent tetrahedra. The parameter a of a mackinawite layer is defined by the shortest distance between the S atoms in the (001) plane. The diagonal of a unit square is close to the period a of tochilinite I, since $a_M \sqrt{2} = 5.16$ Å. The parameter $b = 3a$ chosen normal to this diagonal, is quite close to that of tochilinite I (Fig. 54).

Different sulfide layer thicknesses in tochilinite ($h = 2.523$ Å) and valleriite ($h = 3.14$ Å) resulted in a thinner tochilinite unit (10.72 Å) as compared to that of valleriite (11.37 Å). In hexagonal brucite, $a_{BR} = 3.12$ Å, so that if the tochilinite axis b_T is brought into coincidence with a_{BR}, $b_T = 5a_{BR} = 15.6$ Å and $a_T = a_{BR}\sqrt{3} = 5.40$ Å (Fig. 55).

Geometrical analysis is insufficient for reconstruction of the relative arrangement of brucite and sulfide layers. However, the orthorhombic symmetry of the intensity distribution in tochilinite I point patterns may imply the same symmetry for the structure projection along c. Figures 54 and 55 show that the relevant two-dimensional space group is $C1m$.

The hko reflection intensities have been analyzed to prove the structure model discussed and to explain the observed layer unit-cell parameters, as well as to determine the stacking sequence for different layer types.

Forty-nine independent reflections with $(\sin \vartheta/\lambda)_{max} = 0.64$ have been evaluated. First of all, it was necessary to reveal the nature in the interaction between electrons and the tochilinite crystals and to find the relationship between intensities and structure factors. Thin crystals containing a comparatively large number of atoms in the unit cell as well as the absence of heavy atoms have been in favor of the kinematical approximation. The dependences of Σf_i, $\Sigma \bar{f}_i^2$ and $\bar{I}_{exp}$ on $\sin \vartheta/\lambda$ have also implied the validity of the kinematical approximation and the relationship between intensities and structure factors of the type $I \sim \Phi^2$.

Figure 56a shows the two-dimensional Patterson function. It is seen that the arrangement of maxima corresponds to the sulfide layer model given in Fig. 54. The maxima at $u = 0$, $v = 0$; $u = a/2$, $v = 0$; $u = 0$, $v = b/6$ and $u = a/2$, $v = b/6$ form a square similar to that in Fig. 54 having either S or Fe in the apices. Thus the maxima at $u = 0$ and $u = a/2$ separated by the distance $b/6$ along the axis v represent the interatomic vectors Fe-Fe and S-S. The maxima at the centers of the squares at $u = a/4$ and $3a/4$ with $v = nb/12$ (n is an integer) represent the vectors Fe-S.

The data obtained seemed to prove unambiguously the mackinawite-like structure of sulfide layers in tochilinite. At the same time, the Patterson function contained no peaks indicating the presence of brucite layers. These peaks were revealed in a Patterson map computed by employing the reflections hko in the point pattern from acicular tochilinite. A plate was chosen with low-intensity reflections corresponding to the square network, the number of independent reflections = 60, $(\sin \vartheta/\lambda)_{max} \simeq 0.66$.

Figure 56b shows that there are additional maxima at $u = a/3$, $v = 0$; $u = a/3$, $v = b/5$; $u = a/2$, $v = b/5$. The maxima having the same u and separated by the distance $b/5$ along b may be associated with the presence of brucite layers. As it was mentioned, $b_T = 5a_{BR}$ so that brucite-like layer anions and cations are distributed along the tochilinite axis b with a repeat distance of $b_T/5$. Since the peaks representing the pairs of the heaviest atoms are the strongest in the Patterson maps, one may suppose that the maxima at $u = a/3$, $v = 0$ and $u = a/3$, $v = b/5$ represent the vectors connecting the iron atom belonging to the sulfide layer and the magnesium from the brucite layer. This result permitted to determine the arrangement of the brucite and the sulfide layers within a layer unit. Dashed lines in Fig. 55b represent the tochilinite unit cell for the brucite layer. The origin of this unit cell coincides with the iron atom in the sulfide layer projected along c (Fig. 54).

Comparison of the experimental $\Phi_{exp}(hko)$ values with those calculated in terms of the two dimensional space group $C1m$ utilizing coordinates of 16 independent atoms in idealized sulfide and brucite layer structures yielded $R = 0.42$.

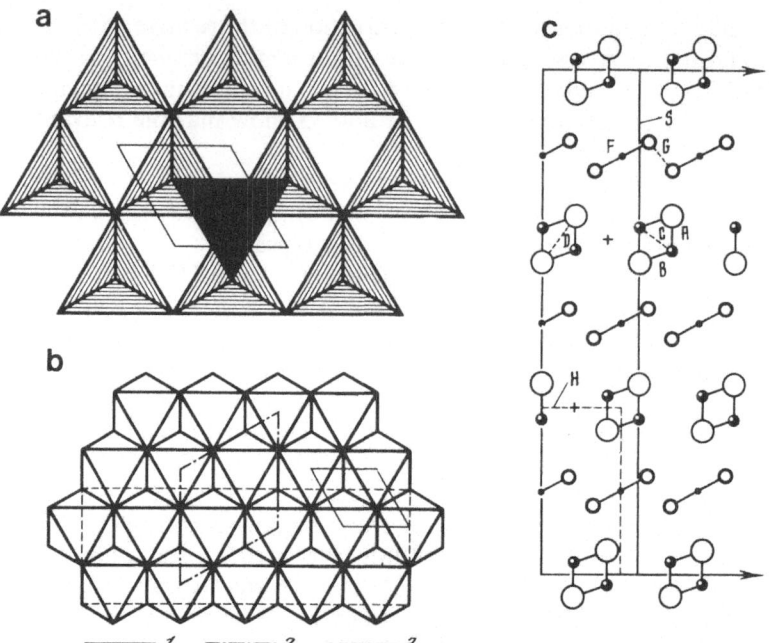

Fig. 55a – c. The valleriite structural pattern: **a** a sulfide-layer structure fragment; **b** the brucite layer: the unit cells are marked for (*1*) disordered and (*2*) ordered cation distribution as well as for tochilinite (*3*); **c** the layer stacking sequence in the valleriite structure (*S* and *H* denote the sulfide and the brucite components, respectively (Evans and Allman 1968)

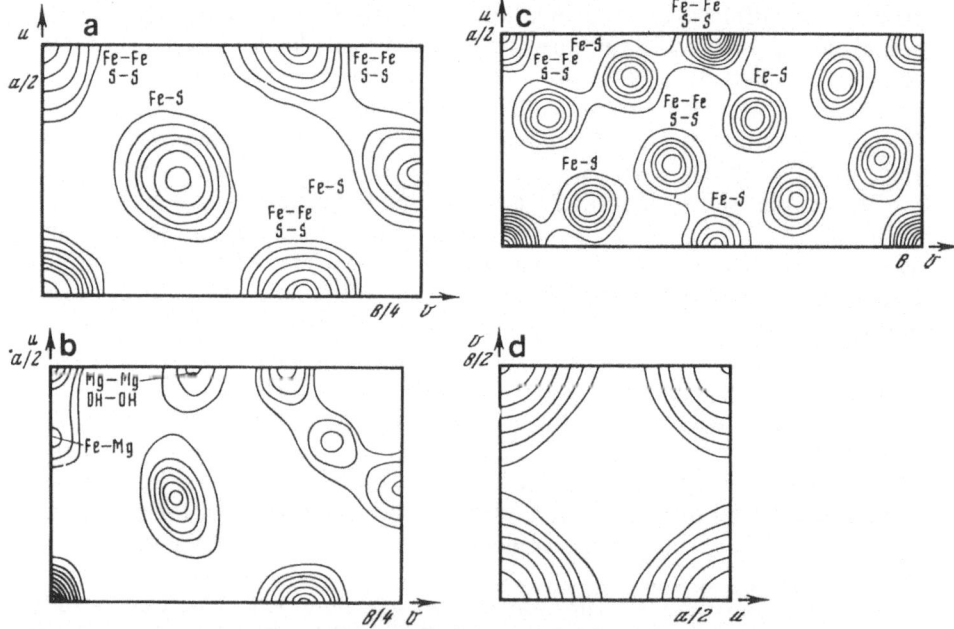

Fig. 56a – d. Functions $\mathscr{S}(uv)$ for **a** isometrical tochilinite; **b** acicular tochilinite I; **c** acicular tochilinite II; and **d** the sulfide component of the phases I and II

Successive Fourier syntheses taking account of peak positions determined from previous potential projections did not reduce this value. Then it was suggested that there was an ordered distribution of vacant tetrahedra, so that tetrahedra located in the symmetry planes (as projected along c) are occupied by Fe cations only partially (Fig. 57). The least-squares refinement yielded the occupancy of these tetrahedra $\mu = 1.3$ and reduced R to 0.12. The final potential projection down c is shown in Fig. 58a.

The validity of the results obtained was confirmed by comparing the tetrahedral and octahedral bond lengths in the corresponding layers in tochilinite and valleriite, since the latter structure had been refined by the conventional X-ray structure analysis (Evans and Allman 1968). As shown in Fig. 55a, the valleriite sulfide layer, as contrasted to that in tochilinite, consists of tetrahedra all of which are oriented with the bases normal to c and have three neighbours each. Thus, there are three shared edges and three unshared ones in each tetrahedron. The average bond lengths in valleriite tetrahedra Fe-S (2.33 Å) and S-S (3.80 Å) are practically the same as those in tochilinite (2.32 Å and 3.80 Å, respectively). In tochilinite tetrahedra, the mean unshared edge length is slightly shorter than the mean shared edge length (3.74 Å vs 3.80 Å), as in valleriite. The brucite layer distortions in tochilinite I are quite similar to those observed in related minerals. The octahedra are flattened so that the shared edge lengths are shorter than the unshared edge lengths. The mean Mg-OH distances are 2.05, 2.09, 2.11 Å for Mg_1-, Mg_2- and Mg_3-octahedra, respectively. The interatomic distances are evaluated with an estimated standard deviation (e, s, d) of 0.02 Å for brucite octahedra.

Figure 59 shows the projection of the tochilinite structure down b. The observed β angle ensures identical mutual arrangement for the neighboring anions having different chemical nature.

The validity of the results obtained with respect to the crystal-chemical features of tochilinite I may be tested by comparing the structure formulae obtained from the refined model and from the chemical analysis. It is seen from Fig. 57 and 55b that the tochilinite unit cell contains 12 sites for the sulfide layer cations and 10 for the brucite layer cations. Taking account of the tetrahedral occupancy of 0.65, we obtain the idealized tochilinite formula 6 [($Fe_{0.94}S$)] · [(Mg, Fe)(OH)$_2$] which, in analogy to valleriite may be written 2 ($Fe_{0.94}S$) · 1.67 · [(Mg, Fe)(OH)$_2$].

If we assume that the sulfide layer negative charge resulting from a cationic deficiency is compensated by tri-valent iron located in the brucite layer, the tochilinite formula becomes 2 ($Fe_{0.94}S$) · 1.67 [(Mg, Fe^{2+})$_{0.85}$ $Fe^{3+}_{0.15}$(OH)$_2$]. The formula calculated from the chemical analysis is 2 ($Fe_{0.91}S$) · 1.67 [($Mg_{0.71}$ $Fe_{0.29}$)(OH)$_2$]. The overall chemical analysis indicated a small quantity of Al_2O_3, but the microprobe analysis revealed no Al in tochilinite particles. On the other hand, among the sample particles there were crystallites having diffraction patterns quite similar to those from Al(OH)$_3$. Thus, a reasonable agreement between the formulae obtained by two independent methods is another argument in favor of the use of point SAED patterns for this structure study.

To interpret the tochilinite II structure, 166 unique nonzero reflections were measured with $(\sin \vartheta / \lambda)_{max} = 0.66$ (Organova et al. 1973b). The situation was

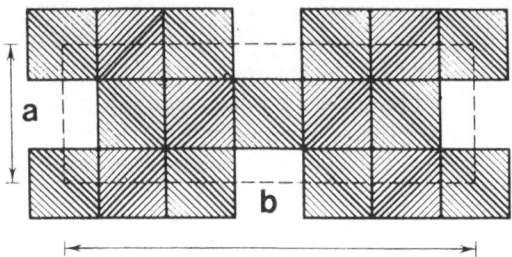

Fig. 57. The distribution of vacancies in the tochilinite I sulfide layer

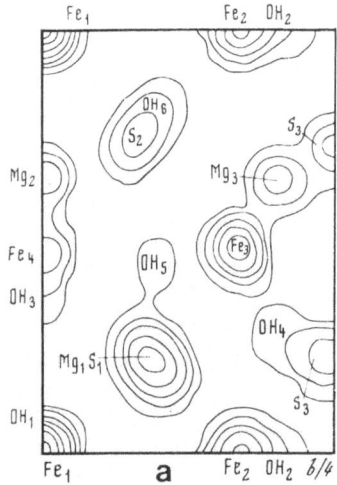

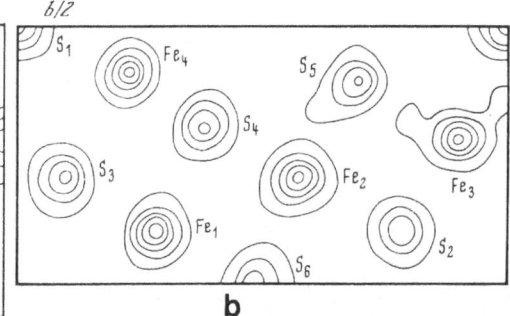

Fig. 58 a, b. Projections of $\varphi(xy)$: **a** for tochilinite I; **b** for the sulfide component of tochilinite II

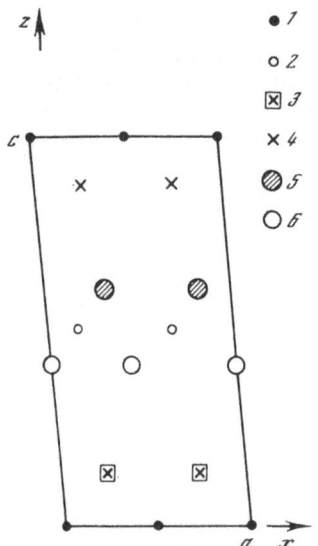

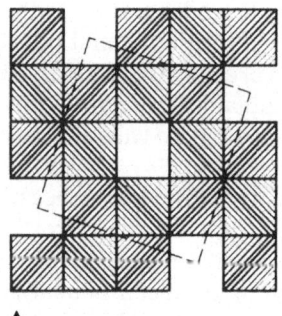

Fig. 60. The distribution of vacancies in the sulfide layer of tochilinite II

◀ **Fig. 59.** Tochilinite structure projection down *b*: *1 and 2* cations in the sulfide and the brucite layers, respectively; *3 and 4* S atoms; *5 and 6* OH groups

complicated by the fact that every fifth reflection in the square network coincided exactly with an orthorhombic network reflection. Therefore, the intensities of 39 coinciding reflections were estimated with an inherent error, since the contribution from the orthorhombic component could be evaluated only approximately. Statistical analysis of intensities again implied the validity of the kinematical approximation and the relationsip $I \sim \Phi^2$. Fig. 56c shows the projected interatomic function having maxima arranged in slightly distorted squares. The pattern in their arrangement corresponds to the atomic pattern in the mackinawite sulfide layer (Fig. 54).

Analysis of the intensities hko of the pseudo-square networks shows that reflections having $h + k = 2n$ are generally stronger than those having $h + k = 2n + 1$. This implies that the cell is pseudo-centered. The possible structure of the sulfide layer is shown in Fig. 60. Reliability factors were calculated for different occupancies of the tetrahedra shown in Fig. 60 as vacant for the two-dimensional space group $P2$. The minimum value $R = 0.18$ was obtained for these completely vacant tetrahedra. The final potential distribution is shown in Fig. 58b. The mean interatomic distances in sulfide layers of tochilinites I and II are quite similar, e.g., Fe-S = 2.34 Å and S-S = 3.82 Å in tochilinite II. Tetrahedral distortions in tochilinite II are also similar to those in tochilinite I. For example, the mean unshared edge (2.78 Å) is shorter than the mean shared one (2.84 Å) (Organova et al. 1973b).

Tochilinite II sulfide layers alternate regularly with brucite layers. This inference is confirmed by a low R value. A different distribution of vacancies in tochilinite II sulfide tetrahedral layers, as compared with tochilinite I, leads to a change in the unit-cell dimensions. Therefore, the brucite-like and the sulfide components in tochilinite II can be described by two interpenetrating sublattices. However, while for valleriite and koenenite the two components shared only the ool reflection in the reciprocal space, for tochilinite II, some of the hko reflections coincide for the brucite-like and the sulfide components. Therefore, the tochilinite structure can be described in terms of a single unit cell with $A = 6a_{sq} = 50.04$ Å and $B = 66_{sq} = 51.2$ Å. However, it is obvious that the structure study is much more simple with separate examinations of reflections for each component. If the relative arrangement of brucite-like and sulfide layers is assumed identical for both tochilinite varieties, the unit-cell parameters for the tochilinite II sulfide component within the space group $P1$ are $a = 8.34$ Å, $b = 8.54$ Å, $c = 10.74$ Å, $\alpha = 87.3°$, $\beta = 94.8°$, $\gamma = 92°$.

According to the structure analysis, the idealized formula for tochilinite II is $2 (Fe_{0.8}S) \cdot 1.67 [Mg_{0.52} Fe^{3+}_{0.48}(OH)_2]$. For acicular tochilinite, which is a mixture of two phases, the chemical analysis gives the overall composition $2 (Fe_{0.88}S) \cdot 1.67 \cdot [Mg_{0.70} Fe^{3+}_{0.30}(OH)_2]$. If the ratio of the contents for tochilinite I and II is assumed 3:2 the average composition calculated from the structural data will be $2 (Fe_{0.88}S) \cdot 1.67 [(Mg_{0.72} Fe^{3+}_{0.28}) (OH)_2]$ which is in agreement with that given by the chemical analysis.

9.4 Structure Analysis of Minerals Related to Tochilinite

Harris and Vaughan (1972) described two fibrous iron sulfides that were considered by them to be new valleriite varieties and therefore called type I and type II valleriite-like minerals. The microprobe analysis gave Fe 50.9%, S 21.5%, and Mg 6.8% for the type II mineral. The X-ray powder pattern was indexed in terms of the valleriite unit cell $a = 3.74$ Å and $c = 32.68$ Å. However, the type II mineral powder pattern appeared to be quite similar to that from tochilinite. A SAED study of the sample kindly provided by D. Harris showed that it is a mixture of phases that are characterized by different point ED patterns (Organova et al. 1974).

Even a visual examination of these SAED patterns indicated almost complete absence of valleriite in the sample. On the other hand, much of the sample was represented by tochilinite I. Figure 61a shows a typical tochilinite SAED point pattern.

As in the case of isometric tochilinite, the crystal elongation direction coincided with the axis b. The sample also contained some plate-like particles characterized by a hexagonal pattern in the distribution of reflections. These particles may be attributed to $Fe(OH)_3$.

In addition to tochilinite, the sample contained two phases having hybride structures and unusual diffraction properties. They were called tochilinite-like phases I and II containing mackinawite-like sulfide layers (Organova et al. 1974).

A point pattern corresponding to the phase I is shown in Fig. 61b. The strongest reflections attributed to the sulfide component form a square network with the same parameter $a = 3.68$ Å as in mackinawite. Relationships between the layer unit-cell parameters for tochilinite and the phase I are given by

$$\begin{pmatrix} A \\ B \end{pmatrix} = \frac{1}{2} \begin{pmatrix} 1 & \frac{1}{3} \\ 1 & -\frac{1}{3} \end{pmatrix} \begin{pmatrix} a_t \\ b_t \end{pmatrix}.$$

A set of weaker reflections that do not coincide with the square network must correspond to the brucite component. Figure 62a sketches the reflections for the sulfide component (open circles) and the brucite-like one (closed circles around the zero point and crosses). It is seen that the "brucite" reflections are located on two circles, so that the radius of the second circle is by a factor of $\sqrt{3}$ greater than that of the first one. The d-spacings corresponding to the reflections lying on these circles are 2.72 Å and 4.71 Å. If the index 100 is ascribed to some reflection on the first circle, then the hexagonal unit-cell parameter will be $a = 3.14$ Å, which is close to the a value for brucite. Note that the phase I point pattern contains a number of additional reflections that are not shown in the scheme (cf. Fig. 61b and Fig. 62a, b).

To determine the sulfide layer structure, intensities were analyzed for 12 unique nonzero reflections. A low general degree of crystallinity as well a fine dispersion of the samples suggested that in this case the interaction between electrons and matter may be treated as kinematical. Owing to an uncertainty in the choice of the right relationship between I and Φ^2, all the calculations were carried out for two versions. Preference was finally given to the formula $I_{hkl} \sim \Phi_{hkl}^2 / d_{hkl}$ implying a comparatively large angular spread in the domain

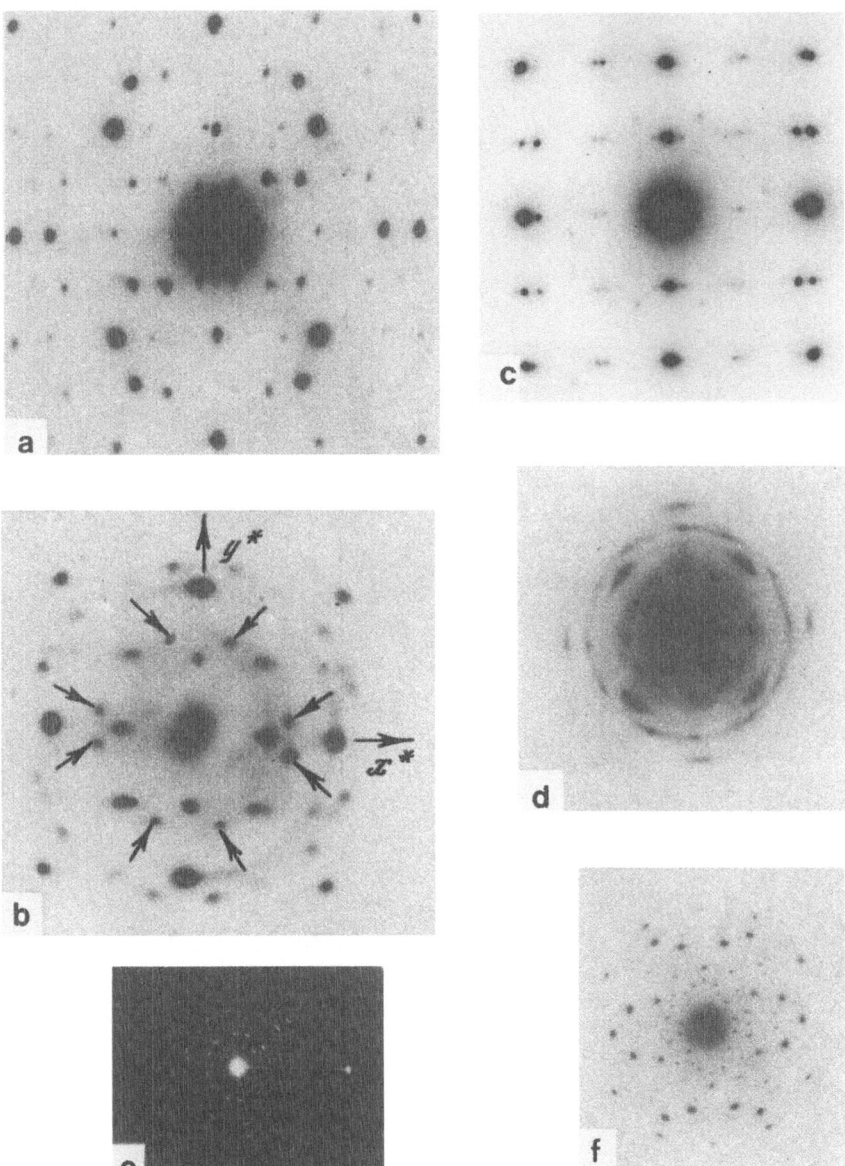

Fig. 61a–f. Point ED patterns from: **a** tochilinite I having incommensurate brucite and sulfide layers; **b** tochilinite-like phase I; **c** tochilinite-like phase II; **d** a phase I twinned crystal (the twins are rotated through 90° with respect to each other); **e** three-component mixed-layer mineral (reflections ool; **f** tochilinite I twins rotated through 120° with respect to each other

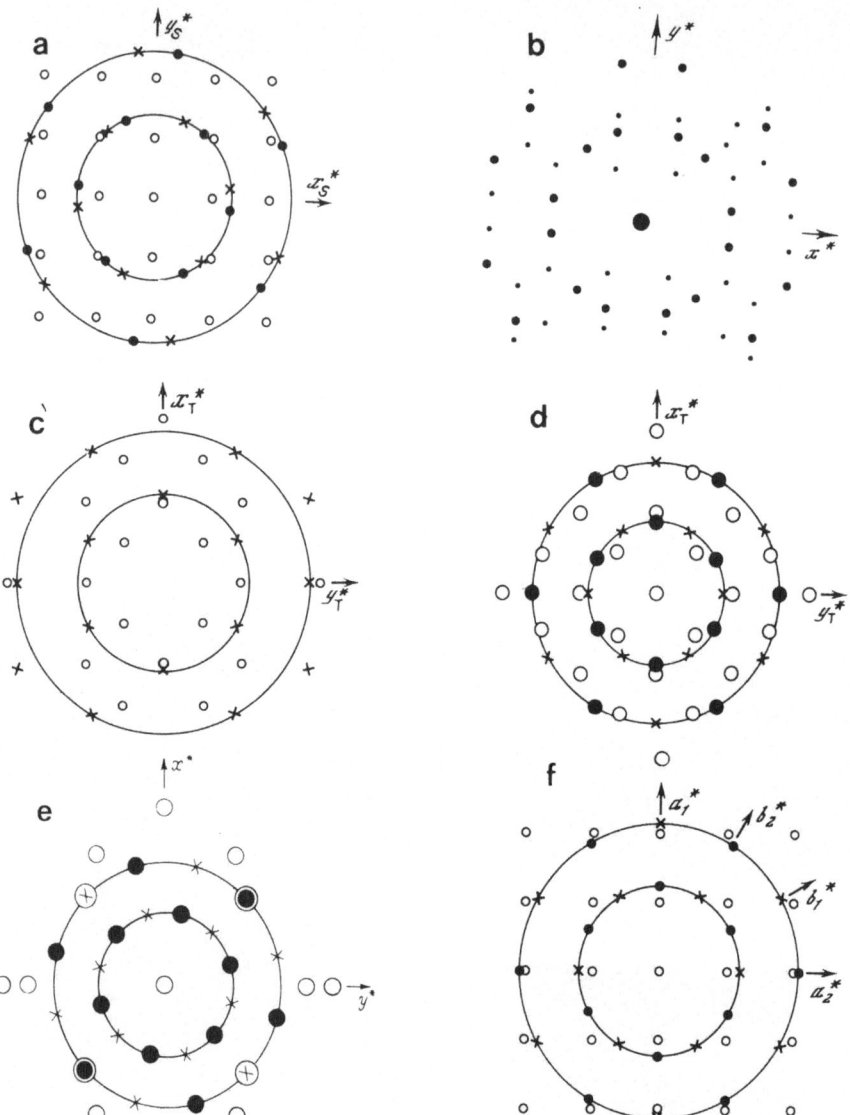

Fig. 62a–f. Reflection distribution in ED patterns from crystallites containing mackinawite-like sulfide layers with a chessboard distribution of vacancies. **a** Idealized distribution of hk reflections obtained from phase I; **b** calculated intensity distribution obtained from the two-layer brucite component of the phase I (the circle diameters are proportional to the corresponding intensities); **c** an example of incommensurability of reflection networks for the sulfide and the brucite components; **d** twinning resulting from a relative rotation of the twins through 90°; **e** reflection networks for the sulfide components of tochilinite and valleriite displayed simultaneously with the tochilinite-like orientation of brucite layers in both components; **f** reflections networks for the sulfide component of phase I and valleriite with the tochilinite-like orientation of brucite layers

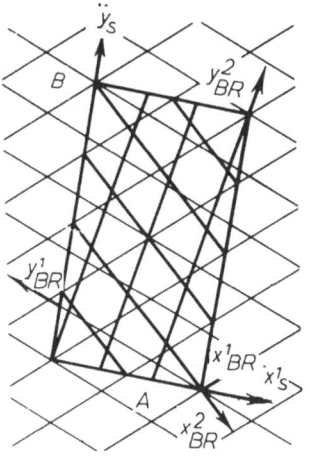

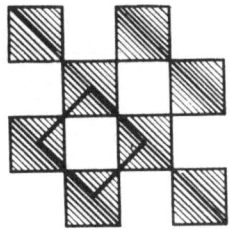

Fig. 63. Distribution of vacancies in sulfide layers in phases I and II

◄ Fig. 64. The arrangement of adjacent brucite layers in the phase I structure

orientations $(3° - 4°)$. This was confirmed by the general intensity distribution in the diffraction pattern.

The two-dimensional Patterson map (Fig. 56d) contained maxima arranged in a square pattern, as should be anticipated for a mackinawite-like sulfide layer. The maxima at $u = 0.5$, $v = 0$ and $u = 0$, $v = 0.5$ represent the vectors Fe-S, while that at $u = 0.5$, $v = 0.5$ corresponds to the vectors Fe-Fe and S-S. The weigtht of the latter peak is appreciably smaller than those of the other two. Hence, it may be inferred that the site corresponding to Fe in the center of the unit cell is partially vacant. The least-squares refinement applied to determine the thermal factor and the occupancy of tetrahedral sites by Fe cations has led to the value $R = 0.14$ for $B = 5$ and $\mu = 0.55$. The sulfide structure prejection down c is shown in Fig. 63. Partially and completely occupied tetrahedra are arranged in chessboard order. The unit cell contains 1.5 Fe atoms per 2 S atoms. Hence the idealized sulfide layer composition for phase I is $2(Fe_{0.75}S)$.

Additional information on the structure of the brucite component can be obtained by comparing the scheme in Fig. 62a with the corresponding point pattern, since the latter contains additional reflections. Organova et al. (1974) suggested that successive brucite layers are alternately rotated with respect to one another, forming a two-layer unit cell. First, this large unit cell explains the presence of reflections that should have been absent in the case of a simple incoherent superposition of two brucite layers, which is represented by Fig. 62a. Second, it leads to distortions of a strictly hexagonal pattern in the intensity distribution characteristic of both "brucite" reflection systems shown in the figure. Note that the additional reflections are of medium intensity, and distortions in hexagonal symmetry are rather weak.

All the brucite reflections can be indexed in terms of a unit cell having its coordinate axes parallel to the axes a and b in the sulfide sublattice and the dimensions $A_{BR} = 8.31$ Å, $B_{BR} = 14.4$ Å, $\gamma = 90°$.

To find the rotation angles for the neighboring brucite layers as well as the relative orientations for the neighboring brucite and sulfide layers, possible versions for the formation of a two-layer brucite unit cell were to be considered. To do this, two identical brucite lattices with a rhomb-shaped unit cell were drawn on two separate sheets of tracing paper. One of these was superimposed on the other and rotated until a large rectangular unit cell common for both lattices was formed so that its parameters allowed indexing for all the brucite reflections (Organova 1972).

Three independent versions were thus obtained. Structural amplitudes $\Phi(hko)$ were calculated for each version. The best agreement between the experimental and the calculated intensities was achieved for the version shown in Fig. 64.

It is seen that the brucite rhomb dimensions are related to the large "brucite" unit-cell parameters by $A_{BR} = a_{BR} \sqrt{7} = 8.31$ Å; $B_{BR} = a_{BR} \sqrt{21} = 14.4$ Å. The relative rotation angle for two neighboring brucite-like layers is 22°. The brucite unit cells are rotated through 41° and 19° with respect to the "sulfide" coordinate system. The above angular values are in agreement with those measured from the point ED pattern (Fig. 61 b; Organova et al. 1974).

The value and the direction for the parameter c cannot be determined from the SAED pattern containing reflections hko. However, the presence in the X-ray powder pattern of only those reflections that can be indexed within the tochilinite unit cell may imply, for a significant quantity of the phase I in the sample, that the interlayer shift for the phase I sulfide layers is the same as in tochilinite I. That is, each sulfide layer is shifted by $-a\sqrt{2}/6$ along the diagonal of the unit square with respect to the preceding layer, where a is the phase I sulfide layer unit-cell parameter. The sulfide sublattice will be then described by a triclinic unit cell $a = b = 3.68$ Å, $c = 10.92$ Å, $\alpha = \beta = 93.5°$, $\gamma = 90°$.

The phase II particles were prevailing in one of the parts of the sample provided by Dr. Harris. The corresponding SAED pattern is hown in Fig. 61 c. The strongest reflections form a square network ($a_{sq} = 3.8$ Å) similar to that of the phase I sulfide reflections.

There was also an orthorhombic reflection system ($a = 4.09$ Å and $b = 3.68$ Å, and $9a = 10a_{sq}$) (Fig. 61 c). The actual parameter a was shown to be $(4.09 \times 2) = 8.18$ Å. Similar intensity distributions in the square networks in SAED patterns from phase I and phase II indicate that their sulfide components are identical. However, the arrangement and the intensities of the remaining reflections are substantially different from what should be expected for brucite layers.

A number of SAED photographs were used to evaluate the intensity for reflections that, by analogy with other hybrid structures, should have been attributed to the hydroxide component. As the oko reflections for different sublattices coincide, the corresponding contribution of the hydroxide component was estimated as the difference between the intensities of reflections oko and hoo, since for the sulfide component $I(oko) = I(hoo)$ for $h = k$. It was again assumed that the kinematical approximation is valid and the formula relating I and $|\Phi|^2$ is the same as for phase I. The Patterson map shows two diffuse maxima at $u = 0.16a$, $v = 0.56b$ and $u = 0.5a$, $v = 0$ (Fig. 65a).

Since the attempts to interpret these maxima in terms of the brucite layer structure failed, other possible structure models had to be sought for the hydroxide component of phase II (Organova et al. 1974). Specifically, it was sug-

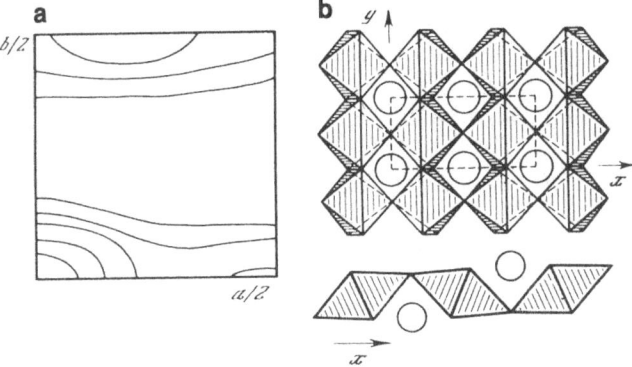

Fig. 65. a The projection of $\mathscr{P}(uv)$ and **b** the structure projection for the hydroxyl component of phase II

gested that the two Patterson peaks corresponded to two oxygen atoms, each of these being connected with a metal atom at the unit-cell origin.

This interpretation leads to a hydroxide layer structure shown in Fig. 65b. Octahedra formed by OH groups and water molecules and filled by Fe cations are connected through shared edges into chains lying along **b**. The neigboring chains are linked together through free apices to form a two-dimensional layer. Shared octahedral edges are formed by OH groups, while the H_2O molecules are at the apices shared by the octahedra belonging to different chains. In addition, water molecules shown in Fig. 65b as large open circles occupy the vacant sites in the layer structure (the introduction of H_2O into these positions improved the agreement between the experimental and the calculated intensities). The presence of weakly bounded water is indirectly confirmed by the formation of lenses or pores in the phase II crystallite volume due to local interlayer swelling observed during EM studies.

Effects of this kind were never observed with particles of other minerals. The best agreement with the experimental data was achieved for the model having a 0.5 probability occupancy of the vacant sites by H_2O molecules. The water content in a microcrystal thus may decrease in the process of illumination by electrons. This may be the reason for variations in the intensity distribution in SAED patterns from phase II particles.

The phase II composition calculated from the structure data is $2(Fe_{0.78}S) \cdot 0.9[Fe(OH)_2\ 3/2\ H_2O]$.

The structure described above is an approximate model and should be treated with care.

9.5 The Crystal Structure of Valleriite

The valleriite structure has been determined by Evans and Allman (1968) and is shown as two projections in Fig. 55c. Two sublattices are required to describe it. Let us denote the positions of S atoms in a close packing as *a, b, c* and those of Fe atoms as α, β, γ. The OH positions in the brucite sublattice will be labeled by *A* and B, and the cation sites by the letter Γ. The valleriite structure is then represented by the formula

$$\overline{a\beta\,abA\Gamma Bb\,y\beta cA\,\Gamma Bc\,\alpha\gamma a\,A\,\Gamma Ba\beta}\,\underline{\alpha b}\,.$$

The upper square bracket marks the repeat distance for the sulfide sublattice and the lower one marks that for the brucite component. The symmetry of a one-layer brucite component is $P\bar{3}m1$ and that of the sulfide component is described by the space group $R\bar{3}m$. This variety is labeled $3R$ according to the sulfide layer stacking.

Since valleriite samples are usually dispersed, it was of interest to carry out a SAED structure study combined with X-ray powder diffraction. Two valleriite samples were chosen (Organova et al. 1973a). The compositions calculated from the chemical analysis data were the following

$$[Cu_{0.81}Fe_{1.19}S_2] \cdot 1.56\,[(Mg_{0.83}Fe_{0.17})\,(OH)_2] \qquad\qquad\qquad \text{(sample 1)}$$

and

$$[Cu_{0.91}Fe_{1.09}S_2] \cdot 1.78\,[(Mg_{0.70}Fe_{0.30})\,(OH)_2] \qquad\qquad\qquad \text{(sample 2)}.$$

The indexing of X-ray powder patterns performed in accordance with the published data led to the unit-cell dimensions $a = 3.787$ Å, $c = 11.37 \times 3 = 34.11$ Å (sample 1) and $a = 3.788$ Å, $c = 11.396 \times 3 = 34.19$ Å (sample 2).

The sample 2 X-ray powder pattern contained more maxima than that from sample 1, while the latter showed a number of wide lines absent in the sample 1 pattern. A large number of point SAED patterns from different valleriite crystals were obtained. Typical valleriite SAED patterns always contain two different but similarly oriented hexagonal hko reflection systems (the incident beam is normal to the microcrystal plane. See Figs. 66, 67). The inner of the two hexagons that are nearest to the center contains stronger reflections and represents the sulfide component. The outer hexagon contains weaker reflections and corresponds to the brucite component.

Even a visual examination of SAED patterns demonstrates considerable advantages of SAED for the study of hybrid minerals. Whereas an X-ray powder pattern of one mineral may be erroneously indexed in terms of the unit cell of another mineral, SAED patterns allow ready and unambiguous identification of valleriite or tochilinite. Moreover, SAED ensures revealing various structural variations for these minerals, while X-ray powder patterns of hybrid minerals never show the contribution of the brucite component.

Reflection intensities for the sulfide component proved to differ substantially for samples 1 and 2. This is seen in the corresponding SAED photographs (Figs. 66a and 67a). For example, the hoo intensity decreases with h for sample 1 and is constant for all h for sample 2. An adequate explanation is given in (Organova et al. 1973a). It has been suggested that in the sample 1 structure, the sulfide sublattice is also described by a one-layer unit cell and labeled $1T$, so that the structure is described by

$$\overline{a\beta\,abA\,\Gamma Ba}\,\underline{\beta ab}\,A\,\Gamma B\,.$$

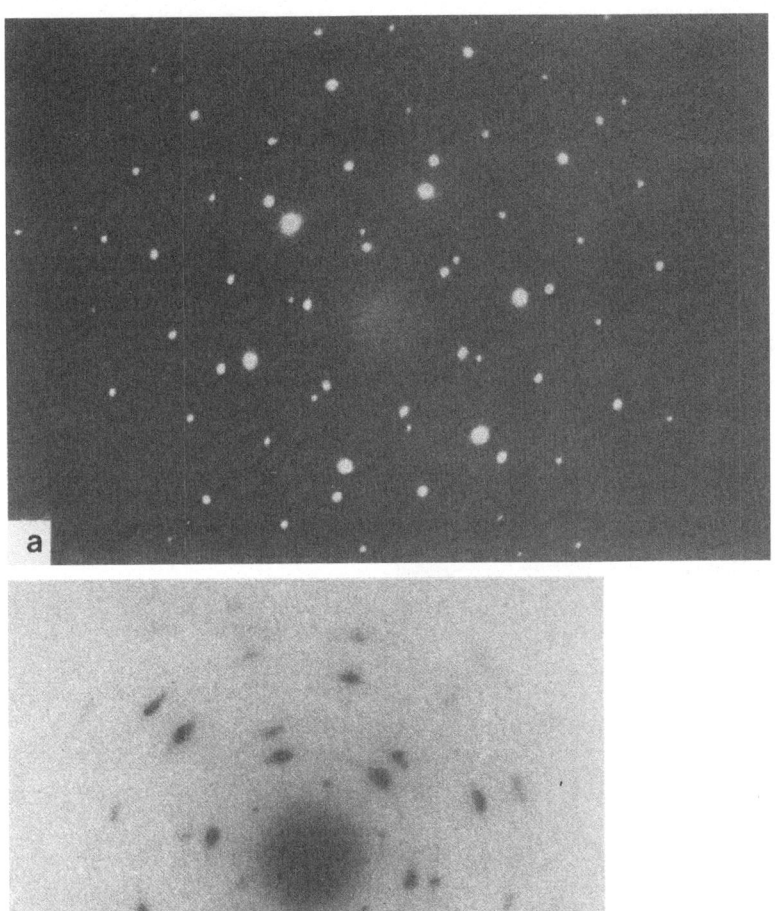

Fig. 66a, b. Point ED patterns from valleriite: **a** sample 2; **b** with an ordered distribution of cations in brucite layers

The notations are the same as in the case of valleriite described above. The unit-cell dimensions are $a_s = 3.79$, $a_{BR} = 3.08$, $c = 11.87$ Å, space group $P\bar{3}m\bar{1}$.

The idealized formula obtained from the structure data is $(FeCu)S_2 \cdot 1.51$ $[(MgFe)(OH)_2]$. A minor deviation of the factor 1.51 from that calculated from the chemical analysis (see above) may be associated with the presence of a small quantity of $Fe(OH)_3$.

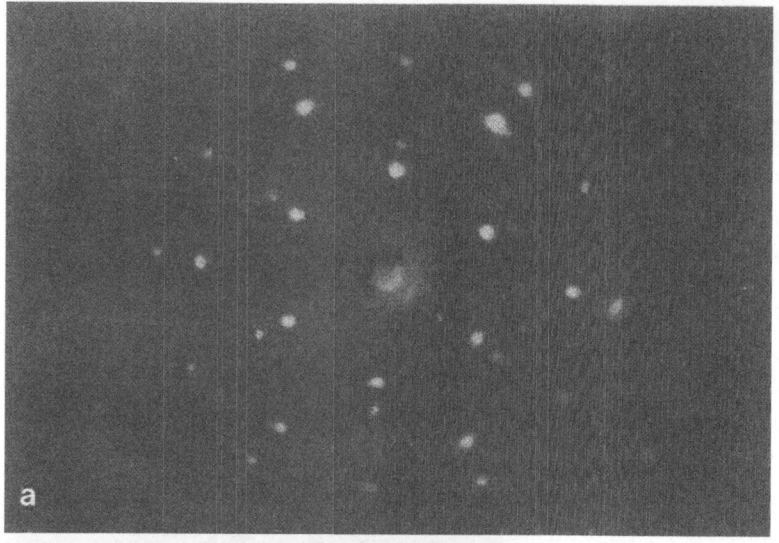

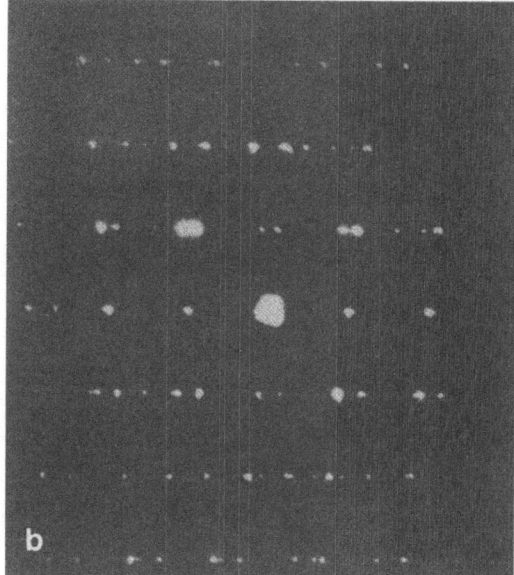

Fig. 67. Point ED patterns from: **a** valleriite from Sweden; **b** twinned F-cupfferite

The inference of a one-layer sulfide component unit cell in sample 1 based on the comparison of the experimental and the calculated hko intensities ($R = 0.14$) was confirmed by a repeated examination of the X-ray powder pattern. It was proved possible to index it in terms of a one-layer unit cell. The presence of a number of wide lines in the pattern may be associated with sulfide layer stacking faults (Organova et al. 1973a).

Another peculiar valleriite variety is identified by the SAED pattern shown in Fig. 66b. The intensity distribution for the sulfide component indicates one-layer periodicity. The pattern contains additional weak reflections corresponding to the brucite-like component ($a_{BR} \sqrt{3} = 3.08 \sqrt{3} = 5.33$ Å). The most reasonable explanation seems the assumption of an ordered distribution of Mg and Fe cations over octahedral sites in brucite layers.

9.6 A Three-Component Hybrid Mineral Containing Brucite-Like, Sulfide and Silicate Layers

SAED patterns from all the valleriite varieties studied contain single systems of basal reflections ool corresponding to $d(001)$ [or to $d(003)$ for rhombohedral valleriite] equal to the sum of brucite and sulfide layer thicknesses. This means that none of the valleriites involved is an epitaxial aggregation of homogeneous crystallites consisting of either brucite or sulfide layers. In SAED patterns obtained from bent edges there are distinct reflections 001 corresponding to a minimum periodicity of 11.37 Å. At the same time, sample 1 contained particles that seemed to give typical valleriite-like diffraction patterns since they contained two incommensurable hexagonal systems of reflections (Fig. 68d). This pattern, as distinct from those obtained from valleriites having ordered cation distribution in brucite layers, showed anomalously strong reflections corresponding to the nonsulfide component. As seen in Fig. 68a, intensities of reflections having identical hk but belonging to different sublattices are quite similar.

The microprobe analysis applied to sample 1 revealed the presence of Si in some of the particles, which may be associated with silicate layers. A more reliable information on the composition and the structure of a hybrid mineral containing silicate layers was obtained by studying a sample provided by V. V. Distler et al. (Organova et al. 1971a). Microprobe analysis indicated the presence of copper, iron, magnesium and aluminum as well as a considerable amount of silicon.

SAED revealed the presence of two phases, i.e., lizardite and a new hybride mineral having a diffraction pattern identical to the unusual sample 1 pattern described above. Diffraction patterns from bent edges containing reflections ool with $d(001) = 19.2$ Å (Fig. 61e) indicate that the crystallites analyzed consist of different layer types alternating regularly. There are several possible versions for the structure of the phase involved. In one of these, the sulfide layers alternate with chlorite packets. One could also imagine a structure where neighboring sulfide layers are separated by pairs of serpentine-like layers. The experimental data available were insufficient for an unambiguous choice of the structural model.

9.7 Forms of Structural Heterogeneity

There are numerous forms of structural heterogeneity in hybrid minerals (Organova 1972; Organova et al. 1971a,b, 1972, 1973a,b, 1974). The basic reason for failures of homogeneity is a relatively weak interaction between layers of different types and their incommensurability. This favors translational and

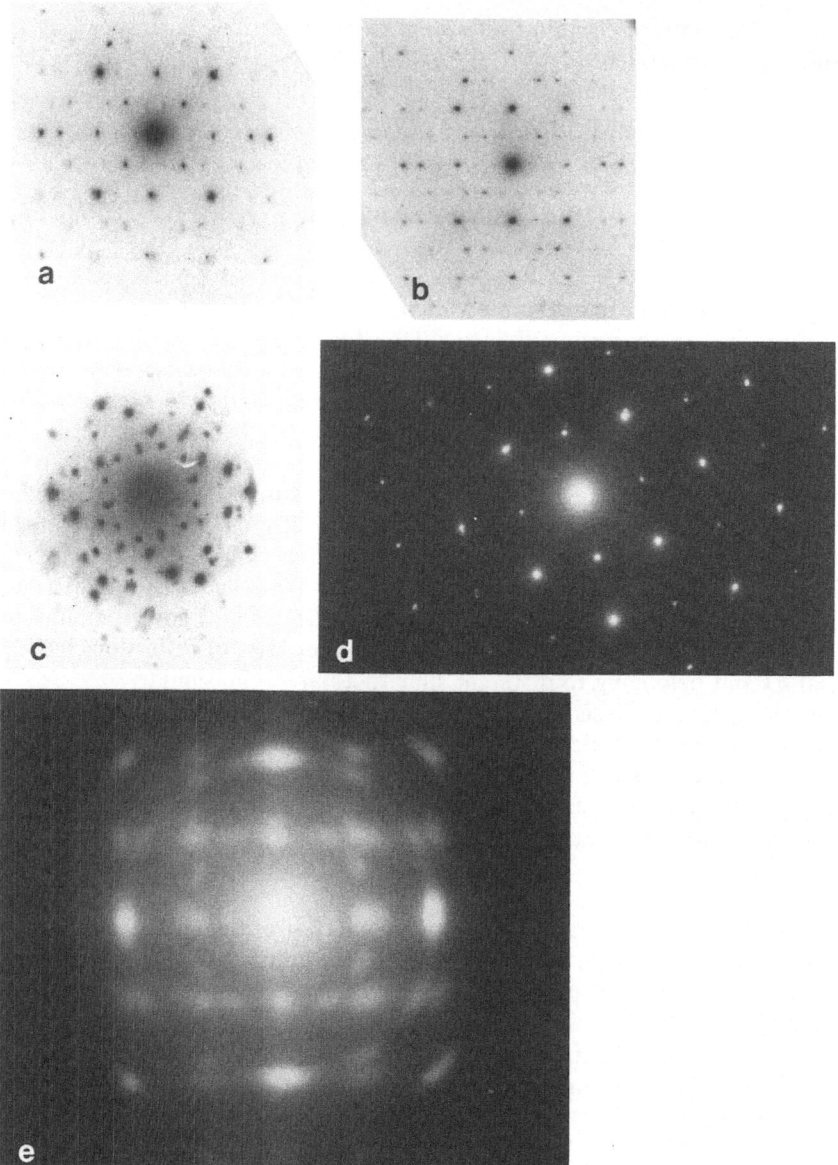

Fig. 68a – e. Point ED patterns from: **a** tochilinite I having a chessboard distribution of vacancies; **b** tochilinite I having a uniform distribution of tetrahedral vacancies in sulfide layers; **c** tochilinite-like crystal containing two sulfide components (tochilinite- and valleriite-like); **d** mackinawite (reflections hko); **e** crystal containing sulfide components corresponding to phase I and valleriite

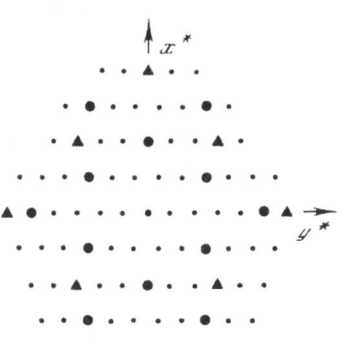

Fig. 69. The *hko* reflection network for tochilinite I

orientational faults in layer stacking. For the same reason, alternation of several layer types is possible in hybrid mineral structures. Moreover, the alternating components being geometrically isolated, fragments differing in structural pattern may exist within the same layer.

SAED proves extremely advantageous in this field. It allows visualizing various types of structure defects in hybride minerals on the basis of geometrical analysis of point patterns and knowledge of the general structure motif.

It has been mentioned that two phases, tochilinite I and II, may form epitaxial aggregations. In addition, a number of other defect types has been found in isometrical tochilinite, which is reflected in the corresponding diffraction effects. To interpret these, peculiarities in the distribution of intensity for tochilinite I reflections *hko* should be considered. Calculation of structure amplitudes show that the reciprocal lattice involved is subdivided into two sublattices. Reflections corresponding to one of these are largely associated with sulfide layers, while those corresponding to the other are associated mainly with brucite-like layers.

Figure 69 sketches the reciprocal lattice points *hko*. The strongest "brucite" reflections are shown by circles (e.g., 350 and 0.10.0) and the "sulfide" ones by triangles (e.g., 260, 0.12.0). These strongest reflections determine the "spider-like" SAED pattern of tochilinite I. On the other hand, it is seen that the strongest brucite reflections form an almost regular hexagon corresponding to $a = 3.13$ Å (for brucite $a = 3.12 - 3.14$ Å), while the "sulfide" reflections form a pseudo-square network corresponding to the unit subcell of the mackinawite sulfide layer with $a_{sq} = 1.87$ Å equal to the projected distance Fe-S in a sulfide tetrahedron (Fig. 54).

Twinning in the plane (001) is the simplest case of structural heterogeneity in tochilinite crystals. Figure 61f shows an SAED photograph containing two tochilinite patterns rotated through 120° with respect to one another. A stacking fault resulting from rotation of the sulfide layer through 120° with respect to the brucite-like one will have little effect on the arrangement of S atoms and OH groups at the adjacent layer surfaces. However, as a result, one part of the structure will be rotated through 120° with respect to the other.

Another type of structural heterogeneity is manifested in variations of distribution of vacancies over the tetrahedral sites in sulfide layers. These defects have been especially distinct in specimens prepared from isometric tochilinite particles kept in water suspension for a month. Diffraction patterns obtained from such preparations differed substantially in intensity distribution from typical tochilinite patterns. For example, a pattern shown in Fig. 68b contains strong reflections forming a large-cell network corresponding to a rectangular subcell in the direct lattice with $a = a_T/2 = 2.68$ and $b = b_T/6 = 2.6$ Å. The arrangement of these reflections is similar to the hko pattern of mackinawite (see Fig. 68d). In the latter structure, all the tetrahedral sites are equivalent, as they are occupied by cations of the same type. Therefore, the intensity redistribution in diffraction patterns from tochilinite-like minerals should imply practically random distribution of vacancies in sulfide layers.

In diffraction patterns from other microcrystals the stronger sulfide reflections are arranged in pseudo-square network with a smaller unit cell than in the previous case. The corresponding subcell in the direct space is a rhomb with $a = 1/2 \times [a_T^2 + b_T^2/9]^{1/2} = 3.74$ Å and $\gamma = 92°$ (Fig. 68a). A square network of "sulfide" reflections with a similar intensity distribution ($a_{sq} = 3.68$ Å) was observed in diffraction patterns from phases I and II. Therefore, one may suppose that a long allowance of the suspension results in a redistribution of vacancies in sulfide layers, so that the chessboard pattern becomes the dominant, although not the only possible, arrangement (Fig. 63). Despite the redistribution of vacancies, the structure still can be described by a single lattice. If the chessboard pattern becomes the only one, the description of all the structure components in terms of a single lattice becomes impossible.

To illustrate this, consider the scheme of the corresponding SAED pattern (Fig. 62c). In spite of the tochilinite-like orientation of the brucite-like layers, the square pattern of sulfide reflections appears incommensurate with the hexagonal pattern of the "brucite-like" lattice points. Therefore, it is impossible to find a common unit cell.

Thus, a distribution of vacancies in sulfide layers differing from that in tochilinite requires two separate sublattices to describe the structure which is essentially the same. Moreover, the chessboard order in the arrangement of vacancies and the resultant change in the sulfide layer symmetry leads to stacking faults different from those in tochilinite.

An SAED pattern indicating twinning in such a tochilinite-like crystal, which results from rotation through 90° of one part of the structure with respect to the other in the plane (001), is shown in Figs. 61d and 62d. The possibility of such a twinning fault is governed by the pseudo-square geometry of sulfide layers, owing to which a 90° rotation of the brucite-like layer does not affect the arrangement of the S atoms and the OH groups in the interlayer. Note that an alternative interpretation for the pattern shown in Fig. 62d is the assumption of random alternation of brucite layers rotated with respect to one another through ±90°. The absence of additional reflections does not permit to employ the idea of two-layer periodicity for brucite-like layers, as it was done for phase I.

The presence of more-than-two-layer types is another defect which is widespread. For example, almost all the diffraction patterns from phase II microcrystals contained weak reflections corresponding to brucite component of phase I (Fig. 61a and b). Thus, the sulfide layers in these structures alternate not only

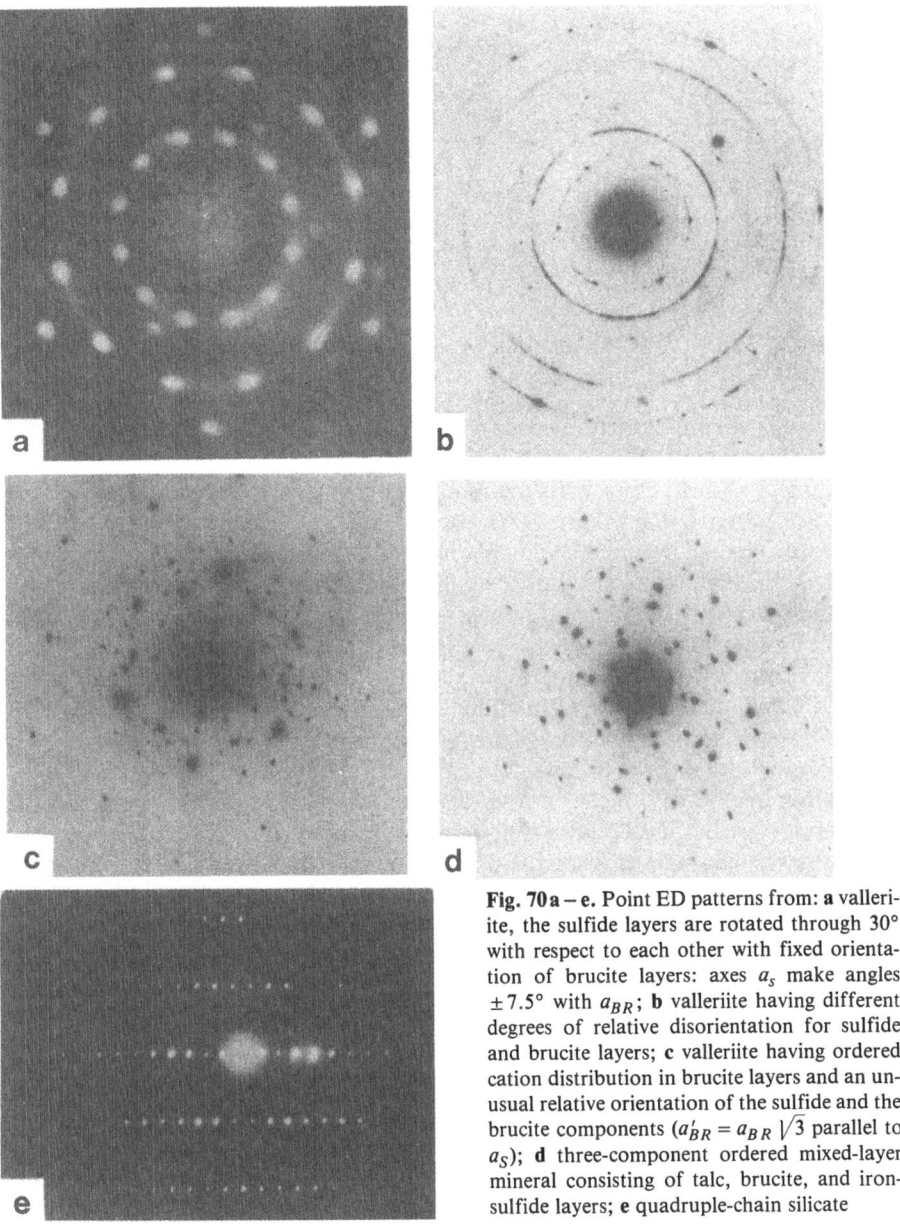

Fig. 70 a – e. Point ED patterns from: **a** valleriite, the sulfide layers are rotated through 30° with respect to each other with fixed orientation of brucite layers: axes a_s make angles $\pm 7.5°$ with a_{BR}; **b** valleriite having different degrees of relative disorientation for sulfide and brucite layers; **c** valleriite having ordered cation distribution in brucite layers and an unusual relative orientation of the sulfide and the brucite components ($a'_{BR} = a_{BR} \sqrt{3}$ parallel to a_S); **d** three-component ordered mixed-layer mineral consisting of talc, brucite, and iron-sulfide layers; **e** quadruple-chain silicate

with octahedral layers sketched in Fig. 65b, but also with "normal" brucite-like layers. The two-layer periodicity of the brucite-like layers allow an alternative interpretation for the patterns involved as resulting from epitaxial intergrowth of phases I and II. Defects in hybrid minerals can be also associated with heterogeneous structure for sulfide layers. The diffraction pattern shown in Fig. 68c was often observed with tochilinite samples of both morphological varieties. The most plausible interpretation is that transitions from mackinawite-like

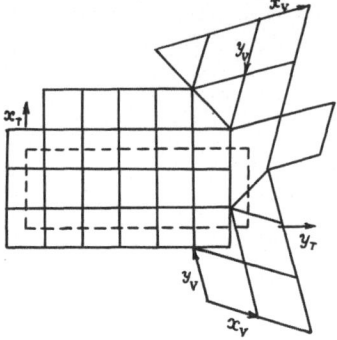

Fig. 71. Transition from tochilinite-like to valleriite-like pattern in the sulfide-layer structure

structure to valleriite structure occur within individual sulfide layers without any failure of continuity.

Figure 71 shows the transformation of square projections of tetrahedra on (001) into triangles (Organova 1972). As a result, the tetrahedral edge forming the square diagonal becomes the small diagonal of the rhomb which represents in Fig. 71 the hexagonal unit cell of the valleriite sulfide layer. Under these conditions there are two, equally probable, orientations for valleriite sulfide layers rotated through 30° with respect to one another (Fig. 71). The existence of relationships between the tochilinite unit cell and the valleriite one is also indicated by the following. The strongest tochilinite reflection 260 associated mainly with sulfide layers corresponds to a d-spacing equal to that of the reflection 110 for the valleriite layers. If the brucite layers are assumed to preserve the "tochilinite" orientation irrespective of the structure of the neighboring sulfide layers, then all the features of the SAED pattern in question can be explained. It contains two reflection systems rotated through 30° that correspond to valleriite-sulfide layers superimposed on a "tochilinite" pattern. Figure 62d shows the "valleriite-sulfide" reflections superimposed on the strongest tochilinite reflections.

The SAED pattern (Fig. 68e) and the scheme in Fig. 62f may serve as another example illustrating the possible existence of two different sulfide-layer types within one microcrystal. The main difference from the previous case in associated with the distribution of vacancies in the mackinawite-like sulfide layer. The valleriite-sulfide pattern and the square network corresponding to mackinawite-like layers with a chessboard distribution of vacancies have no common reflections.

A "wrong" relative orientation of brucite-like and sulfide layers is often met among the samples under study. For example, the SAED pattern shown in Fig. 70c corresponds to valleriite crystallite having brucite layers rotated through 30° with respect to the normal orientation. The directions of the axes of the large hexagonal brucite layer unit cell ($a = a_{BR}\sqrt{3}$) coincide with those of the sulfide layer unit cell.

The "wrong" orientation of brucite layers with respect to sulfide layers sometimes occurs in crystals whose brucite layers have a random distribution of isomorphous cations (Fig. 70a). In some valleriite crystals azimuthal rotations of sulfide layers within a sufficiently wide angular range occur with equal probability. For example, Fig. 70b shows a SAED pattern containing point "brucite" reflections, while the "sulfide" maxima are spread almost uniformly along the arcs.

On the whole, the examples discussed above are also of methodological interest. They demonstrate that conclusions on a structure under study may be drawn by qualitative or even intuitive considerations based on crystal chemistry of related minerals.

9.8 Structure Study of Asbolanes

Our knowledge of Mn-minerals from Fe-Mn nodules is still insufficient despite intense investigation. The main difficulties are associated with their extremely high dispersion, similar chemical compositions and low structural perfection. Besides, the samples are often presented by more than one mineral. X-ray diffraction is often ineffective not only for structure studies but also for reliable phase identification.

The use of SAED and X-ray energy dispersion analysis opened new prospects for structure studies of these minerals. The combination of these methods with X-ray diffraction, thermal and chemical analysis, IR and X-ray photoelectronic spectroscopy has proved especially fruitful, allowing the determination of coordination and valence for cations of different types, the content of molecular water and OH groups, and other important data. Application of the complex of these methods to Fe-Mn nodules has revealed a new world of minerals. Chukhrov and co-authors were the first to reveal the crystal-chemical nature of many of these (Chukhrov et al. 1980a, b, 1982, 1983a, b, 1984). Specifically, this applies to asbolanes that can be regarded as a new group of manganese minerals having unusual structural characteristics. Many asbolanes contain, along with Mn, large quantities of Co and Ni. However, it was not clear whether asbolanes are individual minerals or mixtures of oxides of Mn, Co, and Ni. Systematic investigations by the methods listed above proved the mineralogical individuality of asbolanes and revealed their basic crystal-chemical features.

Asbolanes were found to be hybrid mixed-layer minerals consisting of two alternating layer types having different cationic compositions. Since different cations have different ionic radii, the corresponding layers differ in basic parameters. Therefore, an asbolane structure should be described in terms of two incommensurable hexagonal sublattices which have a common origin, the same orientation, the same period c along the normal to the layers, equal to the sum of thicknesses for two adjacent layers of different types, but differing in the value of a. Therefore, asbolanes, as valleriites, are hybrid minerals with incommensurable layers.

In contrast to other hybrid minerals, in asbolanes layers of one of the types have an island-like structure, while those of the other type are continuous in two dimensions.

At present, asbolanes are classified into two groups (Chukhrov etal. 1980a, b, 1982, 1983a; Drits et al. 1985). In one of these, cations in both layer types are octahedrally coordinated (Chukhrov et al. 1980a, b). In the other, cations belong-

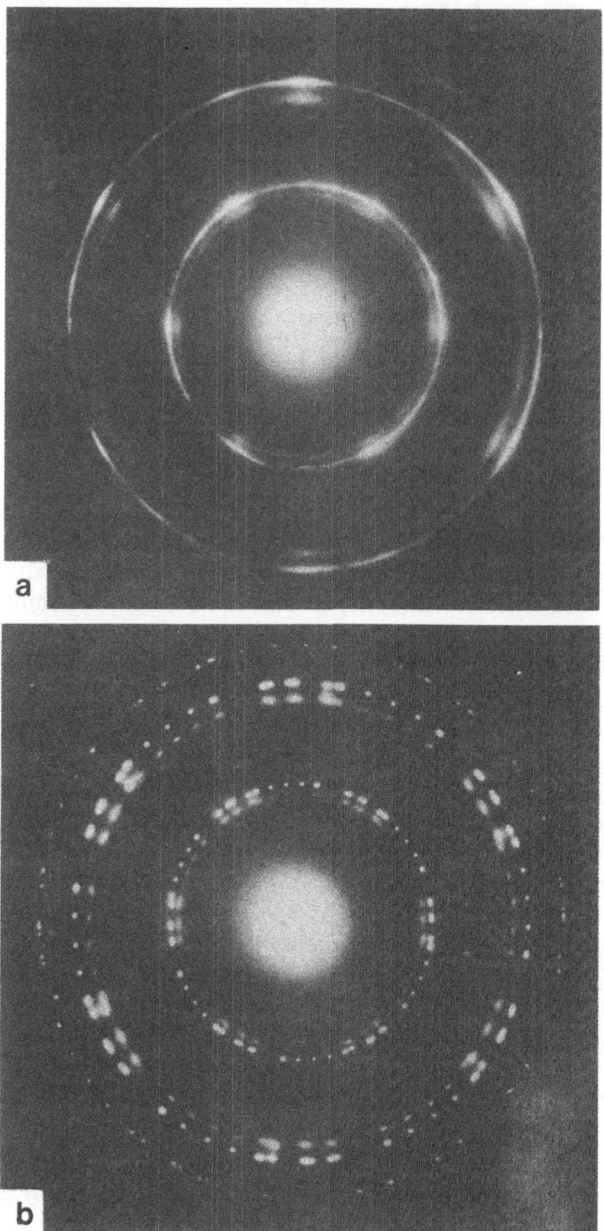

Fig. 72a, b. SAED patterns (**a** and **b**) from Co-Ni asbolane containing hko reflections for two incommensurate sublattices (Chukhrov et al. 1980a)

ing to layers of different types differ in coordination (octahedral and tetrahedral) (Chukhrov et al. 1982, 1983a). Below are discussed the basic structural and diffraction properties of asbolane varieties belonging to each group.

Co-Ni-asbolane (Chukhrov et al. 1980a) belongs to the first group. The X-ray energy dispersion analysis of its particles indicates that the only cations present are Mn, Co, and Ni. Figure 72 shows point SAED patterns from plate-shaped Co-Ni-asbolane crystals. Each of these patterns contains two similarly oriented hexagonal networks of hko reflections. Reflections forming the network situated nearest to the pattern center are several times weaker than those forming the other network. The d-spacings for reflections belonging to different networks and located nearest to the pattern center are 2.445 Å and 2.63 Å. These values correspond to $a = 2.823$ and 3.04 Å, respectively. SAED patterns from bent edges contained a single set of basal reflections ool with $d(001) = 9.34$ Å. SAED patterns from crystals tilted about the incident beam contained the hko reflections and those corresponding to d-spacings of 2.37, 2.17, 1.93 and 1.7 Å. The data obtained indicate that the Co-Ni-asbolane structure can be described by two sublattices: $a = 2.823$ Å, $c = 9.34$ Å (sublattice I) and $a = 3.04$ Å, $c = 9.34$ Å (sublattice II). The space reflections correspond to the sublattice I only. Their indices are 100, 101, 102, 103 and 104 and the calculated d-spacings are 2.445, 2.365, 2.166, 1.923 and 1.69 Å, respectively. The sublattice II is represented by reflections hko and ool only. Figure 73 shows the scheme of the Co-Ni-asbolane structure consisting of two layer types alternating regularly. Layers of one type have the composition $Mn^{4+}O_{1.55}(OH)_{0.45}$. Mn^{4+}-octahedra are linked through shared edges to form layers. The layer charge is positive owing to a partial OH for O substitution. Layers of the second type are also octahedral and have the composition $(Ni_{0.37}Co^{3+}_{0.24})(OH)_2$. Differences in ionic radii for Mn, Co, and Ni lead to a smaller a for Mn layers as compared to that for Co-Ni layers. Octahedra belonging to adjacent layers are of opposite orientations (as in gibbsite). Therefore, in interlayers, anions from different layers are projected approximately onto one another. The interlayer hydrogen bonds in this case are shorter than they should have been if the octahedra in all layers had identical orientations. This is essential since the incommensurability of the adjacent layers in itself should lead to a lengthening of $O \cdots H\text{-}O$ bonds.

Low intensity of reflections hko for the sublattice II (as compared to those for the sublattice I) imply small total volume of coherent scattering domains within Co-Ni layers. This was regarded as an indication of discontinuity, or the island-like structure of Co-Ni layers, which is in agreement with a number of facts: (a) The absence of reflections hkl for the sublattice II implies their low intensity. (b) The scattering powers of Co and Ni are greater than that of Mn. Therefore, if the Co-Ni layers were continuous within the crystal, as are the Mn layers, the ratio of intensities for hko reflections from different sublattices would be inverse to the one observed. (c) The island-like structure of Co-Ni layers favors the formation of stronger interlayer bonds, since the difference in layer lateral dimensions in the case of continuous Co-Ni layers would lead to considerable variations and lengthening in interlayer bonds.

The structural formula for Co-Ni asbolane is

$[(Mn^{4+}O_{1.55}(OH)_{0.45}] \cdot 0.75 \cdot [Ni_{0.37}Co^{3+}_{0.23}Ca_{0.06}\square_{0.34}(OH)_2] \cdot 0.6\ H_2O$,

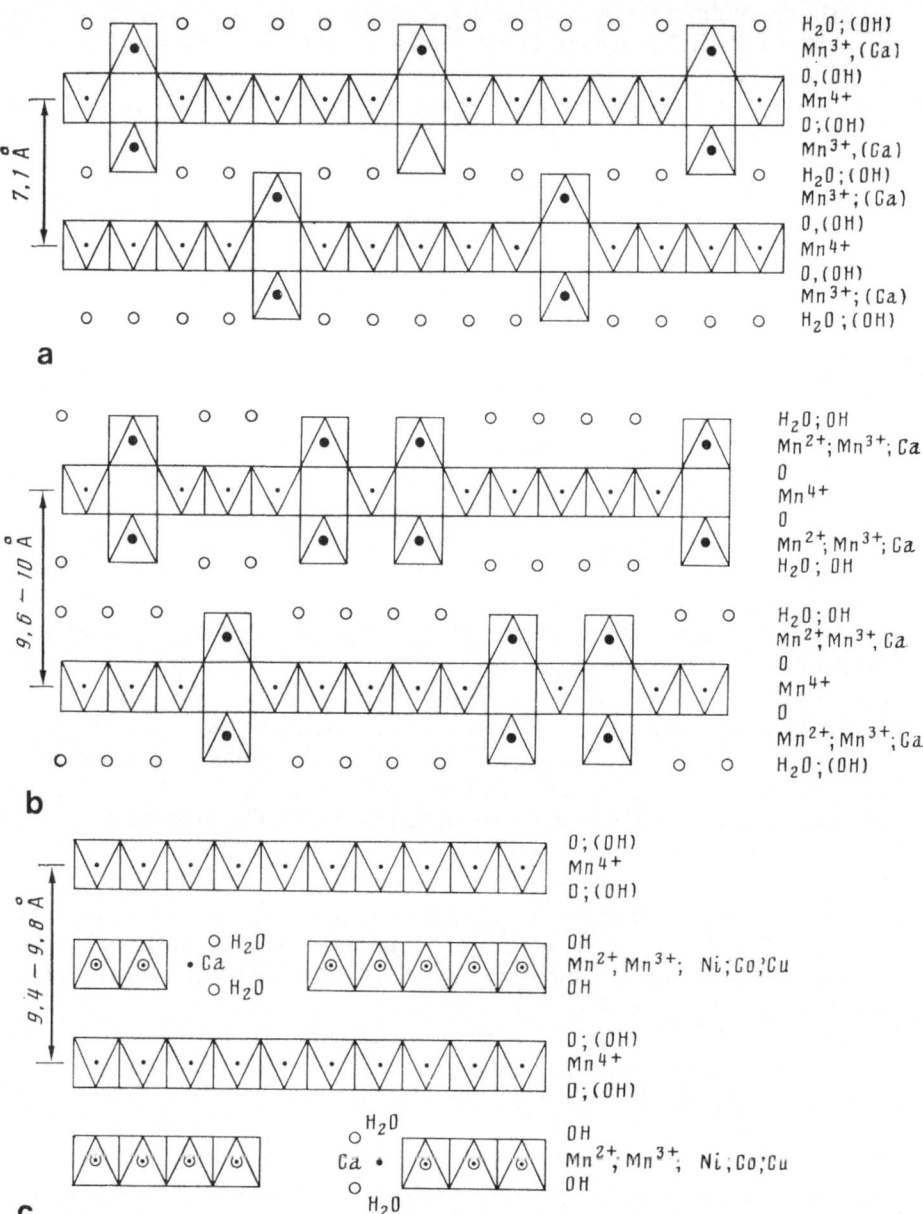

Fig. 73a–c. Structural schemes for Mn-minerals: **a** birnessite; **b** buserite; **c** asbolane with alternating octahedral layers of different type (Chuhrov et al. 1984; Drits et al. 1985)

where $\square$ signifies vacant octahedra. Both hydrogen and electrostatic interlayer bonds should apparently exist.

The Co-Ni-asbolane structure model described is in agreement with the data obtained by various methods and, first of all, with the results of the comparison of experimental intensities for hko and ool reflections with those calculated from the formulae given in Chapter 2. The calculated ratios of intensity for reflections 100, 110, and 200 were 12:20:1 for the Mn component and 4:7:6 for the Co-Ni component. The corresponding ratios for intensities measured on the same scale were 11:21:2 and 3.5:6.5, respectively (Chukhrov et al. 1980a). The $\Phi^2(ool)$ values were also calculated for the same model. The distribution of cations and anions in the layers was assigned in accordance with their contents in the structure. A satisfactory agreement between the observed and the calculated intensity ratios I (001):I(002):I(003):I(004) was observed again, i.e. 15:88:4:0, and 13:88:4:1, respectively.

Ni-asbolane is another variety belonging to the first group (Chukhrov et al. 1980b). The X-ray energy dispersion analysis indicated the presence of only Mn and Ni in the crystallites of the mineral. The Ni-asbolane structure is similar to that of Co-Ni asbolane: Mn^{4+} octahedral layers alternate regularly with brucite-like Ni layers. The Ni-asbolane structure is also described in terms of two hexagonal sublattices: $a = 2.83$ Å, $c = 9.18$ Å (The Mn sublattice) and $a = 5.21$ Å, $c = 9.18$ Å (the Ni sublattice). The difference in parameters for Co-Ni and Ni brucite-like layers is associated with the difference in the distribution of cations over the octahedral sites. In Co-Ni asbolane, Co and Ni occupy octahedral sites with equal probability, whereas in Ni-asbolane, one of the three symmetrically independent octahedra is occupied by Ni with smaller probability than the other two. A number of other asbolane varieties are known at present. Their structures consist of octahedral Mn^{4+} layers alternating with brucite-like layers $R(OH)_2$ (Chukhrov et al. 1980b) where R is Fe and Mn^{2+} (or Mn^{3+}) in some varieties, and only Mn^{2+} and/or Mn^{3+} in others.

Co-asbolane belongs to the second structural group (Chukhrov et al. 1982). The crystallites contain only Mn and Co. A SAED pattern corresponding to the plane (001) of a Co-asbolane crystal was similar to that from Co-Ni asbolane. It contained two identically oriented hexagonal reflection networks differing in dimensions and intensity distribution.

Stronger reflections forming the network which is more distant from the pattern center are characterized by the same intensity distribution and d-spacings ($a = 2.84$ Å, $I_0(100):I_0(110):I_0(200) = 60:100:14$), as found for the Mn^{4+} sublattice in Co-Ni and Ni-asbolanes. Weaker reflections forming the network which is closer to the pattern center represent the Co sublattice: $a = 3.14$ Å. The patterns from bent edges contained the series of basal reflections ool with $d(001) = 9.45$ Å. Weak reflections 101 and 102 with d-spacings 2.36 Å and 2.18 Å respectively, corresponding to the Mn sublattice were obtained from crystals tilted about the incident beam. Low intensity of hko reflections for Co-sublattice, as compared to the corresponding reflections for Mn^{4+} sublattice implies island-like structure for Co layers. However the intensities of reflections 100 and 110 for the Co sublattice were of the same order of magnitude, as compared with Ni- and Co-Ni sublattices in asbolanes described above, where $I_{110} \gg I_{100}$.

This effect was shown to result from tetrahedral coordination of Co (Chukhrov et al. 1982). Figure 75 shows a scheme of the mixed-layer Co-asbolane structure. The Mn^{4+} layers are identical to those in asbolanes discussed above. The Co tetrahedral layers are similar to sulfide layers in valleriite (see Fig. 55). The anionic framework in these layers consists of oxygen atoms.

If the tetrahedral sites are occupied by Co^{3+}, electroneutrality of the layer requires two Co^{3+} cations per three tetrahedra. If regular alternation of Mn^{4+} octahedral and Co^{3+} tetrahedral layers is taken into account, the best agreement between the calculated and the experimental hko reflection intensities is obtained for the model represented by the notation $a\beta abB\Gamma A a\beta abB$ where, as in the case of valleriite, small Latin and Greek letters designate anionic and cationic positions in the close packing of Co-layers, and capital letters denote those in Mn layers. The Co-asbolane structural formula is

$$[Mn^{4+}_{1.0}O_{1.7}(OH)_{0.3}]0.607[Co^{3+}_{1.0} Ca_{0.2}O_2]n H_2O .$$

X-ray diffraction patterns from the asbolanes studied contain, besides the ool reflections, only the reflections hko corresponding to the Mn^{4+} sublattices. Reflections corresponding to the second sublattice, which represents the island-like component, are absent, apparently due to their low intensity. On the whole, an asbolane X-ray pattern is quite similar to those of buserite and poorly crystallized todorackite (see below). Therefore, identification of these minerals based on the analysis of X-ray diffraction patterns is not feasible. SAED is the only simple and reliable method available permitting to reveal asbolanes even when mixed with other minerals. Two, identically oriented, hexagonal reflections networks on a SAED pattern corresponding to the crystal plane (001), and a single series of basal reflections ool with $d(001) = 9.3 - 9.5$ Å indicate unambiguously that the sample under study is an asbolane.

The composition of crystallites obtained from X-ray energy dispersion analysis permits to classify the mineral into one of the asbolane varieties.

9.9 Analysis of Basal Reflection Intensities in SAED Studies of Mixed-Layer Minerals

Along with minerals having regular mixed-layer structures there are natural and synthetic varieties whose crystals contain different layer types alternating irregularly. Such formations are especially abundant among clay minerals (Drits and Sakharov 1976). Recently, random interstratified structures were found among Mn minerals from oceanic Fe-Mn nodules. In mixed-layer crystals, the alternating layers differ in structure and, specifically, in thickness. Therefore, diffraction patterns from irregular mixed-layer minerals may seem extremely unusual. Such a pattern contains an irrational series of basal reflections corresponding to d-spacings that differ from real interplanar distances in the crystals under study. To interpret diffraction effects from mixed-layer structures, special methods have been elaborated, based on the theory of X-ray scattering by quasi-crystalline substances (Brindley and Méring 1951; Kakinoki and Komura 1952; Drits and Sakharov 1976; Sakharov et al. 1982a, b). These methods allow determination of the type, the number and the contents of alternating components as well as of the pattern in their sequence. All these parameters of mixed-layer minerals obtained by X-ray analysis are averaged over all the particles in the sample, consequently, data on the integrated parameters provided by X-ray diffraction may fail to reflect the real structural features for each separate particle in the sample. Therefore, it is of interest to obtain information on the pattern in the sequence of layer types directly in separate particles. SAED provides such a possibility by permitting to obtain basal reflections, using either ultra-thin sections or ion-thinning, or bent edges (Gorshkov 1970). Applications of this method revealed that many particles yielded SAED patterns containing rational series of basal reflections $00l$. This was the case not only with ordered mixed-layer minerals but also with those having layer types alternating at random, as was indicated by X-ray diffraction.

Gorshkov et al. (1975) reported new experimental data on mixed-layer minerals and interpreted the effect observed, making use of conventional methods for structure analysis. Three mixed-layer samples were chosen consisting of layers having substantially different structure factors.

The SAED results for each of the samples are given below.

Corrensite. Rational series of basal reflections present in X-ray patterns obtained from the natural sample, and from those saturated by glycerol and heated to 550 °C (Drits and Sakharov 1976) imply a completely ordered alternation of chlorite and smectite packets. SAED patterns from bent edges contained distinct rational basal reflection series corresponding to a repeat distance of 24 Å along the normal to the layers. The intensity distribution as well as arrangement of basal reflections was found identical for all the particles. In this case, only one ordered sequence of layer types is possible.

The calculated structure factors $\Phi^2(ool)$ were in reasonable agreement with the intensitites $I(ool)$ evaluated visually. The results obtained imply identical structures for all individual particles.

Interstratified Kaolinite-Smectite. X-ray diffraction indicated that the sample is a random interstratification of 75% of kaolinite layers and 25% of smectite layers (Drits and Sakharov 1976).

A set of SAED photographs containing basal reflections was obtained from bent edges of isolated particles. Some of them showed rational or nearly rational series of basal reflections indicating superperiodicity. The repeat distances were different for different particles. The most frequent repeated distances were 48, 54, and 70 Å. This implies, on the one hand, ordered alternation of kaolinite (type A) and smectite (type B) layers within separate particles and, on the other hand, varied relative quantities of layers A and B in different particles.

The best agreement between $\varphi^2_{calc}(ool)$ and $I_{exp}(ool)$ was obtaiend for the following sequences: $ABAAAB$ (48 Å), $AABAAB$ (55 Å) and $ABAAAAAAB$ (70 Å). Thus, layers A and B tend to a relatively uniform distribution within each crystal. In any event, two layers B never occur together and are separated by at least one layer A. SAED patterns from most particles contained no rational basal reflection series.

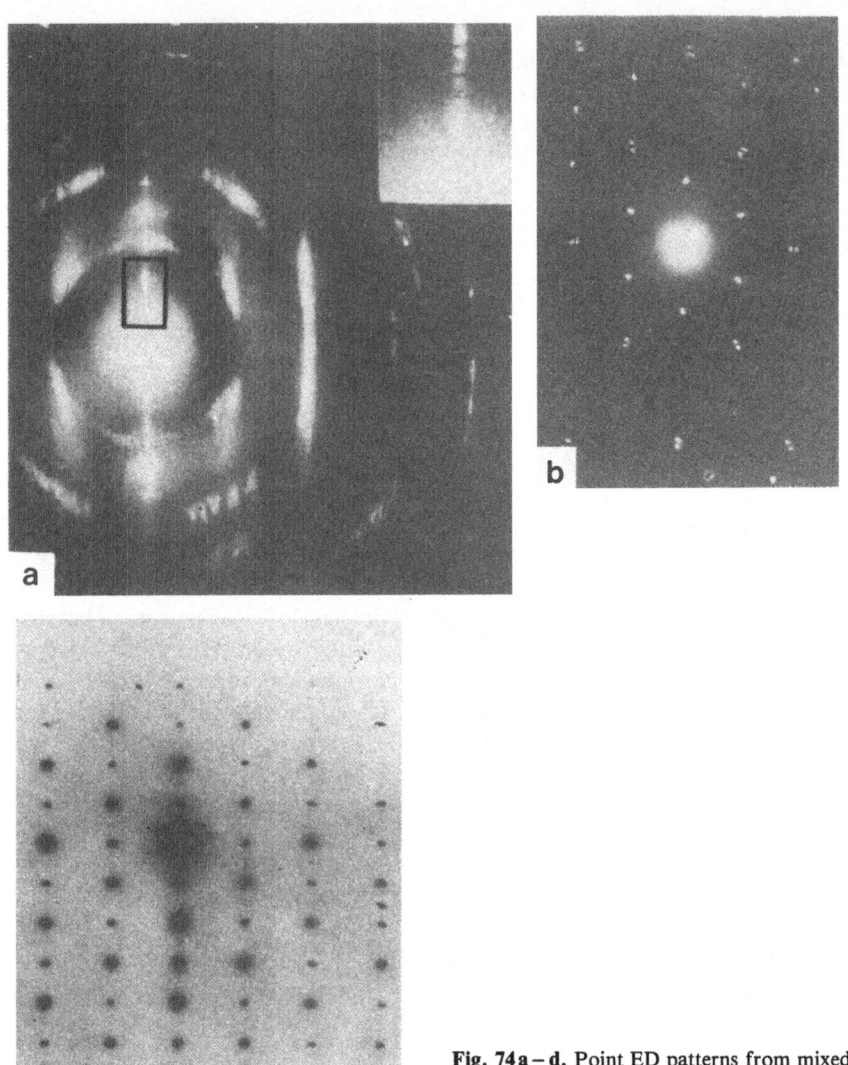

Fig. 74a – d. Point ED patterns from mixed-layer chlorite-swelling chlorite containing reflections **a** ool and **b** hk; **c** okl reflections from F cupfferite

The results obtained showed that the particles of the mineral under study were heterogeneous with respect to the relative quantities of layer types as well as to the pattern in their sequence.

Interstratified Chlorite-Swelling Chlorite. X-ray data indicate that the mixed-layer sample under study consists of 14 Å chlorite packets some of which swell with glycerol to 17.7 Å (Novikov et al. 1973). The portions of 14 Å(A) and 17.7 Å(B) packets are 0.55 and 0.45, respectively. An X-ray pattern from a disoriented sample contained two reflections corresponding to $d(060)$ values of

1.502 and 1.535 Å. Therefore, the sample contained both dioctahedral and trioctahedral layers.

This was confirmed by SAED. Point patterns from a large number of particles (Fig. 74b) contained two hk reflection sets. One of these corresponded to dioctahedral layers ($a = 5.14$ Å) and the other to trioctahedral layers ($a = 5.32$ Å) (Gorshkov et al. 1975).

Along with basal reflections with $d = 14.4$ Å$/n$ (where n is an integer) SAED patterns from bent edges contained weak additional reflections-satellites located near some strong basal reflections (see Fig. 74a). Hence the mineral structure indeed consists of alternating di- and trioctahedral layers differing in structure factors, i.e., it is a mixed-layer structure and not an intergrowth of two mineral species.

Analysis of SAED patterns containing rational or nearly rational ool reflections series showed that the $d(001)$ values are often equal to 99 and 70 Å. The best agreement between Φ^2_{exp} and Φ^2_{calc} was obtained for the sequences $AABBAAB$ (99 Å) and $AAABB$ (70 Å). Thus, in contrast to interstratified kaolinite-smectite, there is a clear tendency for segregation. Note that many of the particles in the sample were found to have random interstratified structures with no superperiodicity.

Among the known silicate mixed-layer minerals, the mineral studied is unique, since it has no strict parameter along c and, at the same time, layers of different types differ in unit-cell parameters in the plane ab.

Interstratified Asbolane-Buserite (Chukhrov et al. 1983b). X-ray diffraction patterns from samples of oceanic Fe-Mn nodules often contain a strong maximum corresponding to a d-spacing of $9.6-10$ Å. Most of the workers attributed this reflection to todorackite. X-ray diffraction patterns containing relatively strong reflections with d-spacings of approximately 9.8, 4.9, and 2.44 Å may also correspond to asbolane, buserite I or buserite II (Drits et al. 1985). The buserite I structure consists of Mn^{4+} octahedral layers (Chukhrov et al. 1984). The presence of vacant octahedra results in a negative layer charge which is compensated by interlayer cations Mn^{3+}, Ca, Na, etc. Interlayer cations have octahedral coordination and are situated exactly below and above the vacancies in Mn^{4+} layers. The parameter c in natural buserite I is 9.8 Å. After dehydration in vacuum or by heating, buserite I is transformed into birnessite having $c = 7$ Å (Fig. 73).

Recent research has shown that the 10 Å manganese phase abundant in oceanic Fe-Mn nodules may be associated with random interstratified asbolane-buserite (Chukhrov et al. 1983b), consisting of asbolane and buserite packets alternating in varied quantities. The mixed-layer asbolane-buserite structure is formed by a set of continuous Mn^{4+} octahedral layers. The interlayers are filled either by island-like layers consisting of octahedra having OH groups at the apices and cations Mn^{2+}, Mn^{3+}, Co^{3+}, Ni etc. in the centers (asbolane interlayers), or by interlayer cations Mn^{2+} and Ca coordinated by water molecules and anions belonging to Mn^{4+} layers (buserite interlayers). The scheme of the asbolane-buserite structure is shown in Fig. 76. Since natural asbolane and buserite packets are of similar thickness, X-ray diffraction patterns from asbolane-buserite contain rational series of basal reflections ($d(001) = 9.6-10.0$ Å).

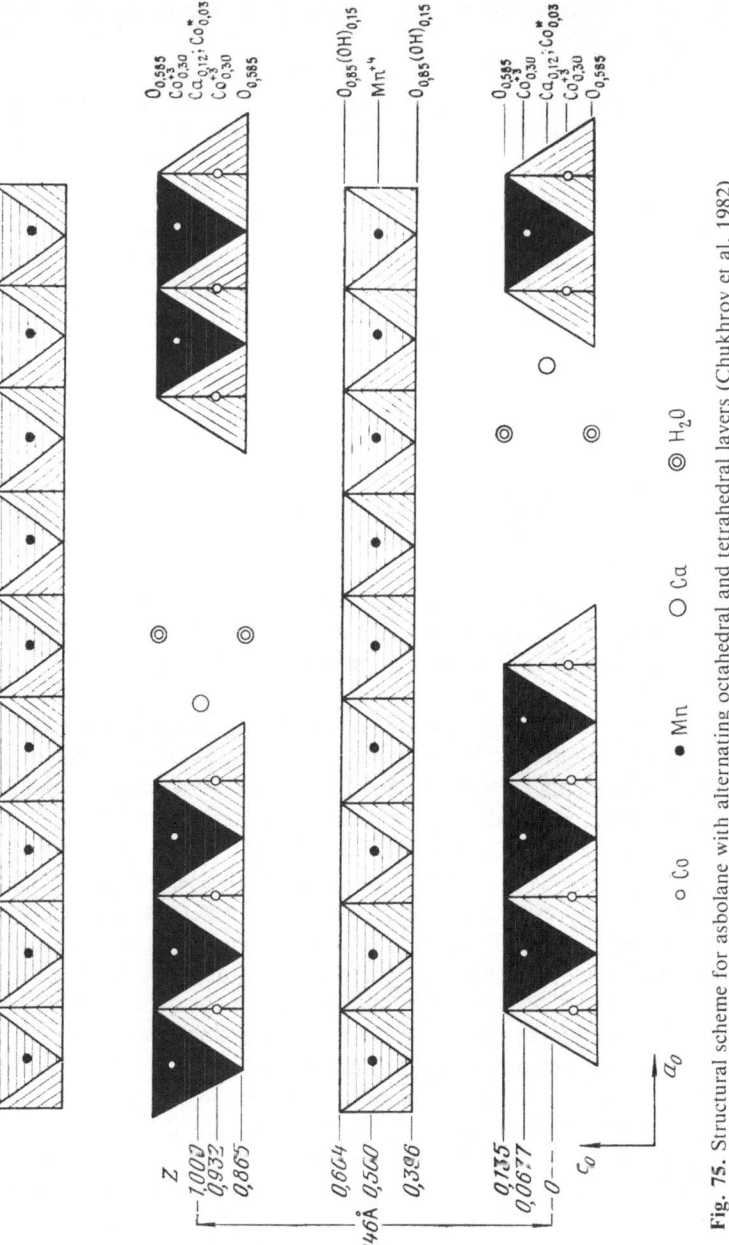

Fig. 75. Structural scheme for asbolane with alternating octahedral and tetrahedral layers (Chukhrov et al. 1982)

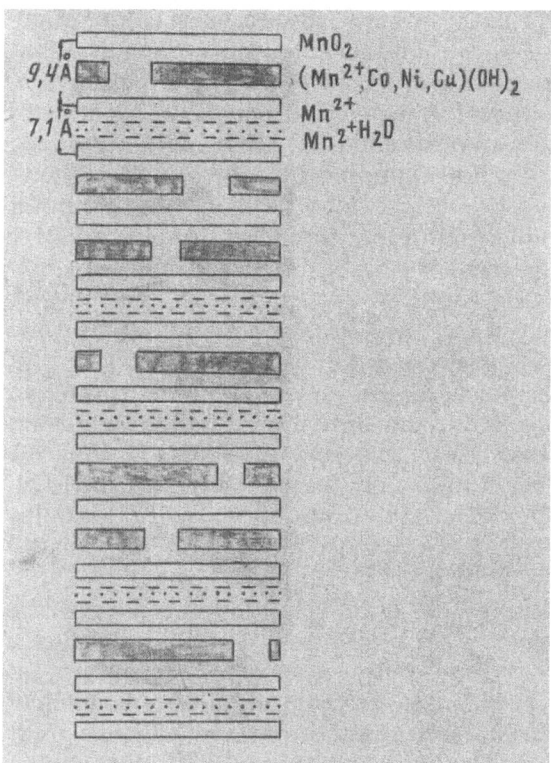

MnO_2
$(Mn^{2+}Co,Ni,Cu)(OH)_2$
Mn^{2+}
$Mn^{2+}H_2O$

9,4 Å

7,1 Å

Fig. 76. Structure of mixed-layer asbolane-birnessite (Chukhrov et al. 1983b)

However, in vacuum buserite packets are transformed into 7 Å birnessite packets, whereas the asbolane thickness remains the same. The resultant alternation of 7.1 Å birnessite packets and 9.4 Å asbolane ones leads to a shift of the first small-angle reflection in SAED pattern into the region of greater ϑ angles. As a result, the basal reflection series becomes irrational, which was distinctly observed in SAED patterns from bent edges. To confirm the structure model proposed, the experimental diffraction characteristics were compared with those calculated. Intensity distributions for mixed-layer structure models differing in the portions of asbolane and birnessite layers W_A and W_B were calculated according to the method of Drits and Sakharov (1976). The calculated intensities I and "apparent" d-spacings determined for basal reflections are in reasonable agreement with the values measured from point SAED patterns. The portions of the alternating components were found to range from 0.6 to 0.7 (asbolane layers) and from 0.4 to 0.3 (birnessite layers).

9.10 Structure Studies of Mixed-Layer Minerals by HREM

The use of HREM in conjunction with SAED is proving effective in structure studies of layer minerals. It allows direct examination of homogeneity/heterogeneity in minerals, the nature of layer types and the pattern in their sequence, and mechanisms for structure transformations.

Initially, clay minerals including mixed-layer phyllosilicates were used as objects for HREM.

Suito et al. (1969) analyzed fringe images from montmorillonite particles containing organic molecules in interlayers. Analysis of the micrographs indicated that layers of thicknesses of 19.26 Å and 40 Å alternate, regularly and at random, in the organo-montomorillonite particles. The authors suppose that the observed local variations in interlayer separations reflect variations in the concentration of absorbed organic molecules and, therefore, the heterogeneity in the charge density distribution in different interlayers.

Yoshida (1973, 1976) used HREM to study interstratified mica-montmorillonite. Analysis of the arrangement of fringes of different widths suggested polar charge distribution at the opposite basal surfaces of a 2:1 layer.

Brown et al. (Brown and Jackson 1973; Brown and Rich 1968; Lee et at. 1975a, b) studied sections 500 – 700 Å thick obtained from oriented preparations with the help of an ultramicrotome. This technique was used to study muscovite (Brown and Jackson 1973), chlorite (Brown and Rich 1968), kaolinite (Lee et al. 1975a), as well as hydrobiotite, glauconite, and interstratified kaolinite-montmorillonite (Lee et al. 1975b). The authors proved that clay mineral particles are highly heterogeneous in respect of structure and composition. They compared the structure of a particle containing 1000 layers with a loosely packed pile of books, each consisting of just 5 – 20 pages. Such a "book" is, in essence, a coherent scattering domain, i.e., it diffracts independently of the "neighbors". It was also found that particles of some minerals can contain variable quantities of layers and interlayers alien to the particular structure involved, although X-ray diffraction indicates monominerality and high-three-dimensional ordering. For example, both continuous and "terminated" micaceous layers were found in kaolinite particles (Lee et al. 1975a).

Interstratified kaolinite-montmorillonite was shown by HREM to contain not only kaolinite and montmorillonite layers but also mica-like layers. Extremely high diversity in the portions of layer types and patterns in their sequence was observed for different particles and for different parts of one particle. Kaolinite segregation areas, areas consisting of regularly alternating kaolinite and montmorillonite layers, as well as three-component systems were found.

In some cases, a mixed layer clay mineral can be regarded as an intermediate stage in a series of solid-phase transformations of one phyllosilicate into another. The possibility for such transformations is largely governed by the layer structure of phyllosilicates, since the interlayer is the "starting point" for the transformation of the mineral. Being most sensitive to changes in the environment, interlayers are transformed rather easily from one structural type to another without any substantial rearrangement of the structure as a whole. This is the reason for the wide abundance of natural mixed-layer minerals containing 2:1 layers and interlayers of different types. HREM proved to be extremely effective for visualization of the mechanisms for solid-phase transformations of phyllosilicates involving intermediate mixed-layer phases. As a result, many of the views on this problem were revised. One of the most important achievements in HREM is the possibility for separate observation of tetrahedral and octahedral sheets and interlayers.

The study of chloritization of biotite is an example illustrating the advantages of HREM in the investigation of mechanisms for solid-phase reactions. It was found that the transformation of biotite into chlorite may proceed according to one of the two fundamentally different mechanisms depending on the specific conditions. Olives and Amouric (1984; Olives et al. 1983) have shown that during metamorphism biotite is partially transformed into chlorite owing to replacement of K-interlayer planes by brucite layers. Crystal fragments have been observed with regular as well as with random alternation of mica-like and chlorite-like layers.

Veblen and Ferry (1983) studied the transformation of biotite into chlorite in granitic rocks. They found another mechanism for chloritizations consisting in the replacement of 2 : 1 mica layers by brucite sheets. The process is accompanied by the formation of amorphous SiO_2 micro-precipitates.

In all the examples discussed, the object structures were known from X-ray data. The development of HREM has allowed a number of direct structure determinations for layer compounds. It was found that interstratification is characteristic not only of phyllosilicates, hybrid minerals and some "mixed" oxides containing the brucite component, but is far more widespread. For example, Horiuchi et al. (1977) studied the mixed-layer structure of $Bi_7Ti_4NbO_{21}$. X-ray powder diffraction indicated that the structure may consist of layer fragments $Bi_4Ti_3O_{12}$ and Bi_3TiNbO_9 alternating regularly. It was also known that the $Bi_4Ti_3O_{12}$ structure contains layers Bi_2O_2 alternating with perovskite-like layers $Bi_2Ti_3O_{10}$. In the latter structure, Bi atoms are in suboctahedral coordination, whereas smaller Ti cations are octahedrally coordinated. It could be supposed by analogy, that Bi_3TiNbO_9 has also a layer structure consisting of alternating layers Bi_2O_2 and $BiTiNbO_7$. However, a detailed X-ray structure analysis was not feasible because of fine dispersion of the sample and a large parameter along c.

A 1 MV electron microscope supplied with a goniometer, with instrumental resolution of 2 Å was used for SAED and HREM studies of the $Bi_7Ti_4NbO_{21}$ structure. Point SAED patterns for the incident beam parallel to the crystallographic directions [110], [100] and [010] were obtained in order to determine the unit-cell parameters ($a = 5.45$ Å, $b = 5.42$ Å, and $c = 58.1$ Å). The possible space groups, as indicated by the systematic absences, were I2cm or Imcm. Then HREM images were obtained for each of the above orientations of the crystal. About 80 diffracted beams contributed to the formation of each image. Analysis of these images, with account taken of the general crystal-chemical considerations, allowed unambiguous determination of the structure. Figure 36,3 shows an image, the interpretation for the contrast observed and the idealized structure model projected down [110]. The image contains fringes of two types differing in thickness separated by darker zigzag fringes corresponding to layers Bi_2O_2. The wider fringes are apparently associated with $Bi_2Ti_3O_{10}$ two-layer perovskite-like packets, while the narrower fringes may correspond to thinner, one-layer $Bi(TiNb)O_7$ perovskite-like packets. The image contrast obtained with other orientations of the crystal was interpreted similarly. It was found that successive layers of the same type are rotated through 180° with respect to one another. The stacking sequence in the structure involved can be thus written as $ABA'B'AB$, where A and B signify the two alternating components $Bi_4Ti_3O_{12}$ and Bi_3TiNbO_9,

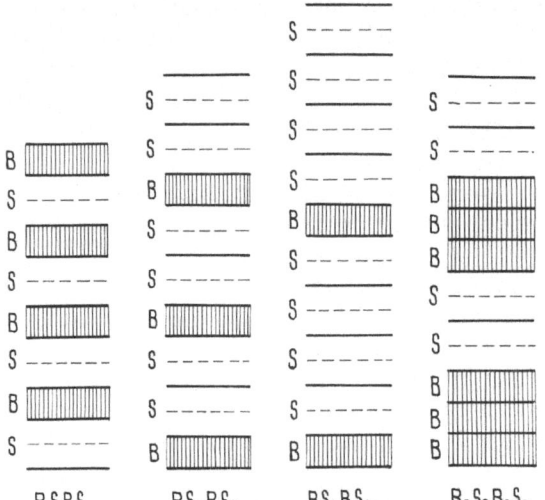

Fig. 77. Structures of different mixed-layer minerals of the type B_nS_m (van Landuyt and Amelinckx 1975)

respectively, and the prime sign indicates the opposite orientation for octahedra in the corresponding layers.

A group of mixed-layer minerals having extremely complex structures and large parameters c belonging to the bastnaesite-synchisite series was found by van Landuyt and Amelinckx (1975). These minerals include bastnaesite $LnFCO_3$, parisite $2LnFCO_3 \cdot CaCO_3$, roentgenine $3LnFCO_3 \cdot 2CaCO_3$ and synchisite $LnFCO_3 \cdot CaCO_3$. One might suppose, juding by the composition, that the parisite and roentgenine structures consist of combinations of bastnaesite and synchisite layers. If the repeated layer sequences in synchisite and bastnaesite are denoted by S and B, respectively, the composition of repeated structural units in mixed-layer minerals will be expressed by B_nS_m.

Analysis of reflections ool revealed that the sample was represented by a set of mixed-layer structures. Along with c values of 23 Å and 28 Å, characteristic of parisite and roentgenine, d-spacings of 32.8, 41.9 and 51.0 Å were found, that were apparently associated with new mixed-layer minerals. Analysis of fringe images from parisite and roentgenine yielded readily the patterns in the sequence of layer types for the new mixed-layer structures (Fig. 77): BS_4BS_4 ($d(001) = 41.92$ Å), $BSBSBS_2BS...$ ($d(001) = 51.03$ Å) and $...B_2SBSB_2S...$ ($d(001) = 32.74$ Å). Many of the crystals are heterogeneous in composition due to combination of domains with different stacking sequences.

It was shown by van Landuyt and Amelinckx (1975) that faults resulting from variations either in the layer sequence (e.g., $BS_4BS_3BS_5BS_4BS_4$) or in composition ($BS_4BS_2BS_4BS_4...$) are frequent in the crystals studied.

CHAPTER 10

SAED and HREM Study of Order/Disorder and Structural Heterogeneity in Layer Minerals

Many minerals exhibit unusual diffraction properties. In addition to strong, sharply defined and widely spaced reflections, corresponding to the substructure, their diffraction patterns contain additional weaker and more closely spaced satellite spots corresponding to a superstructure or a modulated structure. If the spacings of these satellite spots are integral multiples of the substructure spacings, then we are dealing with a superstructure or with a commensurate modulated structure (Cowley et al. 1979). If the substructure spots are not at rational multiples of the satellite spot spacings, the structure is a modulated incommensurate one (Cowley et al. 1979). Satellite spots can result from various structural factors. They are commonly associated with structure modulations resulting from cation ordering and/or positional displacements of atoms. For example, an ordered distribution of isomorphous cations can lead to a superstructure and therefore to satellites in the corresponding diffraction patterns. Modulations may result from the presence of domains separated by out-of-phase boundaries. In this case the periodicity of modulations is determined by the average separation of the boundaries. A survey of modulated mineral structures is given in (Buseck and Cowley 1983).

Experience shows that SAED and HREM are more effective for the study of modulated structures than X-ray single-crystal analyses. Generally, SAED readily reveals satellite spots that are not always observed in X-ray patterns. HREM allows direct determination of the nature of the modulations and reveals the faults perturbing the periodicity.

Some examples of cation order/disorder in phyllosilicates exhibited in diffraction patterns are discussed in the first part of the chapter. Magnesium mica containing Na in interlayers (Drits et al. 1978; Korytkova and Drits 1977) has a completely ordered distribution of isomorphous cations. Its structural features are unusual and require detailed discussion. Vermiculates and micas containing heavy interlayer cations may serve as examples of domain structures (Besson et al. 1974a, b). In the second part of the chapter several examples of structural modulations are discussed, resulting from the lateral misfit of tetrahedral and octahedral sheets in phyllosilicates.

10.1 A New Mica $NaMg_3(Si_{3.5}Mg_{0.5})O_{10}(OH)_2$ Having a Talc-Like Stacking Sequence

Mica of unusual structure and composition was synthesized by treating pure talc by 5% NaOH solution under 450 °C and 500 atm (Korytkova and Drits 1977). The interlayer positions in the structure thus obtained are occupied by Na, while the layer charge results from the presence of Mg in tetrahedra.

An X-ray powder pattern from a disoriented preparation contained only basal reflections ool and two-dimensional diffraction bands typical of disordered phyllosilicates. Thus, on the basis of X-ray data, the crystal lattice of the mica involved could be described only by the layer unit-cell parameters $a = 5.322$ Å and $b = 9.234$ Å, and the minimum repeat distance along the normal to the layers $d(001) = 9.73$ Å.

The structure formula for Na-mica is

$$Na_{0.84}Mg_{3.00}(Si_{3.41}Mg_{0.50}Fe_{0.09})O_{9.75}(OH)_{2.25} .$$

If the distribution of isomorphous Si and Mg within a tetrahedral sheet were random, then there would be finite probabilities determined by the Mg contents, for the occurrence of combinations where each given Mg-tetrahedron could be adjacent to one, two or even three Mg-tetrahedra. If this were the case, there would be a considerable positive charge deficiency at the bridging oxygen atoms, since they would be surrounded only by Mg and Na.

Thus, it is highly probable that tetrahedral Si and Mg are distributed so that each Mg-tetrahedron is surrounded by Si-tetrahedra. SAED confirms the ordered distribution of Si and Mg. Point patterns obtained from Na-mica crystallites always contain additional reflections (Fig. 78a), indicating a C-centered supercell with $a_s = b = 9.234$ Å and $b_s = 3a = 16.00$ Å, where a and b are the conventional unit-cell parameters similar for all phyllosilicates.

Figure 79 shows the model for the tetrahedral sheet. Mg-tetrahedra are centered at the lattice points. The remaining tetrahedral sites in the unit cell are occupied by Si. Different dimensions of Si- and Mg-tetrahedra lead to considerable variations in tetrahedral rotation. This is displayed in appreciable intensity of superreflections, in spite of similar scattering powers for Si and Mg. The Mg distribution shown in Fig. 79 is the optimum one since there is one charge-deficient bridging anion $(Mg - O_{br} - Si)$ per ditrigonal ring. This favors a more local charge compensation by univalent Na cations. Korytkova and Drits (1977) have shown that each Na is shifted from the interlayer center into a hexagonal cavity inside the tetrahedral sheet.

In the light of these data, the results of an OTED study of the Na-mica, which has revealed an unusual stacking sequence, do not seem unexpected. The intensity distribution in the OTED pattern from this mica is far more similar to that for talc than for, e.g., phlogopite. Measured structure factors have been compared with those calculated for two models. The first model assumes a talc-like stacking sequence (Drits and Aleksandrova 1975). The Na cations are arranged in such a way that each Na has at least four nearest anions with Na-O = 2.4 – 2.5 Å (Fig. 79). There are two equally probable interlayer positions for each Na, one inside the hexagonal cavity in the upper layer and the other inside the nearest ring in the

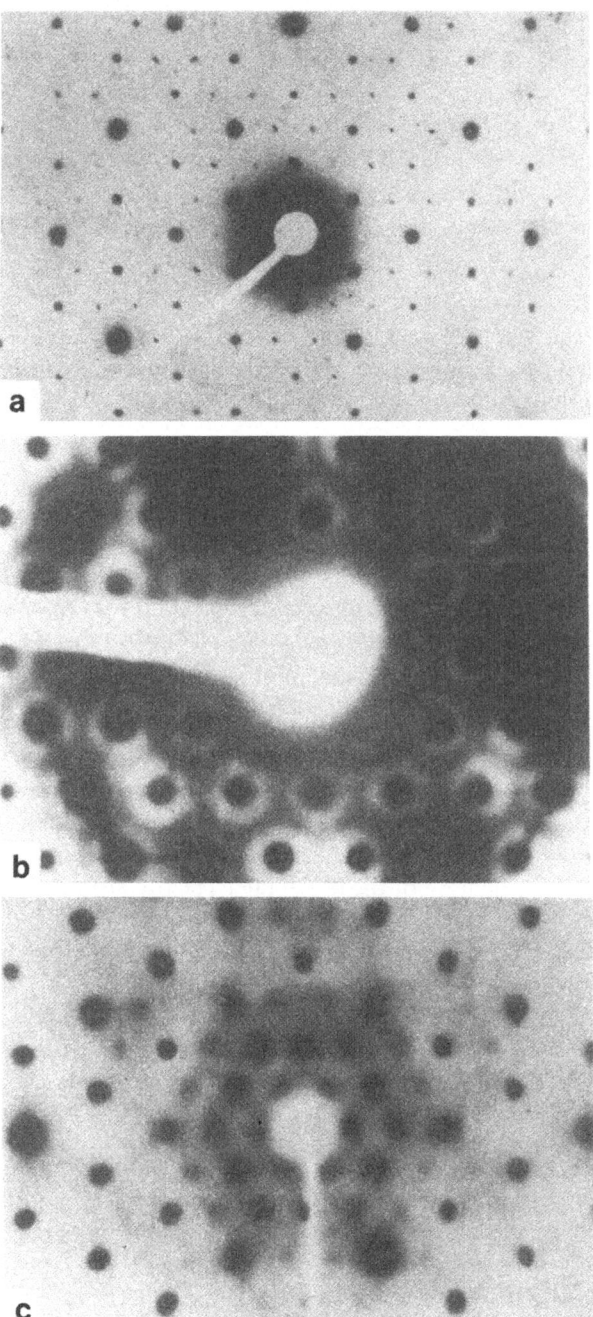

Fig. 78a – c. ED patterns from **a** Na-mica with superstructure reflections; **b** Ba-illite, with strong diffuse "honeycomb" scattering (Besson et al. 1974a); **c** vermiculite, with superstructure spots (Alcover et al. 1973)

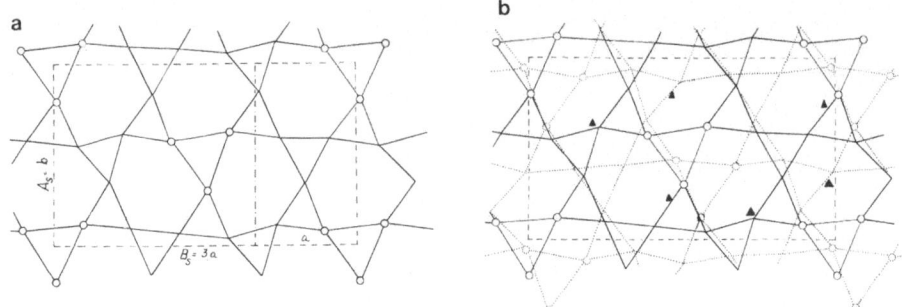

Fig. 79a–b. Structural model for Na-mica: **a** Tetrahedral sheet with ordered distribution of Si and Mg; the supercell is shown by *dashed lines*; the average structure unit cell is shown by *dash-dotted lines*; **b** The talc-like stacking of tetrahedral sheets in adjacent 2:1 layers; *black triangles* mark the possible positions for Na

adjacent lower layer (Figs. 79 and 80). The second model implies a mica-like stacking sequence.

Considerable discrepancy between Φ^2_{exp} and Φ^2_{calc} has been found for the second model. Thus, large lateral dimensions of tetrahedral and octahedral sheets in the 2:1 layers hamper such tetrahedral rotation that could ensure normal Na-O bond lengths in the case of a mica-like stacking sequence.

As the majority of natural talcs, the Na-mica structure studied is semi-random owing to random shifts of adjacent 2:1 layers by distances $\pm b/3$ along b. The possibility for these shifts results from the interlayer cations being "depressed" into the hexagonal tetrahedral cavities.

To conclude, it should be noted that the synthesis of the Na-mica may be associated with a solid-phase transformation of the initial talc. The transformation process must have consisted in introduction of Na cations into interlayers and partial Mg for Si substitution, with the orientation of the 2:1 layers and their stacking sequence being preserved.

10.2 Diffraction Effects from Layer Structures Having Partially Ordered Cation Distribution

In the presence of isomorphous substitutions, the results of structure analysis are adequate to the real mineral structure only for two limiting cases: (1) The distribution of isomorphous cations over the three-dimensional volume of the structure is completely ordered. Diffraction patterns contain, along with the reflections corresponding to the average structure, the superstructure reflections. The arrangement of these depends on the periodicity in the cation distribution. The real structure of the mineral can be reconstructed using the intensities of principal and superstructure reflections. (2) The isomorphous cations are distributed at random and, besides Bragg reflections, continuous diffuse scattering is observed in the diffraction space. Nevertheless, the intensities of Bragg reflections can be

Fig. 80. The supposed distribution of Na cations (*circles*) in the interlayers of the Na-mica

used to determine the average structure. The three-dimensional random cation distribution can be then obtained with the help of simulation methods.

Structures with a partially ordered cation distribution are an important intermediate case. The existence of some correlation in the atomic distribution leads to the so-called anomalous scattering localized within certain domains in the reciprocal space and observed in diffraction patterns between the Bragg reflections.

There are various approaches to the interpretation for diffraction patterns from crystals that are homogeneous and periodic only on average (Cowley 1975b; Wilson 1949). The Patterson method has proved convenient for the solution of many problems. Suppose that the real crystal structures can be described, to a certain approximation, by a mean crystal lattice. In other words, the real structure of the object is assumed to be rather close to that of an ideal crystal. Then the real electrostatic potential distribution in the volume of the object $\varphi(r)$ can be written

$$\varphi(r) = \overline{\varphi(r)} + \Delta\varphi(r), \tag{10.1}$$

where $\overline{\varphi(r)}$ is the average electrostatic potential distribution that can be described by a periodic lattice, and $\Delta\varphi(r)$ is the deviation from the average structure.

The Patterson function for (10.1) is given by

$$\mathscr{P}(r) = \int \varphi\overline{(r')}\,\overline{\varphi(r-r')}dV_{r'} + \int \Delta\varphi(r')\Delta\varphi(r-r')dV_{r'}$$
$$= \overline{\varphi(r)} * \overline{\varphi(-r)} + \Delta\varphi(r) * \Delta\varphi(-r). \tag{10.2}$$

Since the Fourier transform of the Patterson function is the function of the intensity distribution in the diffraction space H,

$$|\Phi(H)|^2 = |\overline{\Phi(H)}|^2 + |\Delta\Phi(H)|^2. \tag{10.3}$$

The expression (10.3) means that the contribution to the intensity for electrons scattered by a real crystal consists of two coherently independent parts. The first is the contribution from the average structure. The average structure can be described by a periodic lattice with the same scattering power for all the lattice

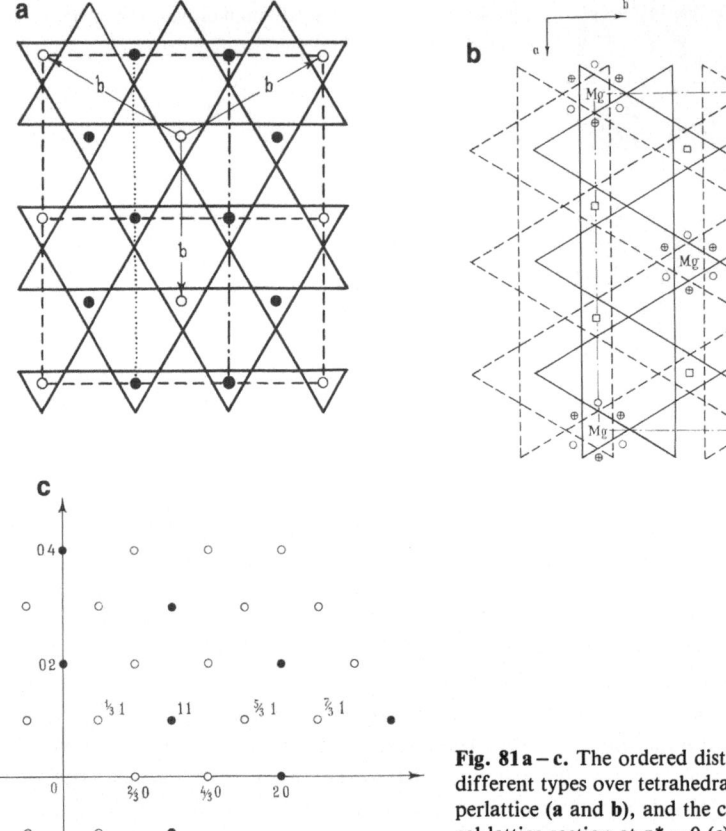

Fig. 81a – c. The ordered distribution of cations of different types over tetrahedral sites leading to a superlattice (**a** and **b**), and the corresponding reciprocal lattice section at $z^* = 0$ (**c**) (Suquet 1978)

points. Therefore, the Fourier transform of the lattice component of the real structure is also a lattice function representing the reciprocal lattice. The corresponding lattice points responsible for the Bragg diffraction are represented by a set of point reflections in the diffraction pattern. The second member in (10.3) is independent of the scattering by the average structure and is determined by the function $\Delta \varphi(r)$. If $\Delta \varphi(r)$ is aperiodic, $|\Delta \Phi(H)|^2$ will be also an aperiodic function describing the diffuse scattering.

To apply (10.2) to the study of diffuse scattering from layer minerals, consider the distribution of Si and Mg over tetrahedra in the 2 : 1 layers in Na-mica discussed above (Fig. 81a). Cations A and B are distributed over the tetrahedral sites so that a C-centered superlattice is formed with $a_s = 3 a_0$ and $b_s = b_0$, where a_0 and b_0 are the unit-cell parameters in the average structure. The superlattice points are occupied by cations A. Assume that the ordering in the cation distribution is of local nature, so that domains separated by anti-phase boundaries exist even within separate layers.

Figure 82a shows three domains having the same pattern in the distribution of cations A and B with $a_s = 3 a_0$ and $b_s = b_0$. These domains differ in orientation and are separated by out-of phase boundaries. If the mean domain dimensions

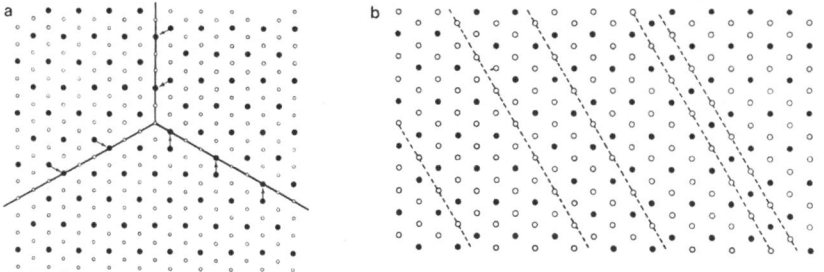

Fig. 82a, b. Examples of domain structures: **a** domains having identical cation distribution separated by anti-phase boundaries; **b** domains differing in the pattern of ordered distribution of cations (including vacancies)

are small and domains of different orientations occur with equal probability, then, after averaging the cation distribution within the layer, the average structure will be described by a conventional unit cell (a_0, b_0). The result will be the same if we suppose that cations are regularly distributed within separate layers, sheets or interlayers without any correlation in cation distribution for neighboring layers, sheets or interlayers. In other words, two-dimensional ordered domains are coherently independent. Crystal structures of this kind are said to have short-range order, as distinct from structures having long-range order.

It is obvious that, in the example discussed, $\Delta\varphi(r)$ will be periodic within a coherent scattering domain, and so will be its self-convolution. Therefore, the diffuse $|\Delta\Phi|^2$ scattering intensity will be a modulated function.

We are interested in scattering from a two-dimensional object, i.e., a planar domain in the tetrahedral sheet. Hence $\varphi(r)$ in (10.1) is the two-dimensional electrostatic potential distribution in a domain with an ordered cation distribution, and $\overline{\varphi(r)}$ is the potential distribution for the same domain for a completely disordered cation distribution.

Assume that A and B have similar sizes but differ in scattering powers, e.g., $f_A > f_B$. Then the superstructure will differ from the average structure only in the distribution of scattering powers, while the atomic patterns within the domain are the same. Therefore, from Fig. 81a

$$\varphi(r) = \varphi_A(r) * \delta(r) + \varphi_A(r) * \delta\left(r - \frac{a_s + b_s}{2}\right)$$

$$+ \varphi_B(r) * \delta\left(r - \frac{a_s}{6} - \frac{b_s}{2}\right) + \varphi_B(r) * \delta\left(r + \frac{a_s}{6} + \frac{b_s}{2}\right)$$

$$+ \varphi_B(r) * \delta\left(r - \frac{a_s}{3}\right) + \varphi_B(r) * \delta\left(r + \frac{a_s}{3}\right), \tag{10.4}$$

$$\overline{\varphi(r)} = \frac{1}{3}[\varphi_A(r) + 2\varphi_B(r)] * \left\{ \delta(r) + \delta\left(r - \frac{a_s + b_s}{2}\right) + \delta\left(r - \frac{a_s}{6} - \frac{b_s}{2}\right)\right.$$

$$\left. + \delta\left(r + \frac{a_s}{6} + \frac{b_s}{2}\right) + \delta\left(r - \frac{a_s}{3}\right) + \delta\left(r + \frac{a_s}{3}\right)\right\}, \tag{10.5}$$

$$\Delta\varphi(r) = \frac{1}{3}[\varphi_A(r) - \varphi_B(r)] * \left\{ 2\delta(r) + 2\delta\left(r - \frac{a_s + b_s}{2}\right) - \delta\left(r - \frac{a_s}{6} - \frac{b_s}{2}\right)\right.$$

$$\left. - \delta\left(r + \frac{a_s}{6} + \frac{b_s}{2}\right) - \delta\left(r - \frac{a_s}{3}\right) - \delta\left(r + \frac{a_s}{3}\right)\right\}. \tag{10.6}$$

The Fourier transform for (10.6) is

$$\Delta\Phi(h_s k_s) = \frac{1}{3}(f_A - f_B)\left\{ 2 + 2\exp 2\pi i \frac{h_s + k_s}{2} - \exp 2\pi i\left(\frac{h_s}{6} + \frac{k_s}{2}\right)\right.$$

$$\left. - \exp\left[-2\pi i\left(\frac{h_s}{6} + \frac{k_s}{2}\right)\right] - \exp 2\pi i\frac{h_s}{3}\left(-\exp - 2\pi i\frac{h_s}{3}\right)\right\}.$$

Multiplying $\Delta\Phi(h_s k_s)$ by the complex conjugate we have

$$|\Delta\Phi(h_s k_s)|^2 = \frac{4}{9}(f_A - f_B)^2 \left\{ 2[1 + \cos\pi i(h_s + k_s)]\right.$$

$$\times\left[1 - \cos 2\pi\left(\frac{h_s}{6} + \frac{k_s}{2}\right) - \cos 2\pi\frac{h_s}{3}\right]$$

$$\left. + \left[\cos 2\pi\left(\frac{h_s}{6} + \frac{k_s}{2}\right) + \cos 2\pi\frac{h_s}{3}\right]^2\right\}. \tag{10.7}$$

Therefore

$$|\Delta\Phi(h_s k_s)|^2 = 0 \quad \text{for} \quad h_s + k_s = 2n + 1,$$

$$|\Delta\Phi(h_s k_s)|^2 = \frac{16}{9}(f_A - f_B)^2\left(1 - \cos 2\pi\frac{h}{3}\right)^2 \quad \text{for} \quad h_s + k_s = 2n,$$

$|\Delta\Phi(h_s k_s)|^2 = \frac{4}{9}(f_A - f_B)^2$, if $h_s = 3h' \pm 1$ and $h_s + k_s = 2n$ simultaneously, i.e., for $3h' + k = 2n \pm 1$, but $|\Delta\Phi(h_s k_s)| = 0$ for $h_s = 3h$ when $3h + k_s = 2n$.

If the layer stacking sequence is a regular one, the average structure is described by a periodic reciprocal lattice having point nodes. The diffused scattering intensity is localized in the reciprocal space in the form of a two-dimensional set of rods parallel to c^* and passing through superlattice points for which $h_s + k_s = 2n$ and $h_s = 3n \pm 1$ simultaneously. If we proceed to the average structure, taking into account that $h_s = 3h$ and $k_s = k$, the contributions of diffused scattering to the Bragg reflection intensity will be zero, since, for Bragg reflections, $h + k = 2n$ or $h_s/3 + k = 2n$, which is possible only for $h_s = 3n$ when $|\Delta\Phi| = 0$. The distribution of the principal and the superstructure nodes in the

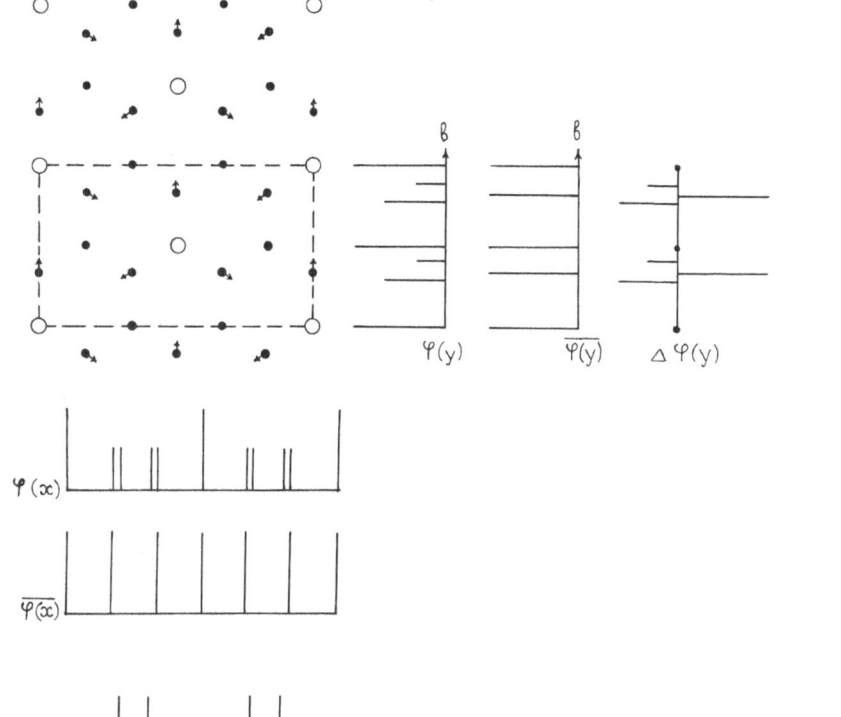

$\varphi(y)$ $\overline{\varphi(y)}$ $\triangle \varphi(y)$

$\varphi(x)$

$\overline{\varphi(x)}$

$\triangle \varphi(x)$

Fig. 83. Distortions of the tetrahedral sheet with an ordered distribution of cations A and B $(r_A > r_B)$. Below and on the right: projections of the distorted "real" $\bar{\varphi}$ and average φ down a and b, respectively, and $\varphi - \bar{\varphi}$

$\left| \longleftarrow \frac{3}{2}a \longrightarrow \right| \longleftarrow \frac{3}{2}a \longrightarrow \right|$

$z = 0$ section of the reciprocal lattice is shown schematically in Fig. 81c. The indices for the diffuse spots are fractional, as they are expressed in terms of fractions of h, k indices of the principal reciprocal lattice points. The rod diameter, i.e., the degree of diffusion in the plane a^*b^* is determined by the inverse mean domain diameter.

Thus, if the basic structure contains coherently independent two-dimensional domains having an ordered cation distribution, the arrangement of diffuse spots in the plane $(001)^*$ at $z^* = 0$ can be used to determine the distribution of cations differing in scattering power. The expression (10.7) implies that the intensity of diffuse spots in the plane $(001)^*$ at $z^* = 0$ decreases as $(f_A - f_B)^2$ with $|H|$. This fact can be used as a diffraction criterion. The important particular case is that of $f_B = 0$, i.e., B is a vacancy. The diffuse scattering intensity increases sharply since it is proportional to f_A^2. Therefore, the greater f_A, the stronger the diffuse scattering.

Assume that $f_A \simeq f_B$ but A and B are of different size, e.g., $r_A > r_B$. Figure 83 shows the arrangement of tetrahedral cations for the average structure. Arrows indicate the directions for shifts, Δ, for atoms B in an ordered pattern, with A cations at the nodes of a centered superlattice $a_s = 3a_0$, $b_s = b_0$. Cations A are shown as large open circles. Below and on the right are shown the positions and the relative scattering powers for atoms in the real and the average structures projected on the axes a_s and b_s, as well as the values for $\Delta\varphi(x)$ and $\Delta\varphi(y)$. The heights of the vertical lines are proportional to the scattering powers. It is seen from Fig. 83 that

$$\varphi(r) = \varphi_B(r) * \left[\delta\left(r + \frac{\Delta}{b}\, b\right) + \delta\left(r + \frac{\Delta}{b}\, b - \frac{a_s + b_s}{2}\right) \right.$$

$$+ \delta\left\{ r - \frac{a_s + 3b_s}{6} - \frac{\Delta}{2b}\,(b_s + a_s) \right\} + \delta\left\{ r + \frac{a_s + 3b_s}{6} + \frac{\Delta}{2b}\,(a_s - b_s) \right\}$$

$$\left. + \delta\left\{ r - \frac{a_s}{3} + \frac{\Delta}{2b}\,(a_s - b_s) \right\} + \delta\left\{ r + \frac{a_s}{3} - \frac{\Delta}{2b}\,(a_s + b_s) \right\} \right],$$

$$\overline{\varphi(r)} = \varphi_B(r) * \left[\delta(r) + \delta\left(r - \frac{a_s + b_s}{2}\right) + \delta\left(r - \frac{a_s + 3b_s}{6}\right) + \delta\left(r + \frac{a_s + 3b_s}{6}\right) \right.$$

$$\left. + \delta\left(r - \frac{a_s}{3}\right) + \delta\left(r + \frac{a_s}{3}\right) \right].$$

The Fourier transform for $\Delta\varphi = \varphi(r) - \overline{\varphi(r)}$ is given by

$$\Delta\Phi(H) = f_B\left(\exp\left[-2\pi i\,\frac{\Delta}{b}\, k_s\right] - 1\right)\left(1 + \exp 2\pi i\,\frac{h_s + k_s}{2}\right)$$

$$+ \exp\left(-2\pi i\,\frac{h_s}{3}\right)\left(1 + \exp 2\pi i\,\frac{h_s + k_s}{2}\right)\left(\exp\overset{\circ}{2}\pi i\,\frac{\Delta}{2b}\,(h_s + k_s) - 1\right)$$

$$+ \exp 2\pi i\,\frac{h_s}{3}\left(1 + \exp\left[-2\pi i\,\frac{h_s + k_s}{2}\right]\right)\left(\exp\left[-2\pi i\,\frac{\Delta}{2b}\,(h_s - k_s)\right] - 1\right).$$

The latter expression is zero for $h_s + k_s = 2n + 1$. Two cases, $h_s = 3n$ and $h_s = 3n \pm 1$, are to be considered for $h_s + k_s = 2n$. For $h_s = 3n$ and Δ less than or equal to fractions of 1 Å, $\Delta\Phi(H)$ is practically zero. For $h_s + k_s = 2n$ and $h_s = 3n \pm 1$

$$|\Delta\Phi(H)|^2 = kf_B^2\left\{ 1 + \left[2\cos 2\pi\left(\frac{1}{3} - \frac{\Delta}{2b}\, h_s\right)\right]\left[\frac{1}{3} + \frac{2}{3}\cos 2\pi\,\frac{3\Delta}{2b}\, k\right]\right\}$$

$$\cong kf_B^2\left[1 + 2\cos 2\pi\left(\frac{1}{3} - \frac{\Delta}{2b}\, h_s\right)\right]. \tag{10.8}$$

Hence the diffuse scattering intensity depends on Δ and increases with $h_s k_s$ and $|\Delta\Phi(H)|^2 = 0$ for $\Delta = 0$. To evaluate $|\Delta\Phi(H)|^2$, assume $\Delta = 0.1$ Å and $b_0 = a_0\sqrt{3} = 9$ Å. Then, while for the spot $h_s = 1$, $k_s = 1$ $|\Delta\Phi(H)|^2$ is practically zero, $|\Delta\Phi(H)|^2 = 0.55 f^2$ for $h_s = 8$ and $k_s = 4$. Thus the arrangement of diffuse rods resulting from modulations in atomic coordinates is the same as that for $f_A > f_B$ and $r_A = r_B$. However, the intensity variation is substantially different. Unfortunately, owing to the fall of f-curves with $|H|$, this is a minor effect.

Let us consider in the light of the above discussion the reported data available on the study of diffuse scattering from layer structures.

It has been mentioned that the distribution of Si and Mg in the Na-mica structure is most favorable for the local charge balance for the interlayer cations. It might be anticipated that, in the case of Al for Si substitution, Al cations would also avoid forming bonds $Al^{IV} - O_{br} - Al^{IV}$. At the same time, different minerals could be expected to differ in the degree of cation ordering.

The problem could be solved by studying the diffuse scattering intensity distribution. For finely dispersed minerals, SAED is the only method available for visualization of diffuse scattering.

Alcover et al. (1973) studied X-ray diffuse scattering from vermiculite from Kenya with Mg or Ni in interlayers. Figure 78c shows a diffraction pattern that contains strong Bragg reflections corresponding to lattice parameters a_0, b_0, as well as weaker diffuse superstructure spots corresponding to the unit cell $a_s = 3a_0$, $b = b_0$ in the direct space. The information on vermiculite average structure and composition was used for the interpretation of the effects observed. The structural formula for the sample was

$$(Mg_{2.47}Fe^{2+}_{0.08}Fe^{3+}_{0.40}Ti_{0.05})(Si_{2.76}Al_{1.24})O_{10}(OH)_2(Mg_{0.35}Ca_{0.02})\,n\,H_2O \, .$$

The average vermiculite structure consists of 2:1 layers, identically oriented and shifted with respect to one another by $\pm b/3$, which results in a two-layer periodicity along c.

The scheme of the stacking of the two nearest tetrahedral sheets belonging to adjacent layers is shown in Fig. 81b. Suppose that interlayer Mg cations surrounded by H_2O molecules are located at the nodes of a superlattice $a_s = 3a_0$, $b_s = b_0$, and the ordered distribution of Mg and vacancies in each interlayer are independent of those in the neighboring interlayers (or even within the same interlayer, Fig. 82a). The situation is then quite similar to that in the above case of cations A and B differing in scattering power. Here B will denote vacancy and $f_B = 0$.

The distribution patterns for A in Fig. 81a and interlayer Mg in Fig. 81b are indeed identical. The vermiculite reciprocal space should contain, along with discrete reflections hkl, a set of continuous diffuse rods parallel to c^* that are cut by the plane (001) at $z^* = 0$ at the nodes having $h_s + k_s = 2n$ and $h_s = 3n \pm 1$, where $h_s = 3h$ and $k_s = k$ for reflections corresponding to the average structure.

It is seen in Fig. 81b that the Mg contents and the charge per half the average unit cell are 0.33 and $+0.66$, respectively. These values are quite close to the number of exchangeable cations and the layer charge in the sample involved (see the structural formula). Two alternative explanations are possible for the local ordering in the distribution of interlayer Mg. (1) One may suppose that the (Si, Al) distribution in tetrahedra is regular, so that the local charge balance requires ordering in the Mg distribution. It is seen in Fig. 81b that Mg is between two tetrahedral bases. If both tetrahedra are occupied by Al leading to two negative unity charges, a Mg cation can compensate this local positive charge deficiency.

(2) Comparatively strong electrostatic repulsion between interlayer cations may lead to the maximum dissociation possible for them. If this is the case, the pattern in the distribution of interlayer cations should be associated with the Mg contents and the total layer charge.

Alcover et al. (1973) studied dehydrated vermiculite and found no local modulations in diffuse scattering. On the one hand, this confirms unambiguously that the effect observed with hydrated vermiculite is indeed determined by local ordering in the distribution of interlayer Mg. After dehydration, the distribution of Mg cations over the hexagonal cavities of the tetrahedral sheets apparently becomes random. On the other hand, complete absence of diffuse scattering all over the reciprocal space seems to imply disordered (Si, Al) distribution in the tetrahedral sheets. Although the scattering powers for Si and Al are similar, the difference in ionic radii should lead to structural distortions resulting in the corresponding diffuse effects for large $|H|$. Nevertheless, the authors suggest that there is local ordering in the distribution of effective negative charges at the basal surfaces of vermiculite 2:1 layers.

Besson et al. (1974a, b) used SAED for a detailed analysis of diffuse scattering from finely dispersed phyllosilicates. As $f_{Si} \simeq f_{Al}$, the authors introduced heavy cations having high scattering powers into interlayers. If the (Si, Al) distribution is partly ordered, it seems reasonable that the interlayer cations should be also distributed more or less regularly.

On the contrary, complete disorder in the distribution of tetrahedral cations should imply random distribution for exchangeable cations. Ba^{2+} cations were used as heavy, strongly scattering atoms. To compensate the layer charge ranging from 0.33 to 1 v.u. Ba^{2+} cations would occupy a relevant part of the interlayer sites available.

Ba-smectites (montmorillonite, beidellite, and saponite), Ba-illites, and Ba-vermiculites were used as objects. Besson et al. (1974a, b) noted that the layer charges resulting from isomorphous substitutions for some beidellite samples were often close to those for illite despite the obvious difference in properties. This was also the case for vermiculite and saponite. For example, it remained unclear why beidellites and illites having similar layer charges differ substantially in expandability.

Fundamental differences were revealed in the intensity distribution for diffuse scattering.

In diffraction patterns from smectites, diffuse scattering is observed in the form of straight or slightly bent lines connecting the Bragg reflections in directions parallel to $[01]^*$, $[11]^*$, and $[\bar{1}1]^*$. The distribution of these lines depends neither on the type of exchangeable cations nor on the presence of any isomorphous substitutions in the structure, since diffraction patterns with a similar distribution of diffuse scattering are also observed both for talc and pyrophyllite. This implies complete disorder in the distribution of Ba over interlayer sites in all smectites irrespective of the composition and the structure. The diffuse lines observed result from atomic thermal vibrations.

In contrast to smectites, the diffuse scattering intensity from Ba-illites and vermiculites formed hexagonal prisms around each hk rod lying along c^* (Besson et al. 1974b). In the SAED patterns representing the plane $(001)^*$ at $z^* = 0$,

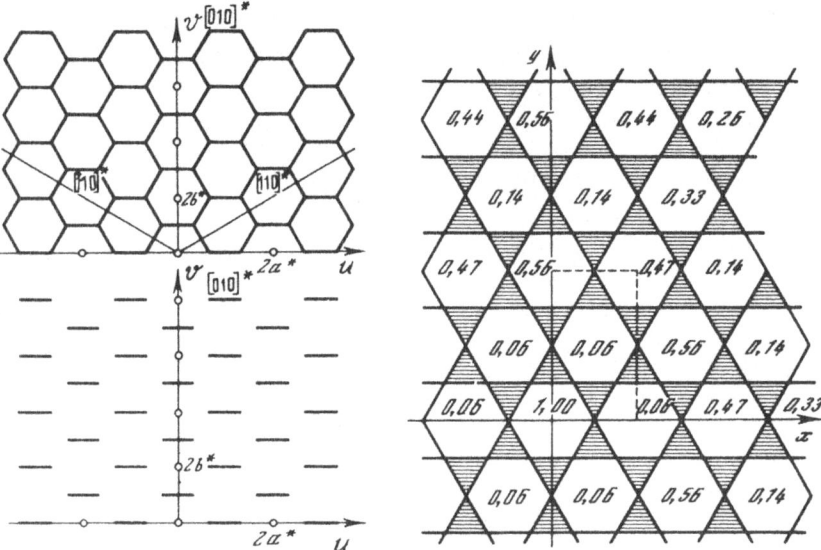

Fig. 84. Diffuse scattering intensity distribution observed in ED patterns from Ba-illite (left) and the pattern of probability distribution for interlayer Ba in Ba-illite (Besson et al. 1974b)

the diffuse scattering intensity is distributed along segments forming a honey-comb hexagonal pattern (Fig. 84). Figure 78b shows the corresponding SAED photograph. The diffuse scattering analyzed was found to have no direct relation to the contents of interlayer cations.

To reveal the distribution of Ba cations in illite and vermiculite, Besson et al. (1974b) computed the Patterson function.

Assume that, although there are two-dimensional domains with ordered cat-ions distribution, there is no correlation in the distribution of cations over neigh-boring layers and interlayers. Therefore, intensity of diffuse scattering is equal to the intensity scattered from one real packet (layer + interlayer) having a certain ordering in the cation distribution minus that for a packet having a random dis-tribution of isomorphous cations including vacancies. In the particular case dis-cussed, it is reasonable enough to assume that the structure of a separate 2:1 lay-er is the same for the real and the average structure.

It follows then that the diffuse scattering intensity for Ba-illite and Ba-ver-miculite is determined only by the arrangement of Ba cations in the interlayer. Therefore the Fourier transform of the function for the distribution of diffuse scattering intensity is the Patterson function representing the probabilities of oc-currence only for vectors Ba-Ba in the layer plane.

To reconstruct the pattern in the distribution of the interlayer Ba, Besson et al. (1974b) used the honeycomb scattered intensity pattern in the SAED patterns from Ba-illite. The scattered intensity was found to vary with $|H|$ from one seg-ments to an other as f_{Ba}^2, as should have been expected, since Ba alternates with vacancies. On the other hand, this indicated constant intensity along each sepa-rate segment. For simplicity, the authors replaced the honeycomb pattern by three sets of parallel segments. One set is normal to $[010]^*$ (Fig. 84), the other two are normal to $[110]^*$ and $[\bar{1}10]^*$, respectively. If the middles of the segments in each set are marked by points, the positions of the points can be described by a lattice function of the type

$$\sum_h \sum_k \delta[H - (ha^* + kb^*)] \cdot \delta(u - a^*).$$

Figure 84 shows that the lattice will be centered if the origin is shifted by a^* along the axis u. The intensity distribution within one separate segment may be described by the function $i(u, v)$, so that

$$
\begin{aligned}
i(u, v) &= 1 \quad \text{for} \quad |u| \leqslant a^*/3, \quad |v| \leqslant b^*/20, \\
i(u, v) &= 0 \quad \text{for} \quad |u| > a^*/3, \quad |v| > b^*/20.
\end{aligned}
$$

Therefore, the intensity distribution within the set of segments normal to $[010]^*$ is the convolution of $i(u, v)$ with the lattice function multiplying the segments according to the lattice law (see Fig. 84)

$$|\Delta \Phi'(H)| = i(u, v) * \sum_h \sum_k \delta[H - (ha^* + kb^*)] \delta(u - a^*) \quad \text{for} \quad h + k = 2n.$$

According to Eq. (4.17) the Fourier transform of the latter expression is given by

$$\mathscr{P}'(n, m) = \frac{1}{ab} \frac{\sin 2\pi x \dfrac{a^*}{3}}{2\pi x \dfrac{a^*}{3}} \frac{\sin 2\pi y \dfrac{b^*}{20}}{2\pi y \dfrac{b^*}{20}} \sum_a \sum_b \delta[r - (na + mb)] \exp 2\pi i x a^*,$$

$$\tag{10.9}$$

where $n + m = 2n'$.

For a mica-like stacking of 2:1 layers in the unit cells of illite and dehydrated vermiculite, only two values for x, y are possible: $x = y = 0$ and $x = y = 0.5$. Proceeding from one unit cell to another, i.e., increasing n and m, the values for $\mathscr{P}'(n, m)$ are calculated. These values are added to those corresponding to two other sets of segments. Figure 84 shows the distribution of Ba-Ba vectors for Ba-vermiculite (Besson et al. 1974b) with account of the real Ba contents in interlayers. The shaded triangles represent tetrahedral bases. A number inside a hexagon signifies the probability to find a Ba cation at the given distance and with the given orientation, if the presence of Ba at the origin is assumed certain. It is seen that the probability to find two Ba cations in the neighboring hexagonal cavities is practically zero. However, the probability for a pair of Ba cations to occur in hexagons separated by one ore two cavities increases up to 0.5.

Further interpretation of Fig. 84 will be given after discussing the results obtained by other authors (Suquet 1978; Kodama 1977; Moret et al. 1976; Ridder et al. 1976, 1977a, b). Suquet (1978) used SAED to study Ba-vermiculites differing in layer charge. Diffuse scattering of appreciable intensity was registered only for some samples.

These data indicate that ordered Ba distribution is not an inherent feature for the dehydrated Ba-vermiculite structure.

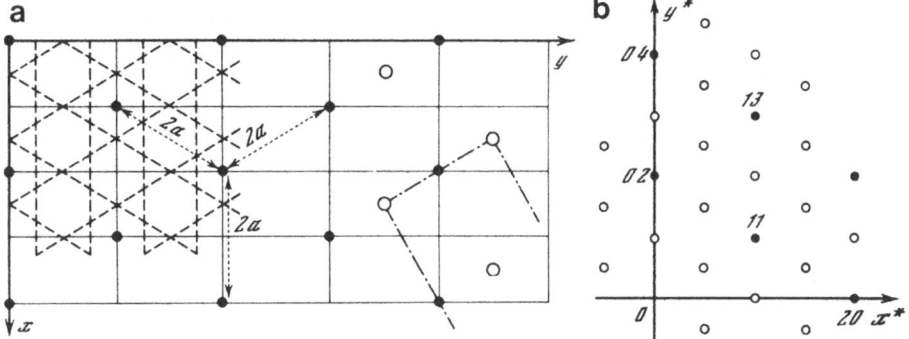

Fig. 85. a The probability distribution for interlayer Ba in vermiculite leading to the superlattice ($a_s = 2a$, $b_s = 2b$); **b** the corresponding reciprocal lattice section at $z^* = 0$ (Suquet 1978)

In some cases, Suquet observed such SAED patterns as sketched in Fig. 85. Additional diffuse spots imply superstructural two-dimensional ordering in the Ba distribution over a centered unit cell $a_s = 2a_0$, $b_s = 2b_0$. The figure shows the distribution of Ba cations based on the assumption that dehydration leads to a mica-like 2:1 layer stacking in the vermiculite structure. The layer charge compensated by Ba is here 0.5 v.u. per half unit cell.

The Ba cations are again equidistant with respect to one another. As it has been noted, the distances between the nearest Ba cations in the unit cell $a_s = 3a_0$, $b_s = b_0$ are equal to b, and the segments connecting each atom with its nearest neighbors make the angles 60° (Fig. 81b). This pattern in the relative arrangement of Ba cations is preserved in the unit cell $a_s = 2a_0$, $b_s = 2b_0$, but the distance between the nearest cations is $2a$ (Fig. 85a).

A similar distribution of diffuse spots was observed for vermiculitized muscovite with residual interlayer K by Kodama (1977) who analyzed X-ray diffuse scattering.

Important results concerning diffuse scattering were obtained by Moret et al. (1976), who studied short-range order in titanium sulfides $Ti_{1+x}S_2$. These structures consist of sulfide layers alternating with completely and partly occupied layers formed by Ti atoms in octahedral coordination.

The stacking sequence in $Ti_{1+x}S_2$ structures is represented by $...STi_1STi_x$ $\square_{1-x} STi_1 STi_x \square_{1-x}S...$, where $\square$ signifies an octahedral vacancy in a partly occupied Ti-layer. $Ti_{1+x}S_2$ compounds were studied with $x = 0.20, 0.23, 0.25$ and 0.35. Patterns in the diffuse scattering intensity distribution in diffraction patterns from these substances were found to be quite similar to those observed for Ba-illite and Ba-vermiculite.

Figure 86 shows geometrical relations for the average-structure reciprocal lattice points and the diffuse scattering in the plane (001)* at $z^* = 0$ for three substances ($x = 0.20, 0.25$ and 0.35). The shaded areas represent the diffuse scattering intensity distribution. For $x = 0.2$, diffuse scattering in the reciprocal space is represented by a set of hexagonal prisms whose axes coincide with hk rods. In the plane (001)* at $z^* = 0$, the diffuse scattering is distributed in a honeycomb pattern (Fig. 86) with a non-uniform intensity distribution along each segment. For

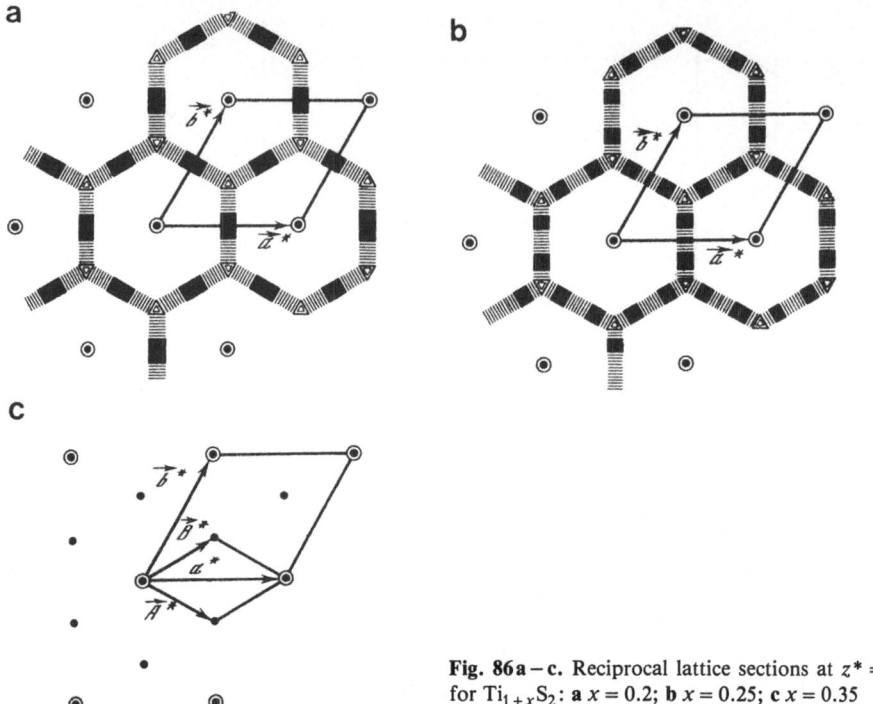

Fig. 86a−c. Reciprocal lattice sections at $z^* = 0$ for $Ti_{1+x}S_2$: **a** $x = 0.2$; **b** $x = 0.25$; **c** $x = 0.35$

$Ti_{1.20}S_2$, increase in intensity (Fig. 86a) is observed at points corresponding to a direct-space superlattice with a centered unit cell ($a_s = 2a_0$ and $b_s = 2b_0$). With the increase in Ti content, the maxima in intensity are shifted toward the hexagon apices, so that the position of one of the maxima in terms of fractional reciprocal lattice coordinates is $u = v = 0.46$ for $x = 0.23$ and $u = v = 0.36$ for $x = 0.25$ (Fig. 86b,c). Finally, for $Ti_{1.35}S_2$, the (001) plane at $z^* = 0$ contains discrete spots indicating local two-dimensional domains with the unit cell $a_s = 3a_0$, $b_s = b_0$ (Fig. 86c).

The results obtained confirm that the pattern in the distribution of Ti-cations is governed by the Ti contents. In other words, the main factor affecting the Ti distribution is electrostatic repulsion leading to the maximum disconnection for cations for each given x.

We shall continue analyzing the probability of occurrence for vectors Ba-Ba (Fig. 84b) in Ba-illite and Da-vermiculite in the light of the above data. It is seen that a center of a hexagonal ring at the distance b from the origin corresponds, with the probability 0.56, to a vector Ba-Ba. Such vectors drawn from a common origin make 60° angles with one another. This pattern in the Ba distribution is described by the sublattice $a_s = 3a_0$, $b_s = b_0$ and the layer charge 0.66 v.u.

The Ba distribution described occurs with the probability close to 0.56. On the other hand, Fig. 84b shows that the probability to find a Ba-Ba vector terminating at a point at the distance $2a_0$ from the origin is 0.46. Such vectors also make angles 60°, which is characteristic of a uniform Ba distribution over a centered supercell ($a_s = 2a_0$, $b_s = 2b_0$) with the layer charge 0.5 v.u. Hence, the value

0.46 can be treated here as the probability for a pattern in the interlayer cation distribution required for the compensation of the charge 0.5 v.u. per half unit cell.

Therefore, we may conclude that the honeycomb diffuse scattering pattern observed in diffraction patterns from some Ba-vermiculites and Ba-illites is mainly associated with the average layer charge that should be close to $(0.66 + 0.5)/2 = 0.58$ v.u. It may also be supposed that the charge distribution is uniform neither within a separate 2:1 layer nor between successive layers. The layer charge distribution governs the pattern in the arrangement of interlayer cations which, in any case, tends to the maximum disconnection. Domains having different patterns in Ba distribution are quite probable to exist in the structure in question. The honeycomb pattern appears in accordance with three, equally probable orientations for domains of each type rotated through $n \cdot 120°$ with respect to one another.

The following conclusions can be made. Certain minimum contents of interlayer cations may imply complete absence of short-range order.

For phyllosilicates having higher layer charges, the diffuse scattering distribution contains information on both the mean layer charge and on the distribution of interlayer cations.

10.3 Structural Modulations Resulting from the Lateral Misfit of Octahedral and Tetrahedral Sheets in Phyllosilicates

In phyllosilicates, tetrahedral and octahedral sheets can be linked together to form 2:1 or 1:1 layers only if they have similar lateral dimensions. However, this is not always the case. The sheets are especially difficult to fit when the lateral dimensions of the octahedral sheet are greater than those of the tetrahedral one. Lateral misfit leads to various structural modulations. For example, modulations in antigorite include tetrahedral tilting with periodic inversion of tetrahedra at four- and eightfold rings to form wavelike structures (Kunze 1956). In stilpnomelane there is a periodic inversion and relinkage of tetrahedral apical directions (Eggleton 1972).

Octahedral sheets containing relatively large Fe^{2+} and Mn^{2+} are especially difficult to fit with the lateral dimensions of the Si-tetrahedral sheets. For this reason, minerals of such octahedral and tetrahedral compositions have complex modulated structures whose nature has been recently revealed by SAED and HREM (Eggleton and Guggenheim 1986; Guggenheim and Bailey 1982; Guggenheim and Eggleton 1986a, b; Guggenheim et al. 1982).

To illustrate this, let us consider the crystal structure of minnesotaite which was previously considered a Fe^{2+}-analog of talc. Guggenheim and Bailey (1982) found that single-crystal X-ray patterns from this mineral contain satellite spots indicating structural modulations. The nature of this was determined by Guggenheim and Eggleton (1986a) by electron-optical investigation. According to the iron content, two structural minnesotaite varieties were revealed. Crystals of the Mg-rich variety are typically characterized by the following unit cell: $a = 5 a_{talc} = 28$ Å; $b = 9.4$ Å; $d_{001} = 9.6$ Å; $c = 12.4$ Å; $\alpha = 101°$; $\beta = 127°$ and

Fig. 87. HREM b-axis image of minnesotaite structure having a C-centered unit cell. Triangles = Si tetrahedra; circles = Fe octahedra (Guggenheim and Eggleton 1986a)

$Z = 1$. Ferroan minnesotaite is dominantly C-centered with $a = 9a_{talc} = 50.6$ Å; $b = 9.4$ Å; $d_{001} = 9.6$ Å; $\alpha = 101°$; $\beta = 127°$; $Z = 2$. The authors revealed complex intergrowths of P and C-centered cells, as well as the presence of diffuse streaking parallel to z^* for reflections okl with $k \neq 6n$ and parallel to x^* and z^* for hol reflections. The nature of structural modulations was determined mainly from the analysis of HREM images that were obtained using reflections hol and hko.

The minnosotaite octahedral sheet is continuous in two dimensions and is linked to two tetrahedral sheets to form a 2:1 layer. However, tetrahedral sheets are continuous along Y and discontinuous along X. They consist of parallel polar tetrahedral ribbons that are four tetrahedra wide. The tetrahedra in these ribbons share apices with the octahedral sheet. The adjacent ribbons are linked together through a single tetrahedral chain which is parallel to Y and located in the inter-layer. The same chain links together the two nearest tetrahedral ribbons from the neighboring 2:1 layer. In the C-centered structure, ribbons that are three tetra-hedra wide alternate regularly with ribbons having the width of four tetrahedra. Figure 87 shows the contrast distribution corresponding to the C-centered struc-ture projection down Y. Dark areas represent the maximum charge density re-

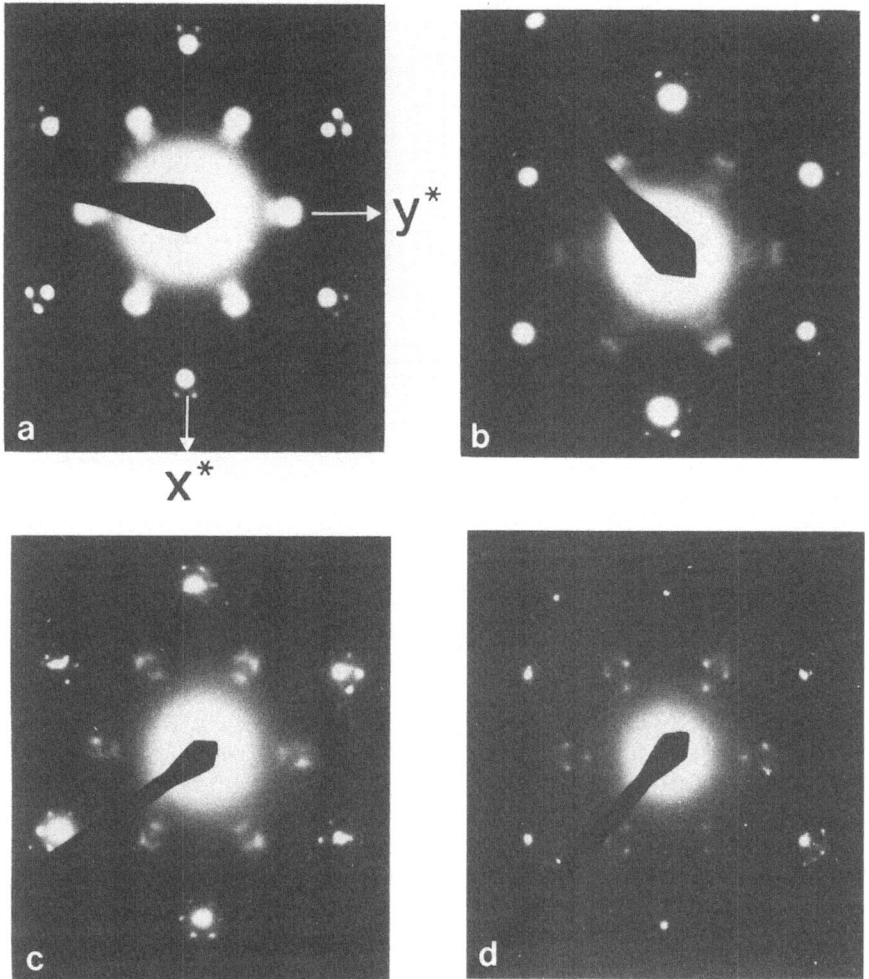

Fig. 88 a – d. ED patterns of diffuse intensity associated with vacant $k \neq 3n$ positions for **a** Cartagena greenalite; **b** Ichinomata caryopilite; **c** and **d** Fallota caryopilite (Guggenheim et al. 1982)

gions. Linear rows of dark points separated by the distance 3 Å apparently correspond to octahedra containing Fe and Mg. Dark bands adjacent to them that are 9 Å and 12 Å wide represent projected tetrahedral ribbons that are 3 and 4 tetrahedra wide, respectively. Every four neighboring dark bands are connected by a narrow dark band 6 Å wide corresponding to the interlayer silicate chain with inverted tetrahedra. Analysis of such HREM images and diffraction data revealed a number of fine features of the mineral associated with various forms of structural imperfections.

Structural modulations of octahedral and tetrahedral sheets were also found in a manganese phyllosilicate ganophyllite, having the idealized formula (K, Na, Ca)$_6$Mn$_{24}$[Si$_{32.5}$Al$_{7.5}$]O$_{96}$(OH)$_{16}$ · 21 H$_2$O (Eggleton and Guggenheim 1986; Guggenheim and Eggleton 1986b). Ganophyllite belongs to space group $A\,2/a$ with

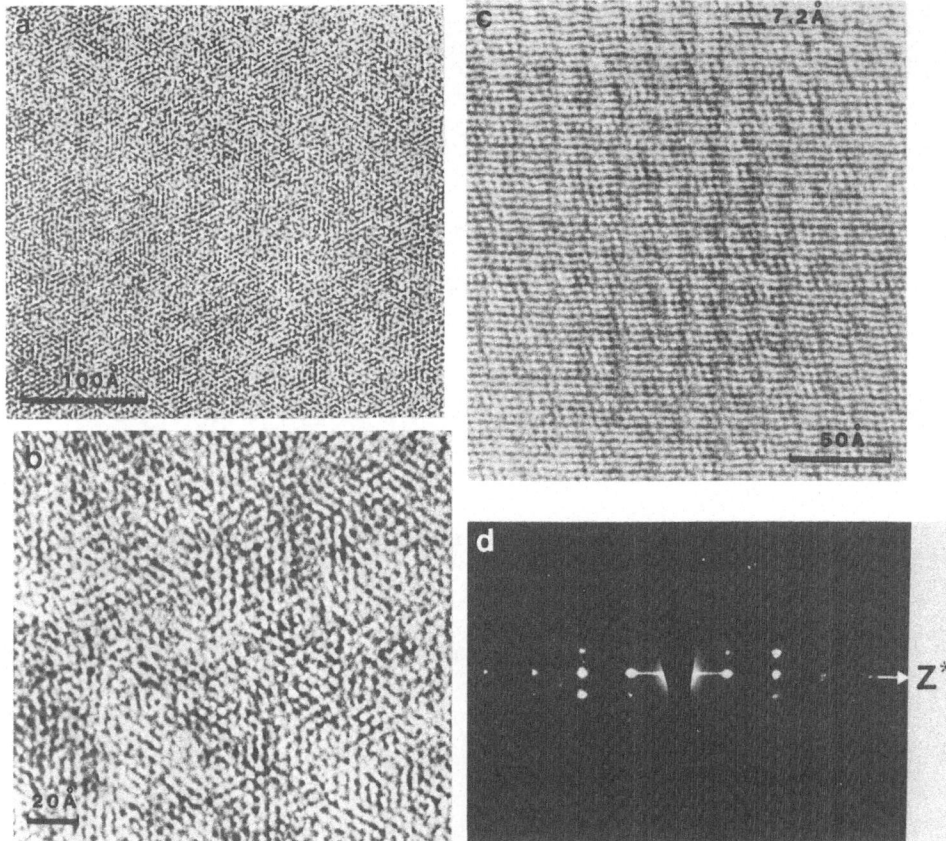

Fig. 89. a and **b** Lattice fringe images of the (010) plane of greenalite from Cartagena showing irregularity in the 4.6Å 020 and 110 lattice fringes; **c** lattice fringe of the 7 Å layers approximately parallel to (001) of caryopilite from Ichinomata; **d** ED pattern of crystal shown in **c** (Guggenheim et al. 1982)

$a = 16.6$ Å; $b = 26.6$ Å; $c = 50.0$ Å and $\beta = 94°$. The structure in question consists of sinusoidal octahedral sheets connected with triple tetrahedral chains parallel to X to form narrow 2:1 layers. The ribbons are linked together through double pairs of inverted tetrahedra forming four-member rings parallel to (010) in the interlayer. The ordered arrangement of inverted tetrahedral pairs leads to a superlattice. It should be stressed that the authors obtained a good agreement between the experimental hko intensities obtained by SAED and those calculated for the model discussed above.

SAED patterns were used by Guggenheim and Eggleton (1986b) to prove that ganophyllite and eggletonite are isostructural.

The efficiency of SAED in combination with HREM was demonstrated by Crawford et al. (1977), who studied polytipism of stilpnomelane $K_2Fe_{42}Si_{70}Al_2O_{180}(OH)_{48}$, which is a complex phyllosilicate with modulated tetrahedral sheets. The stacking sequences for three new polytypes were found from the comparison of the measured intensities with those calculated for dif-

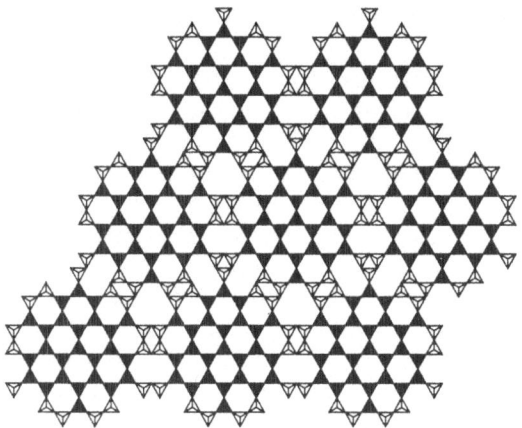

Fig. 90. Idealized tetrahedral sheet for greenalite (Guggenheim et al. 1982)

ferent polytypes with specified number of layers per repeat distance. HREM revealed the twins of one-layer triclinic polytype as well as the coherent intergrowth of 9-and 14-layer polytypes.

The structure studies of greenalite and caryopilite (Guggenheim et al. 1982) illustrate the role of SAED and HREM in revealing the nature of incommensurate modulated structures. Structure modulations in these finely dispersed, serpentine-like minerals having Fe^{2+} and Mn^{2+} in octahedra, respectively, result from the lateral misfit of tetrahedral and octahedral sheets. Figure 88 shows the following main features in hko SAED patterns from greenalite and caryopilite particles. The patterns contain strong and sharp sublattice reflections with $k = 3n$ each surrounded by six satellites forming a hexagon along three pseudo-hexagonal axes y^*. In all cases, two satellites in each hexagon are considerably stronger than the other four. Satellite spacings indicate regular but irrational subcell modulations, since the modulation periodicity varies from $b = 2.4\,b_0 = 23$ Å in greenalite to $b = 1.7\,b_0$ in caryophyllite. Sublattice points with $k \neq 3n$ have zero intensity. Outside these points diffuse intensity distribution is observed that varies from two or three weak blobs to three relatively sharp satellite spots.

HREM images obtained using hko reflections contained three sets of intersecting fringes normal to pseudoaxes Y^* (Fig. 89). The contrast distribution in the images can be associated with the presence of domains containing small numbers of tetrahedral rings. Domain boundaries are poorly resolved and their dimensions and shape vary for different parts of the images.

Lattice fringe images of 7 Å layers obtained with the incident beam parallel to (001) revealed sinusoidal modulations of these layers (Fig. 89). To account for the diffraction effects observed, the authors considered a model based on saucer-shaped domains four and three tetrahedral rings in diameter, respectively, for greenalite and caryopilite (Guggenheim et al. 1982). The increase in lateral dimensions results from an inserted additional tetrahedron per every eight tetrahedra (Fig. 90). Four-member and three-member rings linking the neighboring islands may be inverted and can coordinate the octahedral sheets belonging to the neighboring layers. It is assumed that the linking of tetrahedral islands with the octahedral sheets leads to dome-shaped saucers separated by boundaries consisting of rows of vacant octahedra.

The wavelike structure may result from the packing of saucers in which the convex upward octahedral saucers fit as caps over the tetrahedral islands (Guggenheim et al. 1982).

The authors consider that satellites around the sublattice reflections with $k = 3n$ result from modulations of the octahedral sheets associated with the periodic doming of the octahedral saucers connected with the tetrahedral islands and rows of octahedral vacancies around each saucer. The absence of reflections with $k \neq 3n$ is associated with the four-member rings in the tetrahedral sheets, which allows random displacements for the neighboring domains in (001) by $a/2$. Tetrahedral islands are connected at random with the octahedral anions in greenalite, and in a more regular fashion in some caryopilite crystals, which explains the varied diffuse intensity distribution around reflections with $k \neq 3n$.

CHAPTER 11

Chain Silicates. New Structural Types: Multiple-Chain and Mixed-Chain Minerals

11.1 New Problems in the Structure Study of Chain Silicates

Until recently, pyroxenes and amphiboles were considered the only natural chain silicates having the repeat distance of about 5 Å along the chain direction.

The basic structure units for these minerals are ribbon-shaped fragments of 2:1 mica- or talc-like layers, called *I*-beams. Pyroxenes and amphiboles are rock-forming minerals widely abundant in various environments and especially in magmatic and metamorphic rocks. Extremely diverse isomorphous substitutions lead to numerous mineral varieties, each having a specific name.

Wide abundance and great geological and mineralogical importance of pyroxenes and amphiboles have required their detailed and comprehensive structural and crystal-chemical study. Intensive research has been carried out for more than 50 years since the first structure determinations of tremolite and diopside by Bragg and Warren (Warren 1930; Warren and Bragg 1928). The appearance of modern X-ray diffractometers on-line with computers has marked a qualitatively new stage in the development of crystal-chemistry of pyroxenes and amphiboles. This has favored the use of these minerals as indicators for certain geological processes and environments.

At the same time, up to 1973 – 1975 the structural data contained no fundamentally new information as compared to the classical works by Warren and Bragg (1928) and Warren (1930). In 1971 Belov (1971) noted that although multiple-chain silicate ribbons were found in synthetic $Ba_2Si_6O_{16}$ (Filippenko et al. 1971; Liebau and Hesse 1975), the existence of structures containing polar multiple-chain ribbons that could be treated as intermediate between micas and amphiboles was still uncertain. Two years later a synthetic silicate of a new structural type was reported (Drits et al. 1973; Goncharov et al. 1973a, b). This structure consists of triple-chain silicate ribbons of the composition Si_6O_{16}. The material being highly dispersed, the structure analysis was carried out using mainly SAED.

The triple-chain silicate structure was described in detail in (Drits and Goncharov 1974; Drits et al. 1974). The results obtained indicated that triple-chain minerals should be anticipated under physico-chemical conditions close to the formation conditions of amphiboles and micas. It is remarkable that in 1975 – 1977 Veblen et al. (1977) described independently four new minerals having unusual chain structures. Further detailed discussion was given in a number of papers (Veblen 1981). Two of these structures, jimthompsonite and clinojimthompsonite, contained three-chain silicate ribbons Si_6O_{16}. The other two minerals had fundamentally new structures consisting of slabs made up of I-beams of different width. These slabs alternate regularly along b.

In 1977, Drits and Semenov (unpublished data) completed a structure study of a new mineral belonging to another structural type unknown previously. On the one hand, the structure in question consists of alternating I-beams differing in width. On the other hand, the stacking sequence for adjacent I-beams is alternately pyroxene-amphibole-like or palygorskite-sepiolite-like. SAED again was the basic method for the determination of this structure. Veblen et al. (1977) remark that it is not such a late discovery of minerals with such unusual structures that it is the most surprising, but the fact that similar minerals were discovered in other deposits soon after the first findings. The reason might be the absence of clear diffraction criteria ensuring reliable identification.

Another essential aspect in the structure study of chain-silicates is associated with the revealing of structure defects. The solution of this problem would elucidate the mechanisms for structural transformations of minerals.

To grasp the new aspects in the structure study of chain-silicates, it seems appropriate to give a short description of the pyroxene and amphibole structures.

11.2 Pyroxenes and Amphiboles: Idealized Structures

The basic structure unit in a pyroxene structure is a continuous tetrahedral chain (Fig. 91a) of the composition SiO_3. Each tetrahedron shares two basal apices with two neighbors. Such chains will be hitherto called *pyroxene chains*. All tetrahedra in a pyroxene chain are identically oriented, so that their bases are on one side of the chain and the "free" apices are on the other. Two pyroxene chains linked together through an octahedral chain lying between them form an I-beam.

Figure 91b shows the octahedral ribbon and the upper pyroxene chain. Adjacent I-beams are linked together into a uniform structural motif, so that each tet-

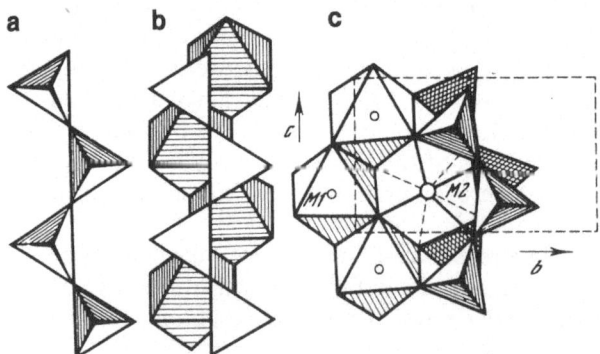

Fig. 91a – c. Pyroxene structure fragments: **a** tetrahedral single-chain; **b** connection of the lower octahedral and the upper tetrahedral single chain; **c** an octahedral ribbon with the adjacent tetrahedral chains from the neighboring I-beams (Papike and Cameron 1976)

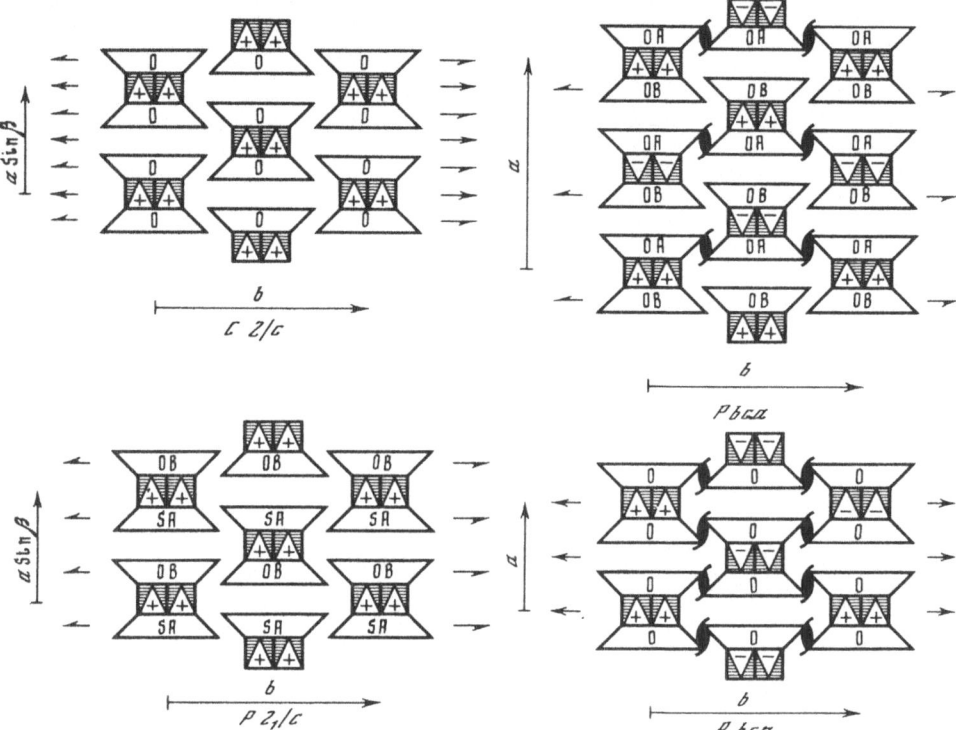

Fig. 92. The main pyroxene structural types (Papike and Cameron 1976)

rahedron shares its third basal apex with an octahedron belonging to the adjacent *I*-beam. Figure 91c shows two pyroxene chains belonging to different *I*-beams and having opposite orientations linked to a central octahedral ribbon. *I*-beams, as pyroxene chains, are supposed to be parallel to *c* which is normal to *b*. Tetrahedral and octahedral bases are parallel to the plane *bc*. Therefore, the packing of adjacent *I*-beams in pyroxene structures proceeds along *a**.

If the upper and the lower pyroxene chains in an *I*-beam are symmetrically independent they are labeled *A* and *B*. If the chains are identical, these symbols are omitted. Figure 91c shows that there are two distinct octahedral positions, *M*1 and *M*2, in an *I*-beam.

Each *I*-beam in the pyroxene structure is described by a definite orientation of octahedra with respect to the crystallographic axes. Assume that the positive orientation denoted by − implies that the octahedral bases are oriented along the positive direction of *c* as viewed down −*a*. The opposite orientation is denoted by +.

Figure 92 shows the main types of *I*-beam packing in pyroxene structures differing in octahedral orientation and silicate chain symmetry (Papike and Cameron 1976).

Figure 93a, b shows the clino- and orthoenstatite $MgSiO_3$ structures projected on the plane (010). It is seen that the two structures consist of similar structural elements, i.e., of two "layer" types parallel to (100) and consisting of chains

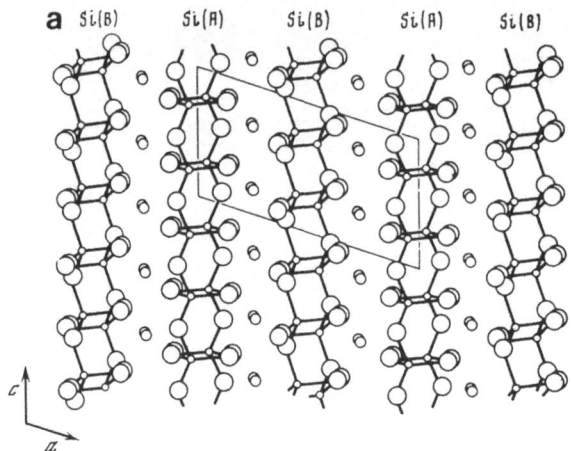

a Si(B)   Si(A)   Si(B)   Si(A)   Si(B)

Fig. 93a–d. Structural schemes for pyroxenes: **a** clinoenstatite structure projection down b; **b** orthoenstatite structure projection down b; **c** twinning of clynopyroxene sketched in the projections down b and c, respectively; **d** twinning on the unit-cell scale (Buseck and Iijima 1975)

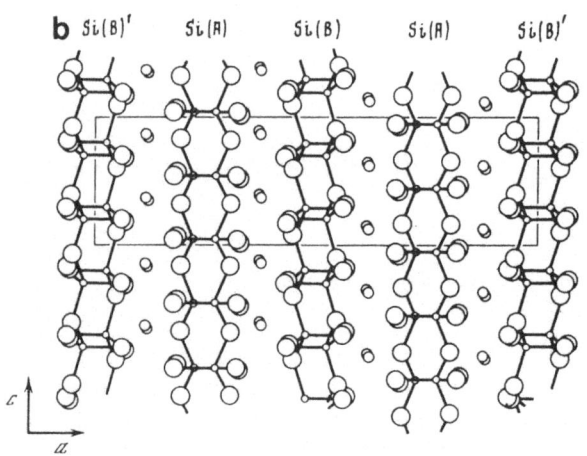

b Si(B)'   Si(A)   Si(B)   Si(A)   Si(B)'

c

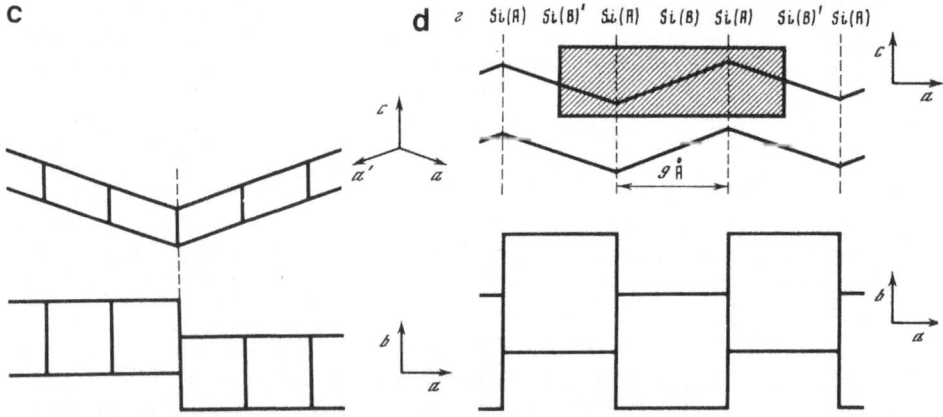

d   z   Si(A)   Si(B)'   Si(A)   Si(B)   Si(A)   Si(B)'   Si(A)

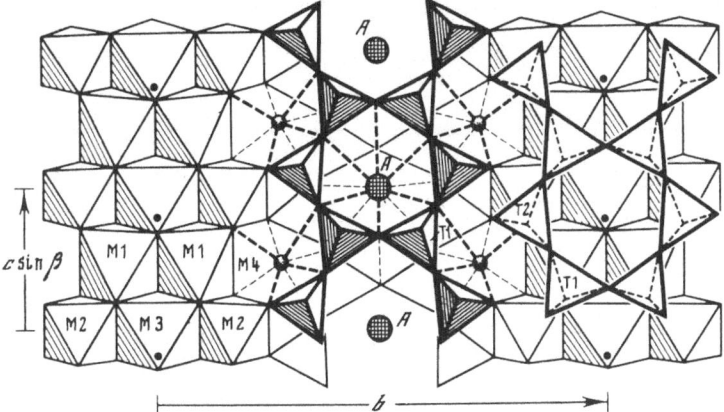

Fig. 94. Amphibole structure fragment (Papike and Cameron 1976)

Si(A) and Si(B), respectively, with adjacent Mg octahedra on the left and on the right. If the plane (100) dividing the Si(A) chain in clinoenstatite in two is brought into coincidence with the glide plane b, then twinning appears. The twinning effect governed by the glide plane is shown in two projections in Fig. 93c, d. If twinning is repeated in every second unit cell, the orthoenstatic structure is formed (Fig. 93d).

Buseck and Iijima (1975), who studied enstatite by HREM, showed that the clinoenstatite structure contains thin coherent intergrowths of orthoenstatite fragments, and vice versa. These effects are readily explained by the mechanism for twinning on the unit-cell scale described above.

To describe micro-twins in clinopyroxene crystals, the two possible unit-cell orientations are denoted by a and b with subscripts to specify the number of unit cells with the given orientation occurring in succession. Thus, the whole diversity of unit-cell sequences can be described easily. For example, Buseck and Iijima (1975) observed a fragment of an enstatite crystal with the following sequence of unit cells differing in orientation: $a(ba)_3 ba_8 ba_2 b_3$.

Twinning can lead to certain sequences of clinoenstatite unit cells alternating regularly with relatively large repeat distances. For example, Buseck and Iijima (1975) observed regular twinning leading of the periodicity along X of 27 Å ($ab_2 ab_2 \ldots$), 36 Å ($ab_3 ab_3 \ldots$) and 54 Å ($ab_3 abab_3 ab \ldots$ or $ab_5 ab_5 \ldots$). Such structure fragments may be treated as certain enstatite polytypes.

The basic structural unit in amphiboles is a double-chain or amphibole ribbon of the composition Si_4O_{11} formed by two condensed pyroxene chains (Fig. 94). The figure shows that amphibole chains are polar. Double-chain tetrahedral ribbons, one on each side, are linked to an octahedral ribbon to form an I-beam which is twice as wide as a pyroxene I-beam. This leads not only to the increase in the number of the cationic structural sites available but also to a qualitative change in the anionic composition which includes, along with oxygen atoms, OH groups or fluorine atoms. The (OH, F) position coincides with the center of the tetrahedral ring in the normal projection on the plane bc.

Table 11. Unit-cell dimensions and shape for some pyroxenes and amphiboles

Mineral	Unit-cell shape	Unit-cell parameters				Space group
		a, Å	b, Å	c, Å	β, °	
Pyroxenes:						
diopside	Monoclinic	9.7	8.9	5.25	100.8	$C2/C$
enstatite	Orthogonal	18.2	8.86	5.20		$Pbca$
Amphiboles:						
tremolite	Monoclinic	9.78	17.8	5.26	106°	$C2/m$
anthophyllite	Orthogonal	18.5	17.9	5.27		$Pnma$

It is seen in Fig. 94 that the outer basal tetrahedral apices in each double-chain ribbon belong, at the same time, to octahedra in the two neighboring (along b) I-beams.

According to the shape of the unit cell, amphiboles are classified into clino- and orthoamphiboles. Clinoamphibole varieties belonging to space group $C2/m$, such as tremolite $Ca_2Mg_5Si_8O_{22}(OH)_2$ are the most abundant. The difference in symmetry between clinoamphiboles and clinopyroxenes results from the difference in the inherent symmetry of I-beams which is $2/c$ for pyroxenes and $2/m$ in amphiboles. Adjacent I-beams related by 2_1 in orthoamphiboles are anti-parallel. Depending on the stacking sequence ($+ - + - \ldots$ or $+ + - - + + - - \ldots$), orthoamphiboles are subdivided into two groups differing in symmetry. For example, antophyllite $Mg_7Si_8O_{22}(OH)_2$ belongs to space group $Pnma$.

Direct relationship between pyroxenes and amphiboles is demonstrated by the comparison of the unit-cell parameters. Table 11 contains cell dimensions for certain clino- and ortho-chain silicates. For unit cells of similar shape the only difference is in the value for b, which is twice greater for amphiboles than for pyroxenes. Zvyagin (1986) has given a general treatment of the main structural features of chain silicates.

Natural amphiboles occur in two morphological forms. Besides usual platy crystals, there are the so-called fibrous asbestiform amphiboles that are used in various industrial fields (Grigorieva et al. 1975; Papike and Ross 1973). Synthesis of asbestiform amphiboles is important in respect of both the study of their formation conditions and the preparation of materials with technologically new properties. Below are given the results of SAED and HREM for synthetic and natural fibrous chain silicates.

SAED studies for these minerals meet a number of difficulties. Fibrous crystallites are elongated along c, so that hko patterns can be obtained only with ultra-fine sections. Diffraction patterns from oriented preparations indicate that (100) and (110) are the most developed faces in monoclinic crystallites. Naturally, it is these faces that are parallel to the support of the specimen holder. However, the incident beam directions normal to the faces (100) and (110) generally do not lead to rational reciprocal lattice sections passing through the origin. To obtain rational reciprocal lattice sections, the orientation of the crystallites is to be varied with the help of a goniometer. Twinning, which is widespread in chain-silicate structures, also complicates the interpretation for point SAED patterns.

Superposition of diffraction patterns from two twins not only imitates the change in the object symmetry and elongates the reciprocal lattice nodes along a^* but also leads to forbidden reflections resulting from double diffraction. Finally, it should be taken into account that chain-silicate crystals having similar morphology may differ in either unit-cell shape or space group. Therefore, geometrical analysis requires a sufficiently representative set of SAED patterns allowing reconstruction of all the details in the reciprocal lattice involved.

SAED being extremely important for identification of asbestiform chain-silicates, the methodological aspects in the interpretation of point patterns will be also discussed.

11.3 Fluorocupfferite Mg$_7$[Si$_8$O$_{22}$F$_2$], a New Amphibole Variety

Among OH-amphiboles, containing neither Ca nor alkaline cations, no purely Fe^{2+}-orthorhombic and Mg-monoclinic varieties were found at present, either in nature or under experimental conditions (Cameron 1975).

Hadji et al. (1978) have described the crystallization conditions as well as crystal-chemical and morphological characteristics for fibrous monoclinic fluoroamphibole having the composition close to Mg$_7$Si$_8$O$_{22}$F$_2$. The purity of this synthetic F-cupfferite has been confirmed by the fact that all the maxima registered in the X-ray powder pattern can be readily indexed in terms of the unit cell $a = 9.512$ Å, $b = 17.98$ Å, $c = 5.278$ Å, $\beta = 102°$.

SAED has also proved that the phase involved should be attributed to F-cupfferite, i.e., monoclinic amphibole with space group $C2/m$. Figure 95 shows the F-cupfferite reciprocal lattice projected down b. If (100) is the most developed crystal face, then the incident beam is initially parallel to a^*. After the crystal is rotated through 12° counter-clockwise with respect to b^*, the incident beam becomes parallel to a. The point SAED pattern will then contain an orthogonal reflection network okl. The lattice being centered, reflections having $k = 2n+1$ are absent (Fig. 74c). If the crystal in the initial position is rotated through 19°

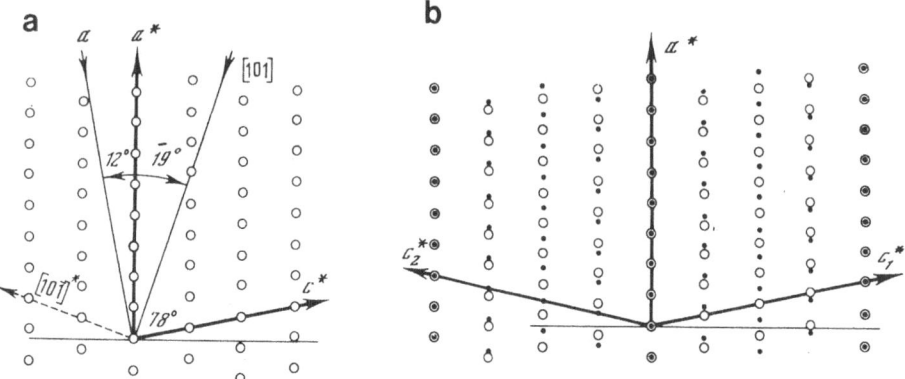

Fig. 95a, b. Reciprocal lattice projection down b for **a** F-cupfferite; **b** a F-cupfferite crystal twinned on (100)

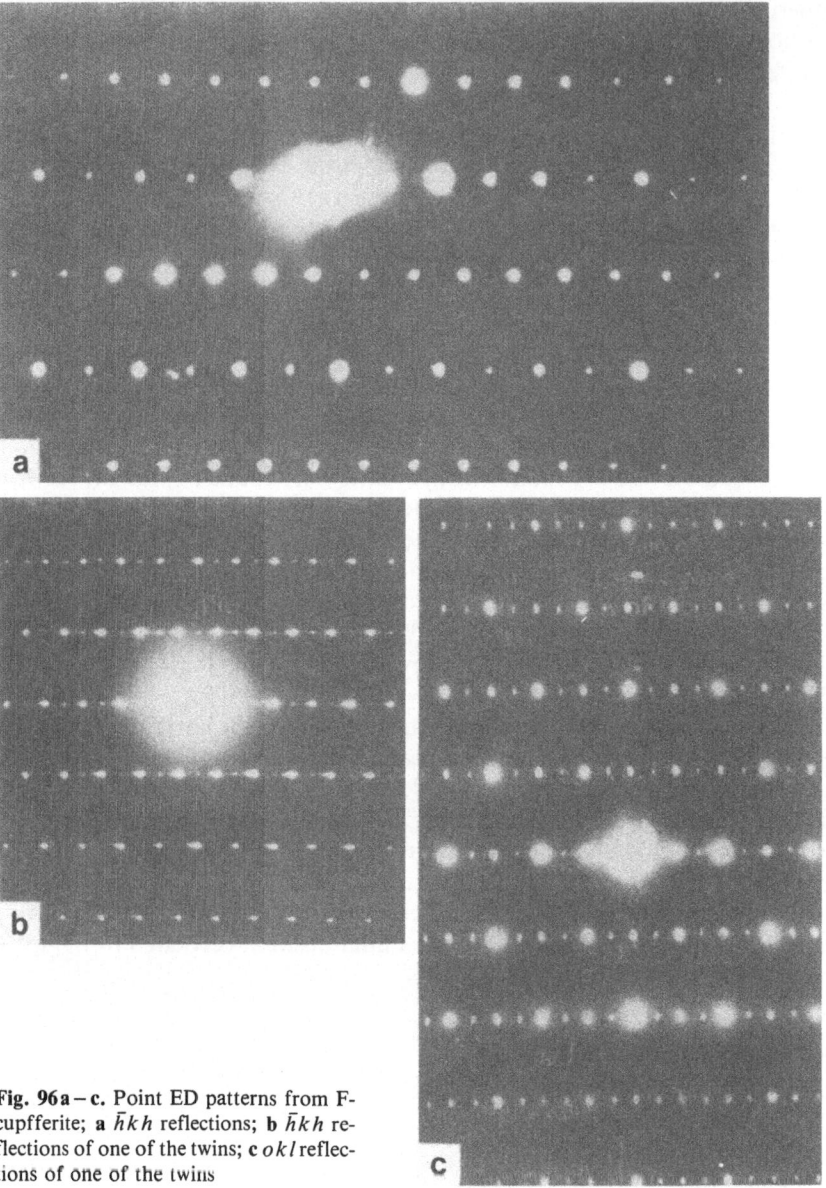

Fig. 96a–c. Point ED patterns from F-cupfferite; **a** $\bar{h}kh$ reflections; **b** $\bar{h}kh$ reflections of one of the twins; **c** okl reflections of one of the twins

counter-clockwise with respect to b^*, the incident beam becomes parallel to the direction [101] and the point pattern will contain reflections $\bar{h}kh$ distributed over a centered pattern, in accordance with space group $C2/m$ (see Fig. 96a). The orthogonal axes for the reflection network are parallel to the reciprocal lattice directions b and $[10\bar{1}]^*$.

The point ED pattern shown in Fig. 96b contains, besides the reflection distribution described, additional weak reflections forbidden for the given space

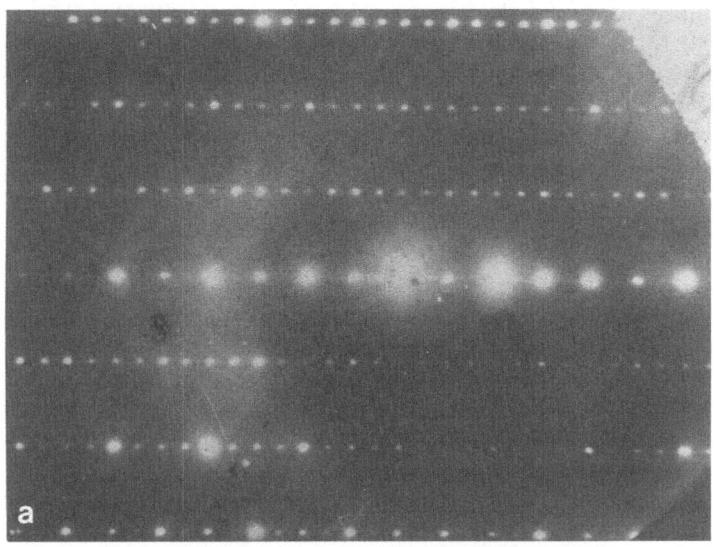

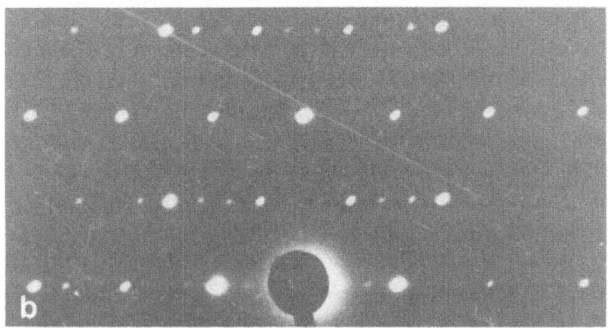

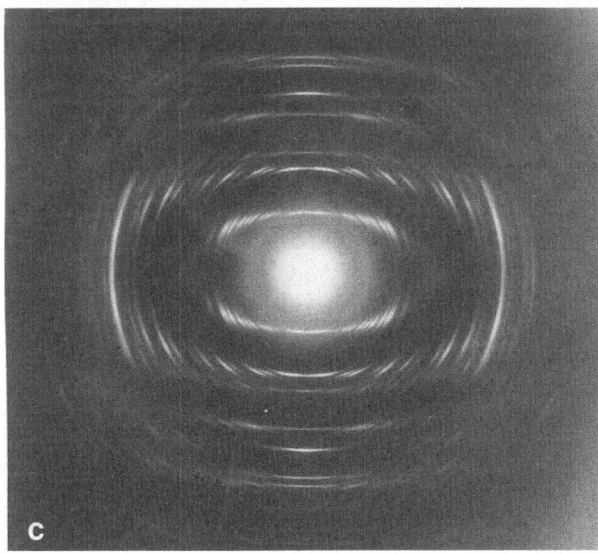

Fig. 97a–c. Point ED patterns from a F-cupfferite, okl reflections; b triple-chain silicate, $(\bar{h}kh)$; c OTED pattern from $2M_1$-muscovite

group. Similar patterns have been described by Hutchison et al. (1975) and Drits et al. (1973, 1976a) for natural and synthetic fibrous amphiboles, respectively. Such diffraction effects result from twinning, where the twins have common axes b and c, while their a axes are related by the mirror plane (see below).

SAED has proved that fibrous amphiboles synthesized at lower temperatures are monoclinic and belong to space group $P2_1/m$. A point ED pattern containing the reflections okl is shown in Fig. 97a. The screw axis 2_1 leads to extinctions only among reflections oko with $k = 2n+1$. The monoclinic unit cell shape with $\beta \simeq 102°$ is clearly seen in the photograph (Figure 113c) representing the reciprocal lattice plane (021)*. According to the space group $P2_1/m$ the zero line contains reflections hoo, the first line contains reflections $h\bar{1}2$ and the fourth contains reflections $h\bar{2}4$ with h even and odd. The reflections are strongly elongated along a^*, which may result from very small dimensions of coherent domains along this directions. As in the case of enstatite this is associated with numerous twinning effects in the plane (100). Clino- and orthoamphibole structures projected on the plane ac are quite similar to the corresponding projections of pyroxene structures (Fig. 93). Therefore, it seems natural that the twinning mechanisms are also similar.

In the clinoamphibole structure (010) projection, lamellae of the width of 4.7 Å can be distinguished passing through, e.g., octahedral centers, or through the middle of the successive tetrahedral ribbons. If two stacking faults related by twinning occur in two successive planes (200), a protoamphibole fragment appears. The parts of the crystal separated by this fragment are incoherent, as the equivalent points are separated by the distance $a \sin \beta$ normal to (001) instead of the vector a. If similar twinning proceeds in (100), an antophyllite fragment is formed, leading to a similar failure of coherence for the parts of the crystal separated by it.

Amphibole crystallites can contain the so-called Wadsley defects associated with chain-width disorder (see below). The zero line in a SAED pattern from such a crystallite contains relatively sharp reflections with spacings of 9 Å/n, while in the other layer lines these reflections merge together to form almost continuous fringes parallel to b.

11.4 Crystal Structures of Triple-Chain Silicates

In connection with the examination of stability fields for asbestiform amphiboles, structures of synthetic products obtained in the system $Na_2O - MgO - SiO_2H_2O$ (system I) and $NaF - MgF_2 - MgO - SiO_2$ (system II) were studied under various thermodynamical conditions. The experimental conditions and the products for each system obtained under different P and T were described by Goncharov et al. (1973a, b). The initial mixture composition was close to that of stoichiometrical $Na - Mg$ richterite.

A pure fibrous phase having diffraction properties differing from all the known $Na - Mg$ silicates was synthesized in system I under $400° - 500°C$ and $250 - 1000$ atm. In system II under $500° - 800°C$ a compound was synthesized whose diffraction properties differed from those of all the known $Na - Mg$ silicates, as well as of the phase obtained in system I.

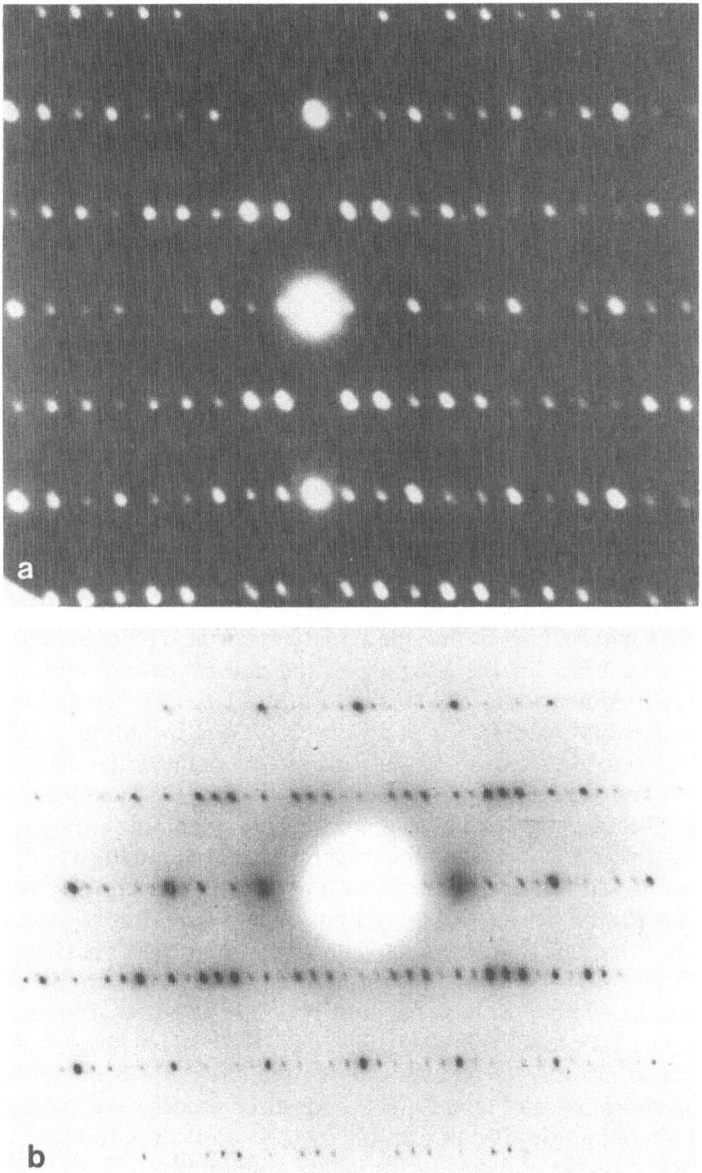

Fig. 98a, b. Point ED patterns from **a** triple-chain silicate (*okl* reflections); **b** twinned triple-chain silicate *okl* and *h̄kh* reflections

The X-ray powder pattern from the phase obtained in system I resembles those from micas rather than from amphiboles, especially as it contains three strong reflections that might be taken for the basal series *ool*. It is not surprising that Franz and Althaus (1974) have indexed a similar powder pattern in terms of a micaceous unit cell, although they have been actually dealing with a chain silicate.

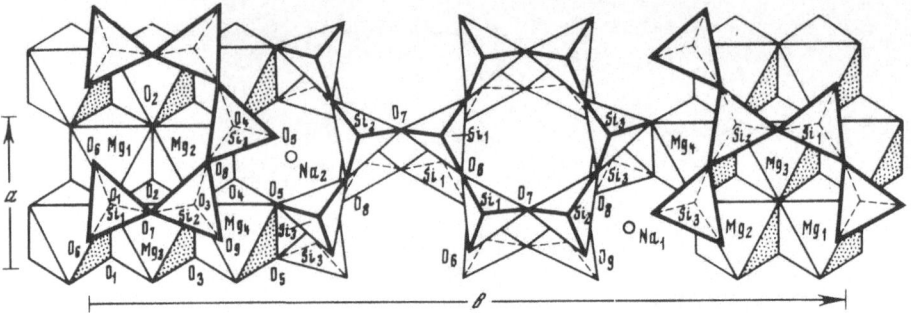

Fig. 99. Triple-chain silicate structure projection on (100)

EM showed that the crystallites in question have the shape of long planar ribbons. A goniometer was used to rotate separate crystallites about the incident beam so that the point SAED patterns contained reflections okl and $\bar{h}kh$ (Figs. 98a and 97b). Geometrical analysis of these patterns indicates that the repeat distance along the fiber axis is 5.25 Å and the period along the direction normal to it is 27 Å. To be more exact, the distance between reflections along the layer lines corresponds to the periodicity of 13.5 Å, but the centered pattern in the reflection distribution in the $\bar{h}kh$ pattern implies that the actual parameter is twice as large.

These data formed a basis for the indexing of the powder pattern and the evaluation of the cell dimensions: $a = 10.132 \pm 0.005$ Å, $b = 27.12 \pm 0.01$ Å, $c = 5.257 \pm 0.005$ Å, $\beta = 106°54' \pm 10'$. The condition $h + k = 2n$ indicates that the unit cell is C-centered. Apart from b, the parameters obtained are close to those for monoclinic pyroxenes and amphiboles (see Table 11). Therefore, a structure model has been devised similar to pyroxene and amphibole structures projected on (010) but with a different arrangement of atoms along the axis b.

The projection of the structure suggested down a^* is shown in Fig. 99. The basic structural unit in this model is a silicate ribbon composed of three pyroxene chains linked together through shared apices. The composition and the charge for such a ribbon is given by the formula $n\,[Si_3O_8]^{4-}$. The ribbons parallel to c are linked to form an I-beam through Mg and Na. The idealized structural formula for this Na, Mg-phase is $NaMg_4[Si_6O_{16}](OH)_2$. By analogy with pyroxenes, the space group $C2/c$ was assumed for the structure in question. Independent atomic coordinates obtained for the idealized model were used to calculate the structure amplitudes for all possible hkl. As a rule, the reflections observed conform to all strong and medium structure amplitudes.

SAED confirmed the validity of the structure model proposed. The methods used for the evaluation of intensity and the determination of the nature of interaction between electrons and matter were the same as in the previous cases described. Projections down [100] and [101] were constructed for the electrostatic potential distribution. Theoretical structure amplitudes were calculated to determine the signs for Φ_{exp}.

For the projections down [101], this direction was assumed to be the main crystallographic axis. In the potential projection along [100], the atoms of the upper and the lower tetrahedral ribbons in I-beams nearly coincide. However,

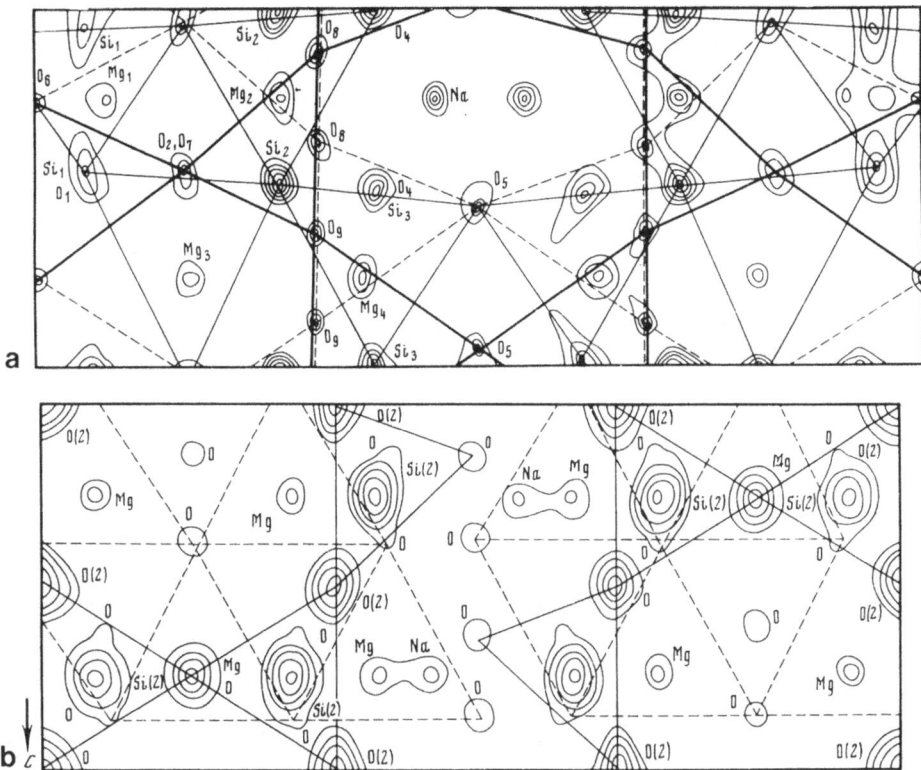

Fig. 100a, b. Electrostatic potential distribution projected down [100] and [101]

the general structural pattern, and especially the three pyroxene chains in a triple-chain ribbon, are clearly seen (see Fig. 100).

The potential projection down [101], which is close to the structure projection down a^* (cf. Fig. 99 and 100), is even more expressive. Note that reasonable agreement between Φ_{exp} and Φ_{calc} with $R_{okl} = 0.22$ ($B = 2$ Å^2) and $R_{hk\bar{h}} = 0.24$ ($B = 1.8$ Å, $h + k = 2n$, $h = \bar{l}$) obtained after a minor change in the initial atomic coordinates.

The X-ray powder pattern from the phase obtained from system II may seem to differ substantially from that of the triple-chain silicate described above. However, the analysis of point SAED patterns indicated unambiguously that the compound in question is also a triple-chain silicate having the unit-cell constants $a = 10.30$ Å, $b = 27.0$ Å, $c = 5.15$ Å, $\beta = 104°$. It is the increase in a and the decrease in β that have lead to the changes in the diffraction patterns. The structural formula for the F-phase calculated on the basis of the silicate composition [Si$_6$O$_{16}$] is Na(Mg$_{4.40}$Na$_{0.60}$)(Si$_6$O$_{16}$)(F$_{0.8}$O$_{0.2}$)$_2$. The experimental data obtained allow deeper insight into the crystallization conditions for micas, amphiboles, and triple-chain silicates of the type described. The latter compounds were found to be stable in a relatively low-temperature region, intermediate between the stability fields for phyllosilicates and amphiboles. It should be mentioned that separate crystals having $b = 35$ Å occur among the synthetic products. These crystals may consist either of quadruple-chain I-beams or of single- and triple-chain I-beams alternating along b (Fig. 70e).

By analogy with pyroxenes and amphiboles, four basic topological types can be distinguished for triple-chain silicates. Figure 92 can be used to represent these if the tetrahedral ribbons are assumed to contain three single chains each.

As it was mentioned, Veblen et al. (1977) carried out a detailed structure study of jimthompsonite and clinojimthompsonite consisting of triple-chain I-beams. The stacking sequence in orthogonal jimthompsonite is described by the notation $+ + - - + + - - ...$, as in orthoenstatite ($Pbca$) and antophyllite ($Pnma$). The jimthompsonite unit-cell parameters $a = 18.63$ Å, $b = 27.23$ Å, $c = 5.30$ Å differ from those for orthopyroxenes and orthoamphiboles only in the value for b. Clinojimthompsonite is isostructural to the synthetic compounds described above but has a different composition, $(MgFe)_5Si_6O_{16}(OH)_2$. The structural formula for jimthompsonite is the same.

11.5 New Minerals Having Regular Mixed-Chain Structures

Veblen et al. (1977) and the present author together with E. I. Semenov and A. L. Dmitrik discovered natural minerals belonging to two, fundamentally new structural types. These minerals consist of I-beams of different widths alternating regularly. Veblen et al. (1977) described chesterite and "clinochesterite" consisting of slabs composed of regularly interstratified triple- and double-chain I-beams. Figure 101 is a sketch of the chesterite structure projected down c. In the chesterite monoclinic variety, the octahedra belonging to I-beams differing in width have identical orientations. The structure projection down b as well as parameters $a = 9.87$ Å, $c = 5.29$ Å, and $\beta = 109.7°$ are quite similar to those of diopside, tremolite, and clinojimthompsonite. However, in clinochesterite $b = 45$ Å, i.e., the sum of the widths for a double- and triple-chain I-beam.

The octahedral orientations in I-beams of different widths packed along a in the chesterite structure are described by $+ + - - + + - - ...$, as in orthoenstatite and antophyllite (Fig. 101).

Fig. 101. Chesterite structure projection down c

All chain and ribbon-layer silicates (sepiolite and palygorskite) are known up to now to have definite a values for each structural type regardless of b. For ribbon-layer structures, $a \approx 13$ Å, while for chain structures a is usually $9.6 - 10$ Å for monoclinic varieties, and $18.2 - 18.7$ Å for orthogonal varieties.

In the present book, a new mineral structure is described consisting of I-beams of different widths. Two patterns in the bonding of I-beams, one typical of chain silicates and the other of ribbon-layer silicates are supposed to be present in the structure.

The mineral in question, a hydrous silicate of Fe, Nb, and Mn, was discovered by Professor Semenov. The name "atlantite" was reserved for the mineral. According to Professor Semenov (personal communication) the density of atlantite is 3.1 and the chemical composition is Nb_2O_5 9.0, TiO_2 2.86, SiO_2 34.90, Al_2O_3 0.47, Fe_2O_3 30.52, FeO 7.95, MnO 4.45, H_2O^- 1.45, H_2O^+ 8.25, the sum of oxides 99.94%.

To determine the unit-cell parameters, SAED and X-ray powder diffraction were applied. Owing to textural peculiarities of the sample, the X-ray powder pattern resembled a rotation pattern. The repeat distance along the "rotation"

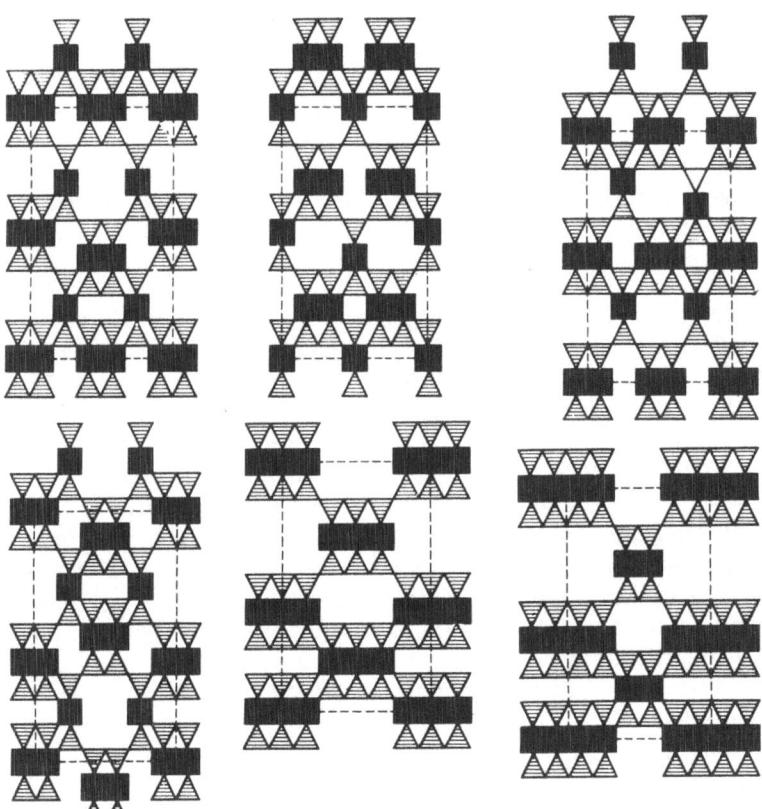

Fig. 102. Possible structures with the unit-cell parameters $a = 23$ Å; $b = 27$ Å; $c = 5.3$ Å

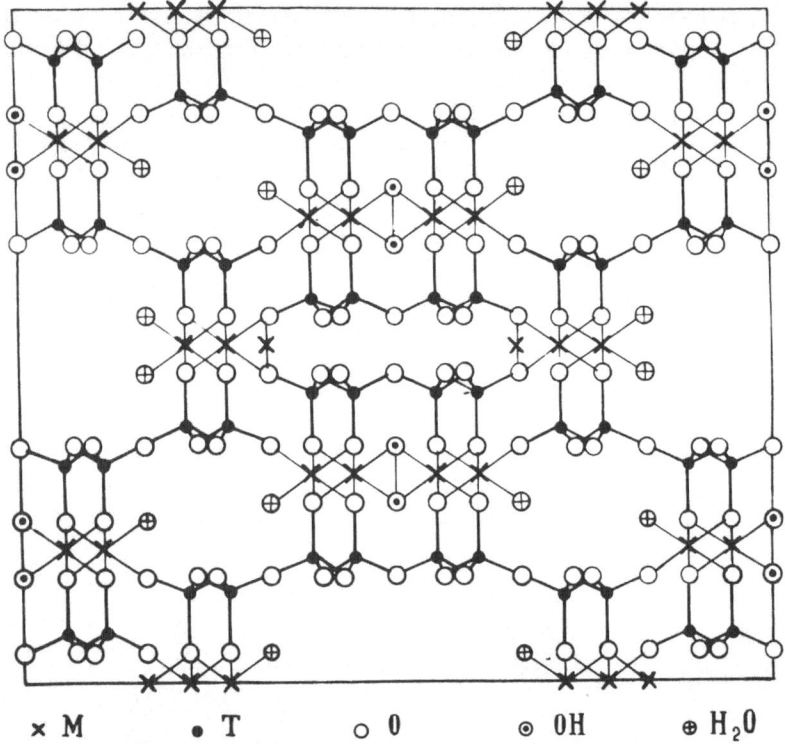

× M   • T   ○ 0   ⊙ OH   ⊕ H$_2$0

Fig. 103. Structural model for atlantite projected down *c*. *M* and *T* are octahedral and tetrahedral sites, respectively

axis was found to be 5.37 Å. The zero layer line was indexed under the assumption that $a \sin\beta = 23.34$, $b = 26.98$ Å. The indexing of reflections in other layer lines suggested a monoclinic unit cell: $a = 23.61$ Å, $b = 26.96$ Å, $c = 5.375$ Å, $\beta = 98°42'$, where $-a\cos\beta = 2c/3$.

A number of reciprocal lattice sections were obtained by SAED. Figure 113a shows a point SAED photograph. The reflections $\bar{h}kh$ are distributed over a centered pattern with $b = 27$ Å and the repeat distance along $[\bar{1}01]^*$ of 5.3 Å. The absence of hol reflections with $l = 2n+1$ implies space group $C2/c$ or Cc. The SAED patterns obtained also show that, for reflections hoo, $h = 2n$. The chemical formula $(Nb_{1.07}Ti_{0.56}Al_{0.16}Fe^{3+}_{6.0}Fe^{2+}_{1.73}Mn_{0.99}Si_{9.14})O_{34} \, nH_2O$, $Z = 4$ reflects only the cationic composition of the unit cell, since the exact quantitative data on different types of structurally bonded water are absent.

The unit-cell parameters of the mineral under study were compared to those of sepiolite, palygorskite, and chain-silicates. The parameters b and c were found to be similar to those of sepiolite and the triple-chain silicate, while $a = 23.6$ Å is close to the sum of the parameters $a \simeq 13.3$ Å for sepiolite, and $a \sim 9.7-10$ Å for amphiboles and pyroxenes. On this basis a number of structural models were considered. In each of these, the two patterns in the bonding of *I*-beams mentioned above were supposed to be present simultaneously.

Figure 102 shows some of the possible structures projected down c that are in agreement with the above unit-cell parameters.

For each model, atomic coordinates were determined and the structure factors were calculated and compared with measured intensities. The best agreement was obtained for the model shown in Fig. 103 projected down c. Single- and double-chain I-beams parallel to c alternate regularly along b. To a certain degree, this structure may be treated as a mixed-chain one. Single-chain I-beams are attached to double-chain ones according to two alternating patterns. In the case of the ribbon-layer silicate pattern, the bases of tetrahedra belonging to different I-beams lie in the same plane, while their apices are in opposite directions with respect to the plane bc. In the case of the chain-silicate pattern, basal anions in tetrahedra belonging to one I-beam are the outer octahedral anions for the adjacent I-beam, and vice versa.

In all the I-beams, the octahedra are identically oriented with respect to c, which leads to monoclinic cell and space group $C2/c$.

The structure factors calculated on the basis of the idealized atomic coordinates for the model described above are only in qualitative agreement with the experimental intensities. At the same time, the agreement is better than for any other structural model.

The structural formulae for the model in question can be calculated irrespective of the mineral density and the unit-cell volume.

There are two possible cases. In the first case, all the "outer" octahedral anions shown as open circles with crosses inside the Fig. 103 are represented by H_2O molecules. Then the anionic framework of the unit cell is $O_{128}(OH)_8 (OH_2)_{16}$ and the formula calculated for $Z = 4$ is $[S_{8.86}Al_{0.14}Fe_{3.00}O_{32}] \cdot [Fe^{3+}_{2.8} Fe^{2+}_{1.67}Mn_{0.95} Ti_{0.55} Nb_{1.04}]_{7.00}(OH)_2 \cdot (H_2O)_4$. It is seen that this is in reasonable agreement with the formula discussed above. We also have to assume that approximately every third tetrahedron is occupied by Fe^{3+}, which ensures a more efficient compensation of high positive charges of octahedral Nb^{5+}, Ti^{4+}, and Fe^{3+}. This could explain the increased value for $c = 5.375$ Å as compared with chain minerals without substitutions in tetrahedra. Furthermore, the trans-octahedra in amphibole- or palygorskite-like I-beams should be assumed vacant. This is acceptable, as the trans-octahedra in the palygorskite structure were found to be vacant even in the presence of octahedral Al having a smaller charge than Nb and Ti (Drits and Sokolova 1971).

Another version for the structural formula can be obtained under the assumption that the anionic composition is $O_{128}(OH)_{16}(OH_2)_8$, i.e., the water molecules are located only in channels that are similar to those in the palygorskite structure. Then we have $[Si_{9.14}Al_{0.14}Fe^{3+}_{2.72}O_{32}] \cdot (Fe^{3+}_{3.27}Fe^{2+}_{1.73}Mn_{0.98}Ti_{0.56}Nb_{1.06}) (OH)_4(OH_2)_2 \cdot nH_2O$. This formula coincides exactly with the one calculated on the basis of the density and the unit-cell volume. Here we have to assume that the trans-octahedra in double-chain I-beams are occupied by cations with a 60% probability. Since the quality of the initial diffraction data was insufficient, it appeared impossible to choose unambiguously between the two versions for the structural formula.

On the whole, the results obtained indicate that there are much closer structural relationships between phyllosilicates, chain- and ribbon-layer silicates than was imagined previously.

11.6 Some Methodological Aspects in the Interpretation for Point SAED Patterns from Chain Silicates

Interpretation for point SAED patterns from amphiboles was discussed by Zvyagin and Gorshkov (1969) and Hutchison et al. (1975). Some peculiarities of diffraction patterns from triple-chain silicates were considered by Drits et al. (1973, 1976a). Below is given a more detailed and systematical treatment of the interpretation for ED point patterns from chain-silicates.

Indexing of the Most Developed Faces of Fibrous Chain-Silicate Crystallites. The crystallites of fibrous silicates being extremely small, SAED is often the only method for reliable determination of indices for the most developed face (or faces). It is sufficient to know the peculiarities in the arrangement of reflections that should be observed for the given structural type depending on the indices of the most developed faces. On the other hand, knowledge of the external form of fibrous crystallites allows the drawing of conclusions on the mechanism of the crystal growth, and mechanical and other properties.

The structural features of asbestiform chain-silicates and their formation conditions lead to the elongation of the crystallites along *c*. Therefore, it seemed natural that the most developed faces should have indices (hko) with small h and k, such as (100), (110) and (010). Let us compare the point SAED patterns calculated for the incident beam assumed normal to the faces (100), (110) and (010) in crystallites of double- and triple-chain silicates with the experimental ones.

The crystallites of the type discussed are usually arranged on the support so that their axes *c* are parallel to the plane. If the incident beam is normal to the support, a pattern from any microcrystal should consist, generally, of a set of the so-called layer lines. Each line contains reflections with $l = \text{const}$ and is separated from the zero line by the distance $r_l = L\lambda lc^* \sin\beta^* = L\lambda(l/c)$ where $l = \pm1, \pm2\ldots$

Assume that (100) is the most developed face in the crystallites of a triple-chain silicate belonging to space group $C2/c$ and having the cell constants $a = 10.0$ Å, $b = 27.1$ Å, $c = 5.28$ Å, $\beta = 110°$. Note that here $2a^* = 3c^* \cos\beta^*$. Inserting the initial data into (1.12), we obtain the ratios for the indices of a reciprocal lattice plane parallel to the crystal face (100)

$$m:n:p = 1:0:c^* \cos\beta^*/a^* = 1:0:2/3 . \tag{11.1}$$

Hence the reciprocal lattice plane containing the "diffracting" nodes is (302)*. According to (1.11), the relationship between the indices for the zone [302] parallel to the incident beam and those for the reflections contained in the SAED pattern is given by

$$3h + 2l = 0 . \tag{11.1a}$$

The zero layer line can contain, according to (11.1a), only the reflections oko. This is natural, as the incident beam is parallel to the axis a^*, and the plane passing through the zero node normal to a^* contains the axis b^* and, therefore, the oko reflections. In accordance with the space group $C2/c$, $k = 2n$ for these reflections.

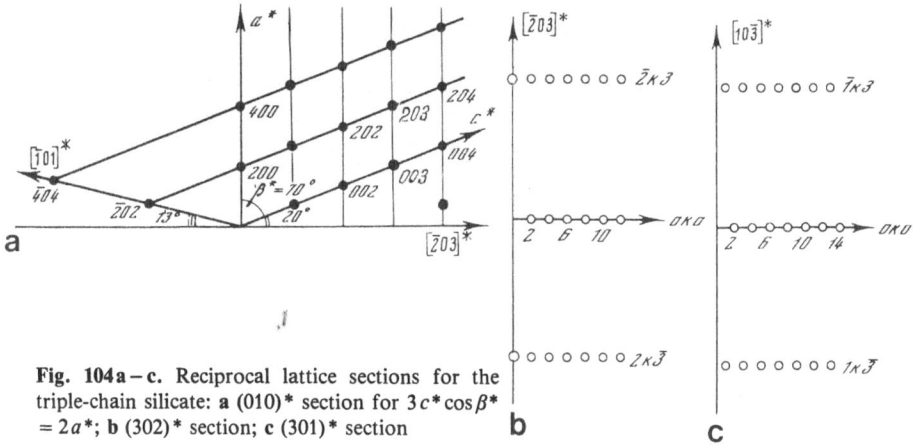

Fig. 104a–c. Reciprocal lattice sections for the triple-chain silicate: **a** (010)* section for $3c^* \cos \beta^* = 2a^*$; **b** (302)* section; **c** (301)* section

Equation (11.1a) implies that integer h values are possible only for $l = 3n$. For example, for $n = 1$, the third layer line will contain the reflections $\bar{2}k3$ where $k = 2n$, as the cell is C-centered. On the whole, the reflection distribution in the SAED pattern can be described in terms of a primitive orthogonal cell whose dimensions are $2/b$ and $3/c$.

A clear view of the specific features of this reflection distribution can be also obtained by a direct geometrical construction. Figure 104a shows the reciprocal lattice plane section (010)* and the trace of the plane passing through the origin normal to a^*, which practically coincides with the Ewald sphere. It is seen that a^* is perpendicular to the direction $[\bar{2}03]$*. Hence the SAED pattern will contain a set of lattice rows containing reflections oko, $\bar{2}k3$, $\bar{4}k6$, etc. (see Fig. 104b).

The example discussed shows that the distribution of reflections in point SAED patterns obtained with the incident beam normal to (100) is largely governed by the relationship between a^* and $c^* \cos \beta^*$. For example, if $a^* = 3c^* \cos \beta^*$ for the crystals in question, there would be reflections only on the zero layer line and on the third one. However, according to (1.11) $3h + l = 0$, and the third layer line would therefore contain reflections $\bar{1}k3$, where $k = 2n + 1$. Thus, reflections would be distributed over an orthogonal centered pattern with the parameters $2/b$ and $6/c$ (see Fig. 104c).

If $a^* = 2c^* \cos \beta^*$ for a triple-chain silicate, layer lines with l odd will be absent in a SAED pattern obtained with the incident beam normal to (100). The reflection ditribution will be described by a centered cell, since the second layer line contains reflections $\bar{1}k2$ where $k = 2n + 1$, the fourth line contains reflections $\bar{2}k4$ where $k = 2n$, etc. In general, a certain relationship between a^* and $c^* \cos \beta^*$ may lead to reflections present only in the zero layer line.

Analysis of the experimental data obtained for synthetic triple-chain silicates shows that the crystallite shape depends on the chemical composition. For example, the most developed face for Co-triple-chain silicate is (100) with $a^* = (3/2)c^* \cos \beta^*$ (Fig. 105a). Crystals of Na–Mg triple-chain silicate having the most developed face (100) are relatively rare. The corresponding SAED patterns contain reflections only in even layer lines, since $a^* \sim 0.56 \cdot c^* \cos \beta^*$ (Fig. 105b).

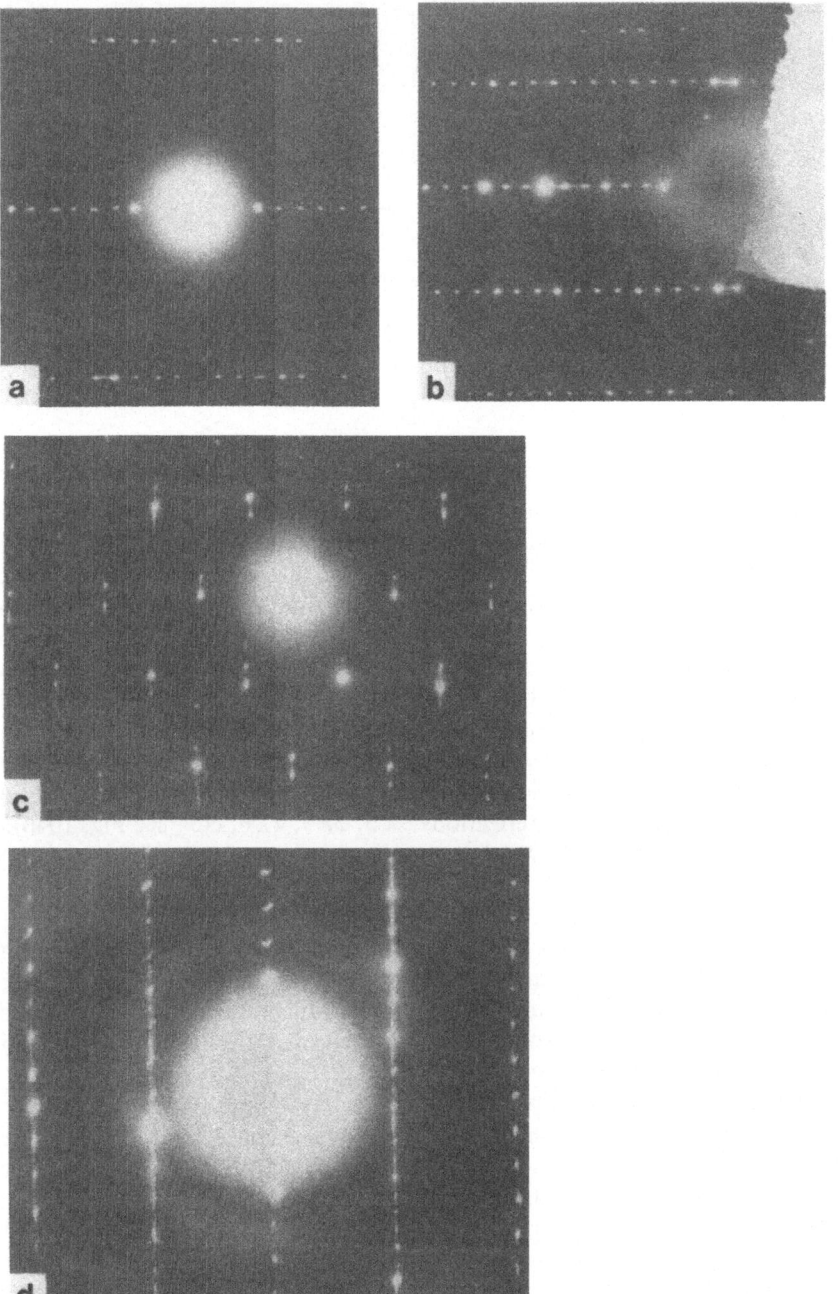

Fig. 105a – d. Point ED patterns representing **a** (302)* of Co-triple-chain silicate; **b** (201)* of Na – Mg-triple-chain silicate; **c** (818)* of Na – Mg-triple-chain silicate; **d** superposition of ED patterns from the twins with the incident beam parallel to b

Table 12. Indices $(mnp)*$ for the reciprocal lattice plane parallel to the crystal face (110) for chain-silicates

Mineral	$\dfrac{c*\cos\beta*}{a*}$	$\left(\dfrac{b*}{a*}\right)^2$	$(mnp)*$	Unit cell parameters			
				a, Å	b, Å	c, Å	β, °
Tremolite	0.47	0.28	(412)*	9.818	18.047	5.273	104.6
F-cupfferite	0.38	0.28	(11.3.4)*	9.512	17.98	5.278	102.2
Grunerite	0.37	0.255	(11.3.4)*	9.564	18.39	5.339	101.9
Anthophyllite	0	1.06	(110)*	18.586	18.065	5.289	90
Na – Mg-triple-chain silicate	0.56	0.128	(16.2.9)*	10.132	27.12	5.257	106.9
Clinojimthompsonite	0.62	0.11	(916)*	9.874	27.24	5.316	109.4
Co-triple-chain silicate	0.625	0.12	(815)*	9.95	27.1	5.29	110
Jimthompsonite	0	0.468	(210)*	18.63	27.23	5.297	90

The general features of the reflection distribution for amphiboles are similar to those described, if $a* = (m/n) \cdot c*\cos\beta*$, where m and n are small integers. The main difference is associated with the periodicity along $b*$, since for amphiboles $b = 18$ Å. Besides, it should be taken into account that the angle β in monoclinic single- and triple-chain silicate structures increases from 105° to 110° with the decrease of the contents of Na and Ca, whereas the amphiboles β decreases from 106° to 101° in a similar situation. Therefore, the most frequent values for $c*\cos\beta*/a*$ range from 1/3 to 1/2 for amphiboles, and from 1/2 to 2/3 for triple-chain silicates. For example, both clinojimthompsonite and fluorocupfferite contain only relatively small cations Mg and Fe. However, $c*\cos\beta*/a*$ is close to 2/3 for the former ($\beta \approx 110°$) and to 1/3 for the latter ($\beta \sim 102°$). It is these differences that determine the reflection distribution in the layer lines with $l = 3n$ in the corresponding SAED patterns.

At the same time, even approximate estimates for the relationships between $a*$ and $c*\cos\beta*$ from SAED patterns obtained with the incident beam normal to (100) should be treated with care. Figure 104a shows that for $c*\cos\beta* = (2/3)a*$, rotation of the crystal through $3° - 5°$ about $b*$ would lead to a set of reflections $ok0$, $\bar{1}k2$, $\bar{2}k4$, etc. distributed over a pattern that should be expected for $a* = 2c*\cos\beta*$. The differences between the SAED patterns are associated with the changes in the spacings between the layer lines which, however, are very difficult to reveal experimentally.

Another face in chain-silicate structures that is often well developed is (110). For monoclinic crystal lattice plane coinciding with the Ewald sphere for the incident beam normal to (110) is defined by

$$m:n:p = (a*)^2:(b*)^2:a*c*\cos\beta* = 1:(b*/a*)^2:c*\cos\beta*/a*. \qquad (11.2)$$

For monoclinic double- and triple-chain silicates of any composition it may be assumed, as a first approximation, that $(b*/a*)^2$ is $0.26 - 0.28$ and $0.11 - 0.13$, respectively. Then m, n, p will be largely determined by $c*\cos\beta*/a*$. The indices $(mnp)*$ for certain chain silicates having different $c*\cos\beta*/a*$ are given in Table 12.

The construction of SAED patterns with the given indices $(mnp)^*$ is discussed for the case of Na − Mg triple-chain silicate where $(mnp)^* = (16.2.9)^*$. The Eq. (1.11) becomes

$$16h + 2k + 9l = 0 .$$ (11.3)

Hence, for hko reflections in the zero layer line the condition $8h + k = 0$ should be satisfied. Therefore, the reflections should have indices $\bar{1}80$, $\bar{2}.16.0$, $\bar{3}.24.0$, etc.

However, reflections with $h + k = 2n + 1$ will be absent, as the cell is centered.

For reflections in the first layer line, $2k = -9 - 16h$, which implies the absence of reflections in this line in the exact Bragg case. However, the values for k, if sufficiently large, can be varied by ± 0.5 or even by 1 for the given h and l, since the neighboring lattice points are closely spaced along the directions parallel to b^*.

Diffraction maxima can appear even if Bragg's Law applies approximately owing to the usual elongation of the reciprocal lattice nodes resulting from the small coherent domain thickness. Therefore, traces of reflections $04\bar{1}$, $\bar{1}31$, $1.\bar{1}\bar{3}.1$, $\bar{2}.12.1$, $2.\bar{2}\bar{0}.1$, $\bar{3}.19.1$, etc. are to be anticipated in the first layer line in a SAED pattern from Na − Mg triple-chain silicate. The second layer line should contain reflection $11\bar{2}$, $1.\bar{1}\bar{7}.2$, $\bar{3}.15.2$ etc.

The values $H(hkl) = 1/d(hkl)$ are calculated for each hkl. A set of parallel lines separated by the distance $1/c$ (on the $L\lambda$ scale) is drawn. The zero layer line is chosen and the pattern center is fixed. Then circles of the radii $H(hkl)$ are drawn marking the intersection points of the zero layer line with circles of the radii $H(\bar{2}.16.0)$, $H(\bar{4}.32.0)$, those of the first line with circles of the radii $H(04\bar{1})$, $H(\bar{1}31)$, ... and those of the second with circles of the radii $H(11\bar{2})$, $H(1.\bar{1}\bar{7}.2)$, etc.

Figure 106a shows the scheme of a SAED pattern that should be expected from Na − Mg triple-chain silicate crystals for the incident beam normal to (110). According to the equation $-h = k/8 + 9l/16$, reflections having equal h are connected by straight lines. Another simple method for simulating SAED patterns in question thus becomes evident.

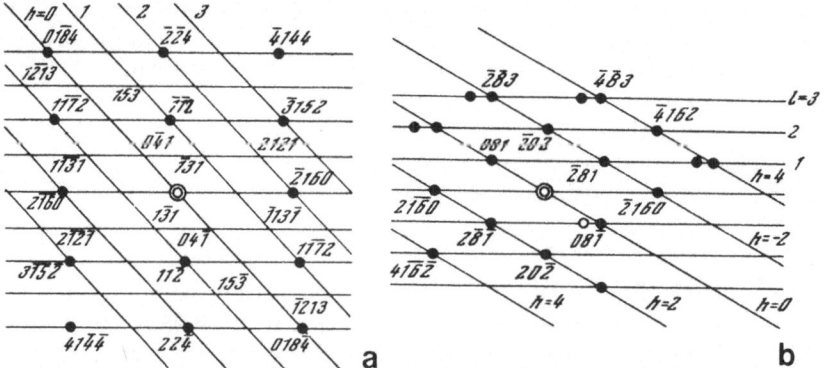

Fig. 106a, b. Reciprocal lattice point distribution for Na − Mg triple-chain silicate; a in the plane $(16.2.9)^*$; b in the plane $(818)^*$

The diagram described differs substantially from SAED patterns from most Na – Mg triple-chain silicate crystals obtained with the incident beam normal to the most developed face (Fig. 105c). This may result from the fact that the most developed crystal face may slightly deviate from (110) owing to the peculiarities in the crystal growth. To see this, let us find the indices for the plane normal to the incident beam in the case of the diffraction pattern shown in Fig. 105c. The diagram representing this pattern with the strongest reflections connected by the straight lines $h = -nk/m - pl/m = $ const is shown in Fig. 106b. Since the parameters are known, the indices for the reciprocal lattice plane $(818)^*$ coinciding with the Ewald sphere are readily obtained from the indices of the strongest reflections. These data are sufficient to find the indices for the plane parallel to the most developed crystal face using the formulae (1.7). The indices being (14.15.2), the plane is nearly parallel to the face (110).

If the incident beam is normal to the face (010), a point SAED pattern contains hol reflections. The arrangement of these permits to distinguish between clino- and ortho-chain silicates, measuring the angle β and drawing certain conclusions on symmetry and structural type. The absence of reflections in layer lines with l odd is associated with glide planes c, typical of chain silicates having an odd number of tetrahedral chains in an I-beam. The absence of hol reflections with $h = 2n+1$ implies a C-centered monoclinic cell, although these reflections may be arranged in orthogonal centered pattern if $a^* = 2c^* \cos\beta^*$. For orthogonal chain-silicates, no limitations are imposed on h for hol reflections.

A number of difficulties may arise in indexing experimental SAED patterns. Parameters a and b in orthoamphiboles have close values. Hence, it should be taken into account that $k + l = 2n$ for reflections okl if the distribution of hol reflections is described by a primitive cell.

For monoclinic triple-chain silicates of a certain cationic composition, the values for $d(110)$ and $d(100)$ may appear quite similar. For example, jimthompsonite containing only Mg and Fe in the M sites has $a \sin\beta = 9.19$ Å, while for Na – Mg triple-chain silicate $d(110) = 9.13$ Å. Therefore, only very careful measurement of d-spacings in the pattern analyzed can decide whether it is $(010)^*$ or $(1\bar{1}1)^*$ that coincides with the Ewald sphere. For example, the reflection distribution shown in Fig. 107a represents the plane $(010)^*$ of a monoclinic triple-chain silicate with space group $P2_1/c$ and unit-cell parameters $a = 9.58$ Å, $c = 5.23$ Å and $\beta = 106°$. Layer lines with $l = 2n+1$ should be absent and there should be no limitation imposed on h. On the other hand, a nearly identical reflection distribution is obtained for a triple-chain silicate crystal belonging to space group $C2/c$ if the Ewald sphere is parallel to $(1\bar{1}1)^*$. Figure 107b shows schematically the distribution of lattice points in this plane calculated for $a = 10.132$ Å, $b = 27.12$ Å, $c = 5.278$ Å, $\beta = 106.9°$.

Determination of Unit Cell Parameters. To determine the unit cell parameters and the structure symmetry, a set of rational reciprocal lattice sections is required. As it was mentioned, a set of SAED patterns having a common lattice row is of practical importance. Figure 105a shows a SAED pattern obtained from a monoclinic triple-chain silicate crystal with the incident beam parallel to a^*. The zero and the third layer lines contain reflections oko and $\bar{2}k3$, respectively.

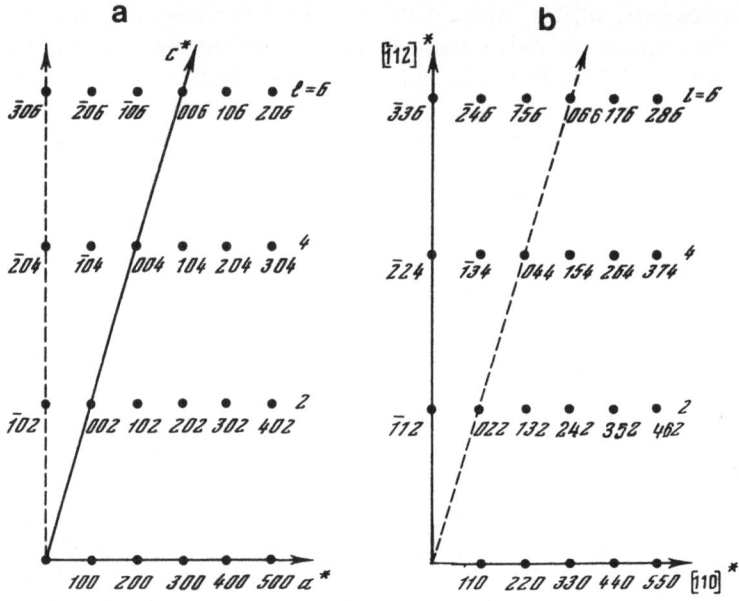

Fig. 107a, b. Reciprocal lattice point distribution for monoclinic chain silicates with different space symmetry: **a** in the plane $(010)^*$, space group $P2_1/c$; **b** in the plane $(1\bar{1}1)^*$, space group $C2/c$

The repeat distance along b^* is $0.074 \, \text{Å}^{-1}$. Rotation of the crystal through $20°$ about y^* has led to a pattern containing an orthogonal reflection network which was assumed to be $(100)^*$ (Fig. 98a). Thus c^* and β^* are readily obtained.

It is possible to proceed from a monoclinic C-centered cell to a body-centered one if the direction [101] in the initial lattice is assumed to be the new axis a, and c direction is reversed (the transformation matrix is $\begin{pmatrix} 101 \\ 010 \\ 00\bar{1} \end{pmatrix}$. Therefore, [101] is an important crystallographic direction which corresponds to the plane $(101)^*$ containing $\bar{h}kh$ reflections. If the crystal is rotated about b^* until the incident beam becomes parallel to [101], the pattern will contain linear reflection rows oko, $\bar{1}k1$, $\bar{2}k2$, etc. in the zero, the first, the second, etc. layer lines, respectively. Since $d(001)$ and $d(\bar{1}01)$ are quite close to each other for all monoclinic chain-silicates, the spacing between the layer lines in the pattern containing okl reflections is practically the same as in the pattern $\bar{h}kh$.

Rotation of the crystal through $13°$ about b^* from the initial position (incident beam parallel to a^*) in the opposite direction with respect to the previous case has resulted in a SAED pattern containing a centered orthogonal reflection network (Fig. 97b). This implies that the true periodicity along b^* is $0.037 \, \text{Å}^{-1}$. The distribution of reciprocal lattice points in the plane $(010)^*$ passing through the origin is obtained with the help of three straight lines drawn from one center. Positions occupied by reflections along the lines $[\bar{2}03]^*$, $[001]^*$, and $[\bar{1}01]^*$ in the three patterns described above are marked in along these lines. Figure 104a shows that the extinctions observed imply a C-centered cell and space group $C2/c$. The data obtained are sufficient to determine the unit-cell parameters.

Since the space symmetry of chain-silicate structures for the given evenness/oddness of the number of tetrahedral chains in silicate ribbons is generally limited to four space groups, it can be determined from a relatively small set of SAED patterns. Specifically, monoclinic chain-silicate varieties can be readily discerned using SAED patterns representing the plane $(100)^*$. Figures 97a and 74c show the SAED photographs from amphiboles belonging to space groups $P2_1/m$ ($k = 2n$ for oko only) and $C2/m$ ($k = 2n$ for all okl).

Twinning Effects: Manifestation in Point SAED Patterns. Various forms of structure defects are inherent in chain-silicates. The nature of defects in pyroxenes and amphiboles was studied by SAED and HREM by a number of workers (Buseck and Iijima 1974; Chisholm 1973; Drits et al. 1976a; Hutchison et al. 1975). Imperfections in synthetic triple-chain silicates crystals displayed in point SAED patterns were treated in Drits et al. (1977, 1976a). Below are discussed various aspects in the study of the real structure of fibrous chain-silicates on the basis of their diffraction patterns. Special attention is given to triple-chain silicate structures.

Figure 105d shows a point ED pattern from a triple-chain silicate crystal obtained with the incident beam parallel to b. Two oblique reflection networks are clearly seen, both of which are described by the following unit cell: $a^* = 1/9.30$ Å; $c^* = 1/5.0$ Å^{-1}, $\beta^* = 70.4°$. These parameters are in agreement with a, c and β for Co triple-chain silicate. Since the condition $c^* \cos\beta^* = 0.625\, a^*$ applies almost exactly, the two networks have common reflections not only in the zero layer line but also in the fourth one. The absence of reflections in layer lines with l odd implies c glide planes. The space group is $P2_1/c$, as there are no extinctions among reflections with $h = 2n+1$. It is obvious that the presence of two reflection sets results from twinning on (100).

The twinning mechanism can be demonstrated by using the structure projections shows in Fig. 108. In the simplest case, the twins are related by the glide plane c passing through the tetrahedral chains belonging to the adjacent I-beams as shown in Fig. 108.

Figure 109 shows the superposed reciprocal lattice projections down b^* for the Na $-$ Mg triple-chain silicate twins. It is seen that the plane $(100)^*$ for one of the twins lies close to $(101)^*$ for the other and vice versa. Therefore, if the incident beam is parallel to the axis a of one of the twins, the diffraction pattern will contain, along with okl, reflections $\bar{h}kh$ from the other twin. If the crystal is rotated about b through $31°$, the Ewald sphere becomes coincident with the plane $(101)^*$ of the first twin and, simultaneously, will be close to $(100)^*$ of the other. As a result, the pattern obtained will be practically identical to the initial one.

Figure 98b shows a SAED photograph from a twinned crystal, containing reflections okl from one twin and $\bar{h}kh$ from the other.

The intensity distribution confirms that the crystal in question is composed of two twins both of which belong to space group $C2/c$. For a triple-chain silicate, reflections having $k = 6n$ in the layer lines with l even, and those having $k \neq 6n$

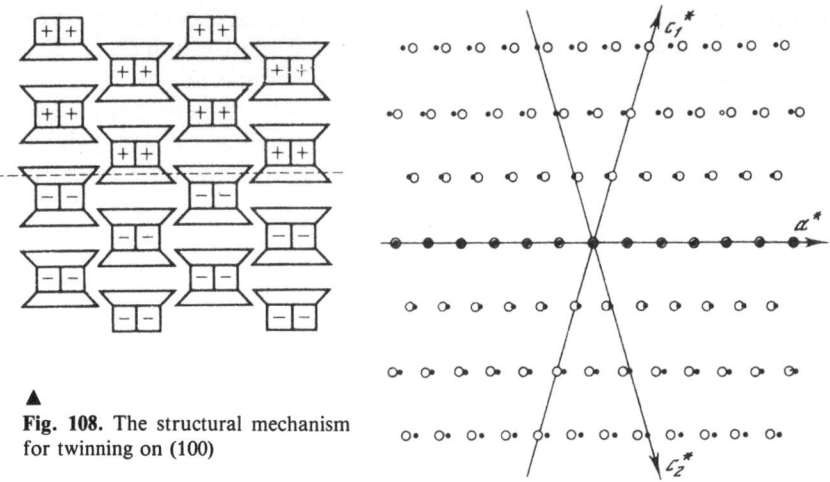

Fig. 108. The structural mechanism for twinning on (100)

Fig. 109. Superposed reciprocal lattice projections down b^* for the twins in Na – Mg triple-chain silicate

in the lines with $l = 2n + 1$, are usually the strongest among the okl reflections (Fig. 98a). In contrast, the strongest $\bar{h}kh$ reflections in the first layer line are those having $k = 6n + 3$. The superposition of diffraction patterns from the twins is in reasonable agreement with the observed pattern, except for reflections oko and $ok2$ with $k = 2n + 1$. The presence of these "forbidden" reflections in the experimental SAED pattern may be associated with secondary diffraction, i.e., the diffracted beam from one of the twins acts as the primary one as it propagates throw the second twin.

Figure 110a and b shows two diagrams representing the point ED patterns from the twins for the incident beam parallel to [100] and [101], respectively. Assume that the incident beam is parallel to the axis a of one of the twins and the strong reflection 021 acts as the primary one for the second one. Then the diffraction pattern from the second twin representing the plane (101)* should be superposed on the pattern from the first twin, so that the reflection 021 should be at the origin of (101)*. The resultant diffraction pattern (Fig. 110c) contains forbidden reflections with $k = 2n + 1$ in the layer lines with l even. Any of the reflections okl may become "initial", so that reflections generated by it in the second twin would lead to additional reflections in the pattern from the first twin. Similarly, strong reflections $\bar{h}kh$ (both h and k odd) of one of the twins may generate reflections okl, which would also lead to "forbidden" reflections. In the case of $c^* \cos^* \beta^* / a^*$ differs substantially from 1/2 there are some other peculiarities in diffraction effects from twinned crystals. Figure 67b shows the SAED photograph from twinned fluorocupferite crystallites ($c^* \cos \beta^* / a^* = 0.38$). Twinning is clearly indicated by two distinct monoclinic reflection networks. The indices for reflections hko, i.e., 130, 260, etc. are readily obtained from the spacing between hko reflections in the zero layer line. In layer lines with $l \neq 0$, reflections belonging to one of the networks have indices $n \cdot 3n \cdot l$, and those belonging to the other have $n - 1 \cdot 3n + 1 \cdot l$. In accordance with the lattice parameters, the angle between [130]* and [001]* is 96°.

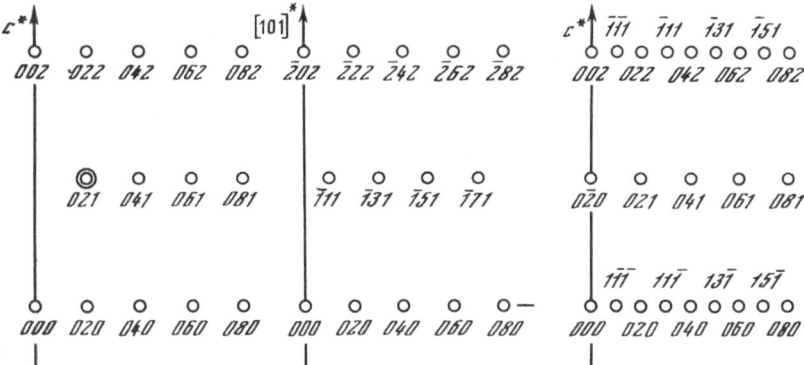

Fig. 110. Forbidden reflections resulting from secondary diffraction in the ED pattern from the twinned crystal

Twinning is also displayed in diffraction effects if the Ewald sphere cuts the planes (100)* or (101)* in one of the twins. Figure 96b, c shows the SAED patterns for twinned fluorocupfferite for two different orientations of the same crystal. The two patterns have the same geometrical arrangement of reflections but differ substantially in intensity distribution. In one of the patterns, strong reflections form an orthogonal network with a primitive cell, which is similar to the $okl(k = 2n)$ reflection distribution for the C-centered clinoamphibole crystal lattice. Weak reflections are also arranged in an orthogonal pattern and may be treated as reflections okl where $k = 2n + 1$. Thus, the incident beam for the given crystal orientation was parallel to a at least for one of the twins. After rotation through 31° about b the incident beam should have been expected to be parallel to [101] for the same twin (see Fig. 95). In the SAED pattern obtained for this orientation, the strong reflections are indeed distributed over a centered orthogonal network so that indices $\bar{h}kh$ can be ascribed to them.

The weak reflections are most probably associated with twinning on (100) (Fig. 108). Figure 95 shows the superposed reciprocal lattice projections for the two twins. It is seen that, for the incident beam parallel to a of one of the twins, the Ewald sphere passes close to the lattice rows $[\bar{1}k1]^*$, $[\bar{2}k3]^*$, $[\bar{3}k4]^*$ of the other twin. The arrangement of reflections $ok3$ in third layer line is identical to that of $\bar{2}k3$ for a C-centered lattice. Therefore, additional reflections should be anticipated only in the first and the fourth layer lines for the given orientation of the crystal.

The experimental photograph, however, contains forbidden reflections in all layer lines including the zero line. This should be associated with secondary diffraction. As in the case of the triple-chain silicate, the strong reflections $ok1$ from one of the twins may generate reflections $\bar{1}k1$ in the other twin. These would appear in the zero and the second layer lines in the diffraction pattern from the first twin, thus simulating reflections $ok0$, $ok2$ where $k = 2n + 1$. If the incident beam is parallel to [101] for one of the twins, the Ewald sphere would pass near the rows $ok1$, $1k3$, $1k4$ in the reciprocal lattice for the other twin (Fig. 95). The corresponding diffraction pattern should contain along with $\bar{h}kh$

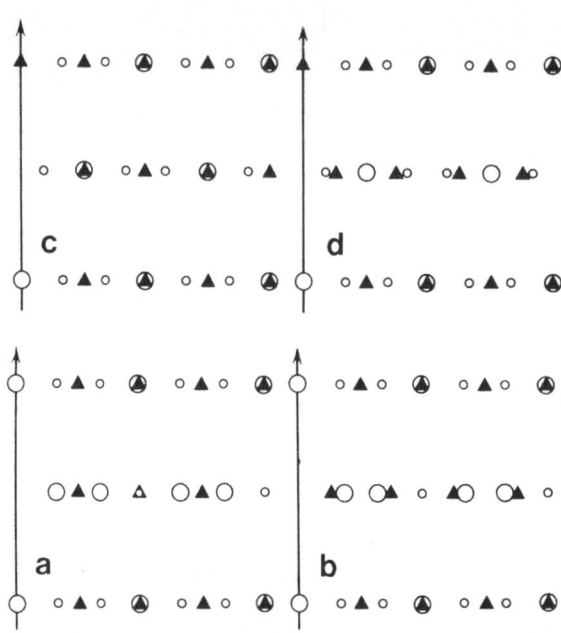

Fig. 111. Schemes for ED patterns from intergrown double- and triple-chain silicates depending on their linkage and the orientation with respect to the incident beam

additional reflections in the first and the fourth layer lines. The combination of reciprocal lattice vectors $[\bar{1}k1]^*$, $[\bar{4}k4]^*$ of one twin with vectors $[ok1]^*$, $[1k4]^*$ of the other should lead to forbidden reflections having $k = 2n+1$ in even layer lines and $k = 2n$ in odd ones.

Thus, twinning is readily identified not only from hol patterns but also from okl and $\bar{h}kh$ reflection distributions. Note that twinning breaks the coherence of the crystal along a^*, which is displayed in the elongation of the reciprocal lattice nodes parallel to a^*.

Intergrowth of Double- and Triple-Chain Silicates and Mixed-Chain Structures: Diffraction Effects. The possibility for intergrowth of amphibole and triple-chain silicates within one crystal was confirmed by us in the study of synthetic and natural asbestos crystals. Figure 111 shows four types of point SAED patterns expected for different patterns in the packing of triple- and double-chain structural fragments and for different orientations of these with respect to the incident beam. If the corresponding axes in two fragments are identically oriented, the patterns shown in Fig. 111a,d should be observed for the incident beam parallel to a and [101], respectively. If the direction [101] in one of the fragments coincides with [100] in the other, SAED patterns of two types are possible differing in the orientation of the intergrowth with respect to the incident beam. The incident beam direction may coincide either with the amphibole a axis (Fig. 111c) or with that of the triple-chain silicate (Fig. 111b).

Systematic analysis of the experimental data shows that intergrowth of double- and triple-chain silicates within one crystal occurs often enough among syn-

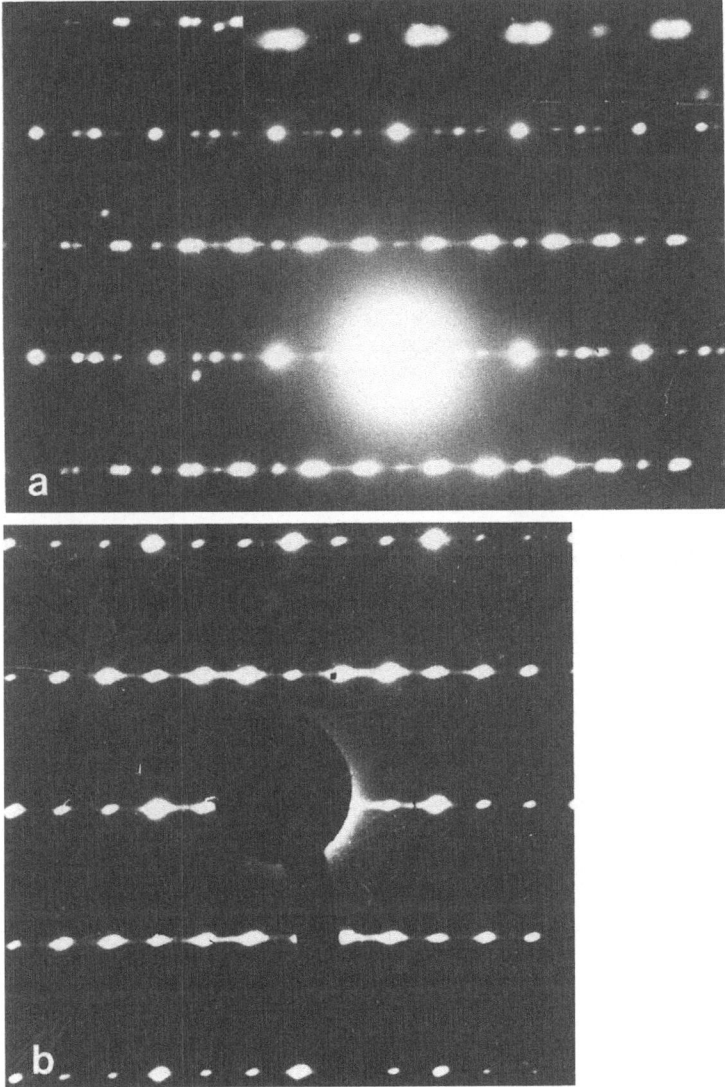

Fig. 112a, b. ED patterns from **a** intergrown double- and triple-chain silicate crystallites; **b** triple-chain silicate with numerous Wadsley defects

thetic and natural asbestoses, especially if $a^* = 2c^* \cos\beta^*$. In all these cases, the direction [100] of one component coincides with [101] of the other.

Typical SAED patterns representing the intergrowth of two fragments belonging to different structural types are shown in Figs. 112a and 113b. Comparison with Fig. 111 indicates that here the incident beam is parallel to the axis a of the triple-chain silicate and the direction [101] of the amphibole. The SAED pattern representing the twinning effect in the triple-chain silicate crystal (Fig. 98b) resembles the reflection distribution that should be observed from the inter-

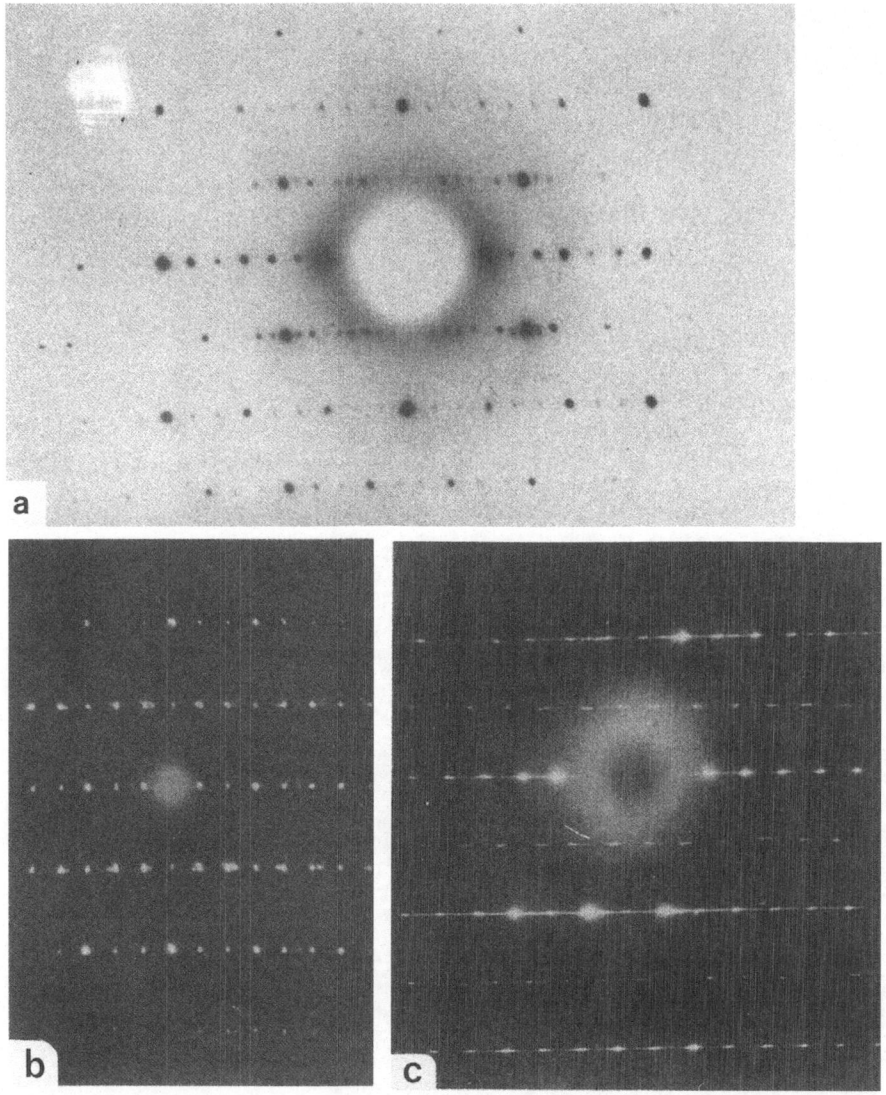

Fig. 113a–c. Point ED patterns from **a** atlantite (hkh reflections); **b** intergrown double- and triple-chain silicates with prevailing triple-chain component; **c** F-cupfferite, (021)*; reflections are elongated along *a* owing to multiple twinning

growth of double- and triple-chain silicates, with the incident beam parallel to the common axis *a* (Fig. 111a). However, the peculiarities in the intensity distribution described above imply twinning. In addition, the SAED pattern obtained after rotating the crystal about *y** to bring the incident beam parallel to [101] of one of the twins was close to the initial one obtained with the incident beam parallel to the axis *a* of the same twin. If the crystal analyzed had consisted of two intergrown fragments of different structural types, the corresponding rotation about *b* would have resulted in the pattern shown in Fig. 111d.

To elucidate the mechanism for intergrowth, one should take account of the following. Firstly, the packing of I-beams to form a uniform structure is the same for all types of chain silicates. Therefore, linking together slabs consisting of double-chain I-beams and those composed of triple-chain ones is essentially similar to the same process in pyroxenes, amphiboles, and triple-chain silicates.

Secondly, the packing leading to identical orientations for octahedra belonging to I-beams of different width is the most natural one for monoclinic structures. At the same time, as it will be shown below, identical orientations for octahedra and the preservation of centered lattices for both components imply that they are coherently linked together in such a way that the amphibole direction [100] coincides with or is close to the triple-chain silicate direction [101] and vice versa.

11.7 Direct HREM Observation of the Structural Motif of Asbestiform Chain Silicates

In the recent decade, numerous works have been published on direct HREM imaging of chain-silicate structures (Cressey et al. 1982; Darling and Zussman 1980; Drits et al. 1979; Hutchison et al. 1975; Jefferson et al. 1978; Millinson 1980; Millinson et al. 1977, 1980; Nakajima and Ribbe 1980; Nissen et al. 1979; Smith 1977; Veblen 1980, 1981; Veblen and Buseck 1979a, b, 1980, 1981; Veblen et al. 1971; Whittaker et al. 1981; Zakharov et al. 1979), etc. The experimental data obtained indicate that images even from relatively thick crystals may be directly interpretable in terms of the chain structure.

Veblen et al. (1977) and Veblen and Buseck (1979a) observed images from various chain minerals, such as anthophyllite, jimthompsonite, chesterite, with the incident beam parallel to a as well as to c.

HREM revealed intergrowth of chain-silicates differing in I-beam width in minerals that had been believed to be structurally uniform. For example, Jefferson et al. (1978) and Millinson et al. (1980) observed extremely long triple-chain silicate fragments lying parallel to b in the amphibole matrix. Nakajima and Ribbe (1980) and Veblen and Buseck (1979a) revealed domains consisting of intergrown clinojimthompsonite and altered augite. Small triple-chain silicate domains in a disordered crystal matrix intermediate between athophyllite and jimthompsonite were described by Nissen et al. (1979).

Veblen and Buseck (1979b) revealed, using HREM, a number of new ordered mixed-chain silicates having unusually large b parameters. If the symbols 2, 3, and 4 are used to designate double-, triple-, and quadruple-chain-I-beams, respectively, sequences of (010) slabs of different widths in the primitive cells of the new structures can be written (2233), (233), (232233), (222333), (2332323), (433323), (2234), etc.

We shall confine our attention to examples of direct HREM observations of amphiboles and triple-chain structures (Zakharov et al. 1979; Drits et al. 1979).

Co-containing double- and triple-chain silicates synthesized from the system $Na_2O - CoO - SiO_2 - H_2O$ under conditions close to those described in (Drits et al. 1976b) were used as objects. The unit-cell parameters were $a = 10.0$, $b = 27.0$,

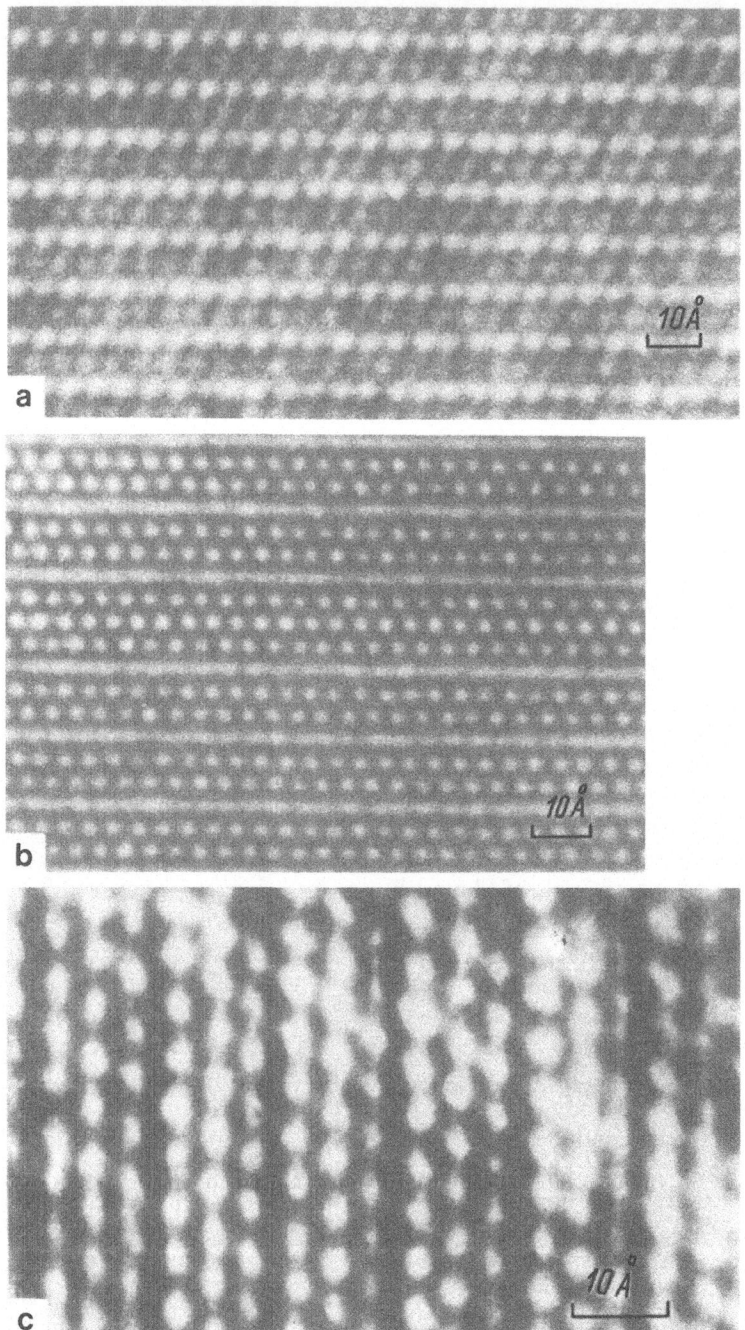

Fig. 114a – c. HREM images of **a** amphibole; the incident beam is parallel to *a*; **b** triple-chain silicate containing a quadruple-chain slab (the incident beam is along *a*); **c** triple-chain silicate; the incident beam is along [101]

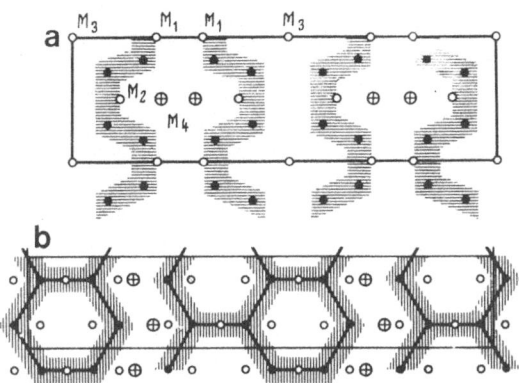

Fig. 115. Structure projections down *a*: **a** for monoclinic amphibole; **b** for monoclinic triple-chain silicate. *Shaded areas* correspond to high charge density; *open and closed circles* represent cations in M_i and tetrahedra, respectively

$c = 5.29$ Å, $\beta = 102°$ (amphiboles); $a = 10.0$, $b = 27.0$, $c = 5.29$ Å, $\beta = 110$ Å (triple-chain silicate). Experimental conditions for obtaining the images are described in (Drits et al. 1979; Zakharov et al. 1979). Let us consider first the degree of correspondence between structure and image, and discuss the possible reasons for failures of two- or one-dimensional contrast periodicity.

Figure 114a shows an image from an amphibole crystal. The image contrast consists of fringes parallel to *c* and separated by the distance 4.5 Å. Each fringe is a linear row of bright spots. Fringes consisting of very bright spots alternate along *b* with those consisting of darker grey spots and are separated by black zig-zag lines. The distance between each two bright (or grey) fringes is 9 Å, i.e., the amphibole ribbon width. The distance between the centers of the neighboring bright (or grey) spots in each fringe is 5.2 Å, i.e., the parameter *c*.

Figure 115a shows the cation distribution in the clinoamphibole structure (space group $C2/m$) projected down *a*. A qualitative relationship between the structure and the image can be obtained if the dark lines are attributed to the structure fragments having the maximum charge density, i.e., containing $2\,Co + 4\,Si + 6\,O$ per unit cell. (The Co cations occupy the sites M_1, M_2 and M_3.)

The problem is how to determine which structural elements lead to bright and grey spots. It is obvious that the contrast should be governed by the composition $2\,M_4 + 4\,O$ per unit cell for fringes of one type, and by $M_3 + 6\,O$ per unit cell for fringes of another type. If there is no strict quantitative relationship between the "density" of dark lines in the image and the value of the projected electrostatic potential or projected charge density, it is difficult to choose unambiguously between the two versions.

Interpretation for the contrast in the images from triple-chain silicate crystals proved more successful owing to the specific features of the structure projection down *a*. It is seen in Fig. 115b that the cation distribution near the boundaries between the adjacent *I*-beam slabs differs substantially from that within the slabs. Double rows of empty channels parallel to *a* are clearly seen within each slab. Traces of these channels are arranged in a chessboard pattern within each triple-chain ribbon.

Qualitatively, HREM images from the triple-chain silicate crystallites (Fig. 116a) can be regarded as directly related to the structure projection if the dark re-

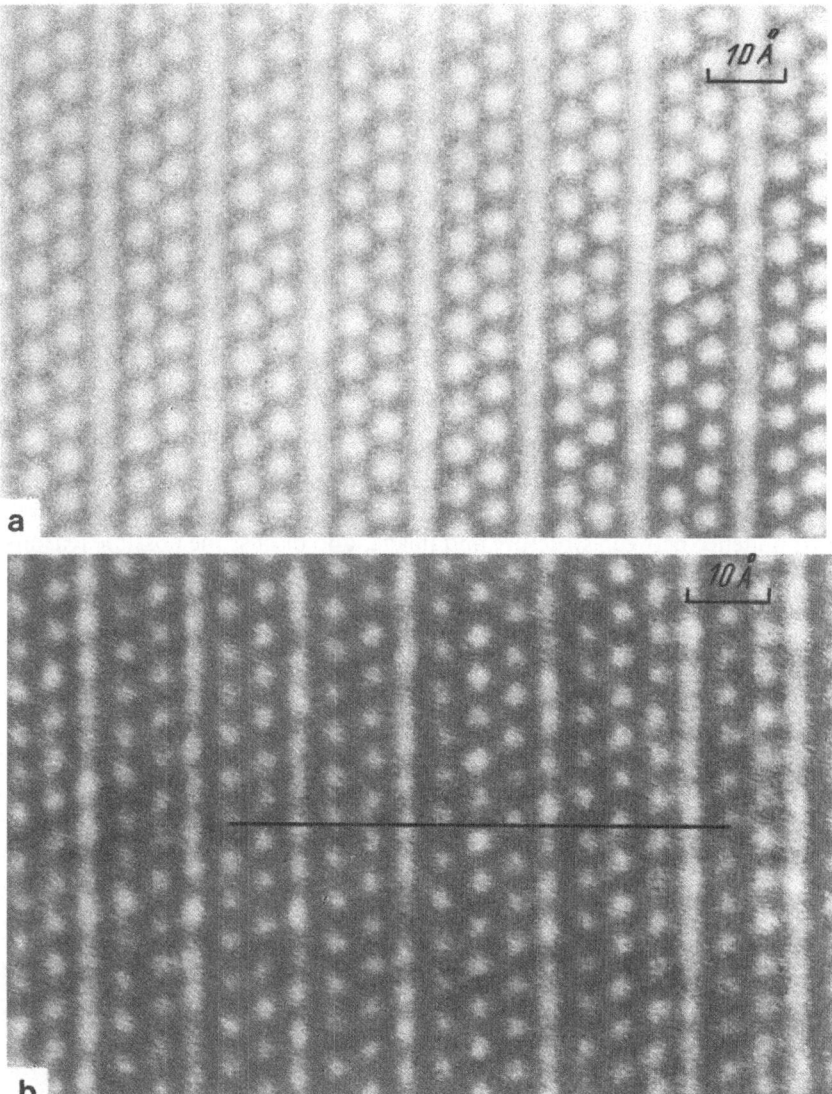

Fig. 116a – b. HREM images of **a** triple-chain silicate; the incident beam is along *a*; **b** triple-chain silicate containing two adjacent quadruple-chain slabs

gions in the image are assumed to correspond to the structure fragments having the maximum charge density (shaded in Fig. 115b). Continuous light fringes in the image correspond to the boundaries between the adjacent *I*-beam slabs.

An extremely loose packing of cations and anions in these boundary structure fragments, as well as the low scattering power of Na occupying the M_5 sites, lead to such a small contribution of diffracted beams to the corresponding regions in the image. Pairs of parallel rows of bright spots shifted with respect to one an-

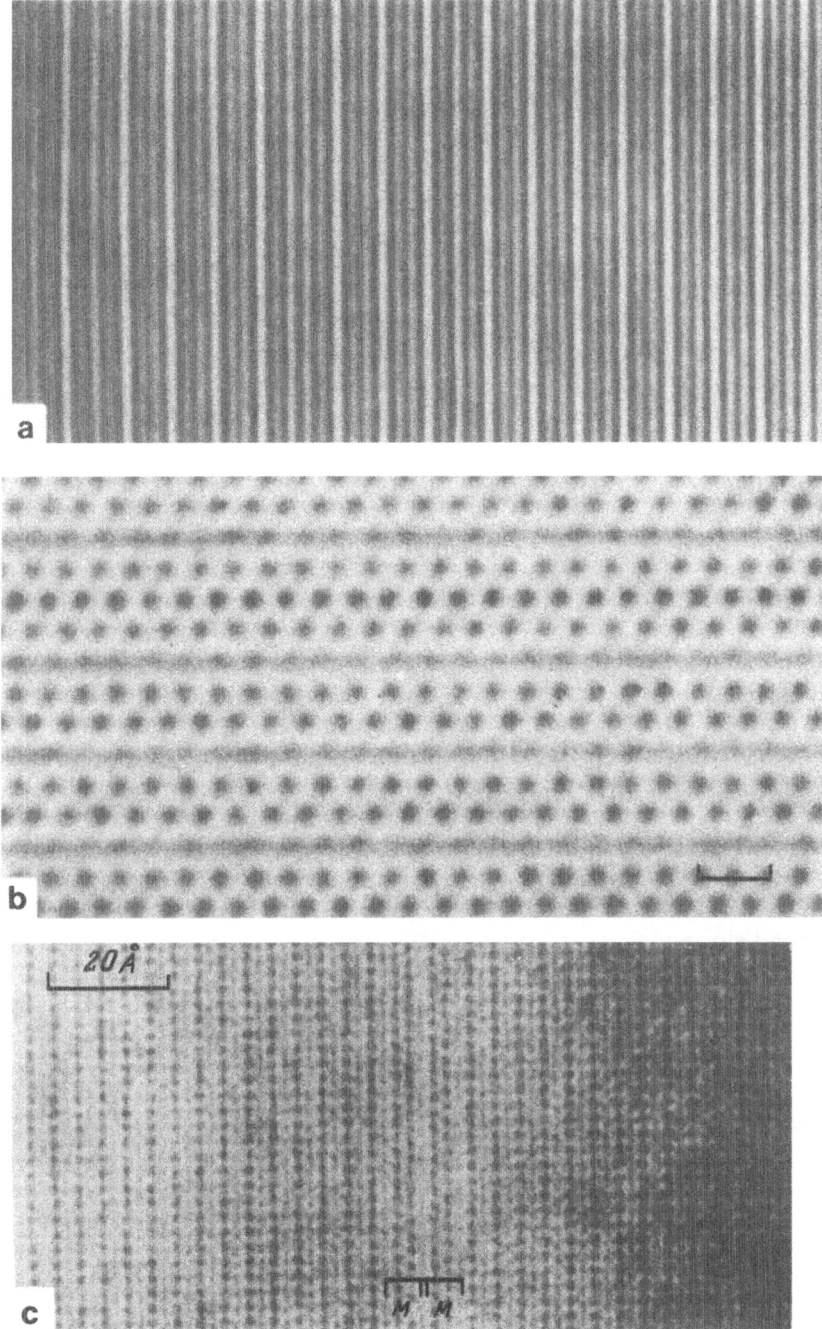

Fig. 117a–c. HREM images of **a** triple-chain silicate containing quadruple- and quintuple-chain slabs (one-dimensional fringe contrast); **b** triple-chain silicate containing a quadruple-chain slab; **c** amphibole (Hutchison et al. 1975)

other by $c/2 = 2.6$ Å between each two neighboring bright continuous fringes correspond to projections of vacant channels distributed similarly within each I-beam.

Fringe images formed by reflections oko only are also readily interpretable. The brightest lines correspond to the boundaries between I-beams, while three black straight lines located between each two bright ones correspond to three tetrahedral chains forming triple-chain silicate ribbons (Fig. 117a).

The true repeat distance along b is distinct in the images from triple-chain silicates for the incident beam close to the direction [101] (Fig. 114c). There are three rows of bright spots arranged in a hexagonal pattern in each band between each two neighboring dark, nearly continuous fringes parallel to c. The arrangement of the spots can be described by an orthogonal centered unit cell ($b = 27$ Å, $c = 5.2$ Å, see Fig. 114c). There is no direct relationship, however, between the image in question and the structure projection down [101]. Thus the periodic contrast distribution in the image resulting from the interference between the diffracted beams is only a rough approximation of the object structure projection, since the crystal is too thick, and the incident beam is not strictly parallel to the direction required.

11.8 Chain-Width Disorder in Chain-Silicates

In addition to structures consisting of (010) slabs of different widths alternating regularly, crystals having chain-width disorder are abundant among natural and synthetic chain-silicates, especially among asbestiform varieties. In such crystals, the basic structure contains (010) slabs of the "wrong" width.

In 1973, Chisholm, relying on the analysis of diffraction patterns, was the first to suggest that chain-width disorder might exist in chain-silicates. Chisholm (1973) found that okl reflections in SAED patterns from fibrous amphibole were elongated along b^*. He explained this by the intrusion of (010) slabs containing pyroxene chains or ribbons consisting of more than two chains. Drits et al. (1973, 1976a) found elongated okl reflections in SAED patterns from triple-chain silicate crystallites which they explained by the intrusion of amphibole (010) slabs into the basic matrix. Hutchison et al. (1975) were the first to prove directly the presence of chain-width defects in asbestiform amphibole structures by means of HREM.

Later, triple-chain defects in asbestiform amphiboles were observed by a number of authors (Alario Franco et al. 1977; Chisholm 1973; Cressey et al. 1982; Darling and Zussman 1980; Veblen 1980; Veblen and Buseck 1979a, 1980; Veblen et al. 1977; Whittaker et al. 1981). Slabs wider than triple-chain ones were also found in these minerals (Cressey et al. 1982; Darling and Zussman 1980; Veblen 1980; Veblen and Buseck 1979a; Veblen et al. 1977). For example, slabs consisting of 4, 5, 6, etc. I-beams were observed. Note that HREM revealed similar structural defects in nephrite (i.e., amphibole having the actinolite composition (Hutchison et al. 1975; Jefferson et al. 1978; Millinson 1980; Millinson et al. 1977, 1980; Drits et al. 1986).

Chain-width disorder was also found in natural and synthetic triple-chain silicates (Drits et al. 1979; Nissen et al. 1979; Veblen and Buseck 1979a; Zakharov et al. 1979), chesterite (Veblen and Buseck 1979a), and pyroxenes (Nakajima and Ribbe 1980; Veblen and Buseck 1981).

One of the most obvious reasons for Wadsley defects is that one and the same pattern in the packing of *I*-beams is in fact realized in all chain silicates. Therefore, alternation of slabs differing in width within a uniform structure is quite feasible without breaking the continuity of the chain-silicate structure. Then the width and the concentration of defect domains may vary within a very wide range. If a chain silicate structure contains only isolated intrusions of "alien" slabs, the uniformity and periodicity of the structure, as well as its belonging to the given mineral species, are unaffected. If (010) slabs differing in width alternate irregularly in a crystal, a mixed-chain structure is formed. Such crystal structures are in fact neither uniform nor periodic, their composition is nonstoichiometric and it is difficult to classify them into one definite mineral group. Therefore, they should be assumed to compose a distinct structural type.

Figure 121b shows an image of the whole area of a crystallite. The fringe contrast distribution implies that the crystal in question is a mixed-chain structure. Coherent scattering domains do not exceed four or five unit cells along *b*. The alternation of slabs of different widths along *b* apparently should be accompanied by diffraction effects similar to those observed from mixed layer structures (Drits and Sakharov 1976). Therefore, the reciprocal lattice nodes for mixed-chain silicates should be elongated along *b* (Fig. 112b). The elongation of reflections and the modulations of their intensity depend on the concentration of defects, their distribution and the *I*-beam widths, and scattering powers.

11.9 Contrast Distribution in *a*-Axis HREM Images for Chain-Silicate Crystals Having Chain-Width Disorder

Analysis of experimental data available has revealed a number of general features in the contrast distribution in the images from chain-silicates with chain-width disorder. The image shown in Fig. 114b corresponds to a triple-chain silicate structure fragment with an intruded quadruple-chain *I*-beam slab.

Four dark zig-zag fringes separated by three rows of bright spots are contained between two fringes corresponding to boundaries between neighboring *I*-beams. The distribution of bright spots on one side of the "defect" region is shifted by *c*/2 with respect to that on the other side.

The contrast distribution of this kind is explained by the following peculiarity in the packing of *I*-beams in monoclinic chain-silicate structures. If the corresponding crystallographic axes are brought into coincidence, octahedra in amphibole *I*-beams (and those in all chain-silicates with an even number of tetrahedral chains in silicate ribbons and space group $C2/m$) and octahedra in pyroxene *I*-beams (and in all odd-number chain-silicates) are of opposite orientation along *c*. Therefore coherent intergrowth of slabs differing in width leads to the fact that the amphibole direction [100] coincides with [101] in the triple-chain slab (and vice versa) for identical octahedral orientations. As the cells are *C*-centered in both chain silicates in question, a centered potential distribution corresponds to the projections of both structures down [101]. Thus several rules can

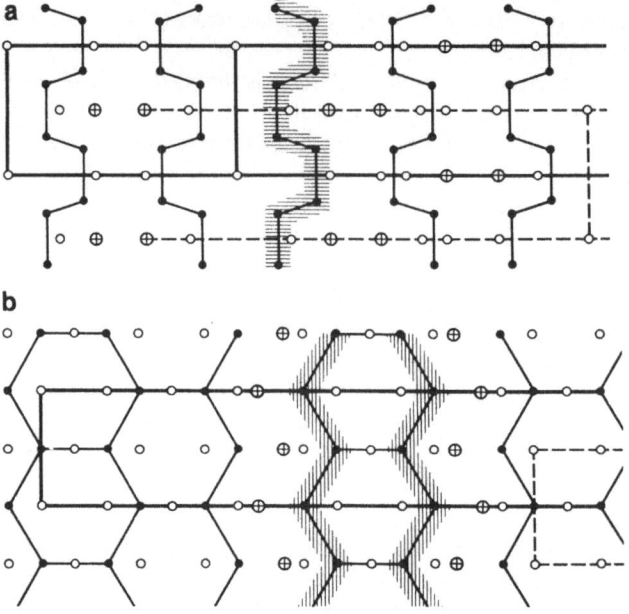

Fig. 118a, b. Shifts of one part of the structure with respect to the other by $c/2$ along c: **a** intrusion of a pyroxene I-beam slab into the amphibole matrix projected down a; **b** intrusion of an amphibole I-beam slab (shaded area) into the triple-chain silicate matrix. Open and closed circles represent M_i and tetrahedral cation sites, respectively

be formulated governing the contrast distribution on both sides of a defect structure fragment depending on the evenness/oddness of the number of silicate chains in the I-beam.

Assume that the monoclinic structure under study consists of I-beams containing an odd number of tetrahedral chains. If an I-beam slab containing an even number of silicate chains is intruded into the basic structure matrix, it is seen in Fig. 118 that the two structure fragments separated by the defect region are shifted by $c/2$ along c with respect to one another. This results from the shift component determined by the body-centered potential distribution in the defect slab (the incident beam is parallel to a). The identical points on both sides of the defect slab are related by translations $(b+c)/2$.

If the intruded slab differs from the basic ones only in the I-beam width and has the same evenness/oddness, the lattice of the crystal is preserved after removing the defect and connecting the two fragments by a parallel shift. For example, Fig. 119a shows that intrusion of a nine-chain I-beam slab in a triple-chain silicate structure preserves translational equivalence of the structure fragments on both sides of the defect region.

If the basic structure matrix contains a pair of adjacent slabs containing an even number of silicate chains, the structure fragments on different sides of the defect are again identical translationally. The numbers of silicate chains in the adjacent I-beam slabs may be different. The bright spots corresponding to the

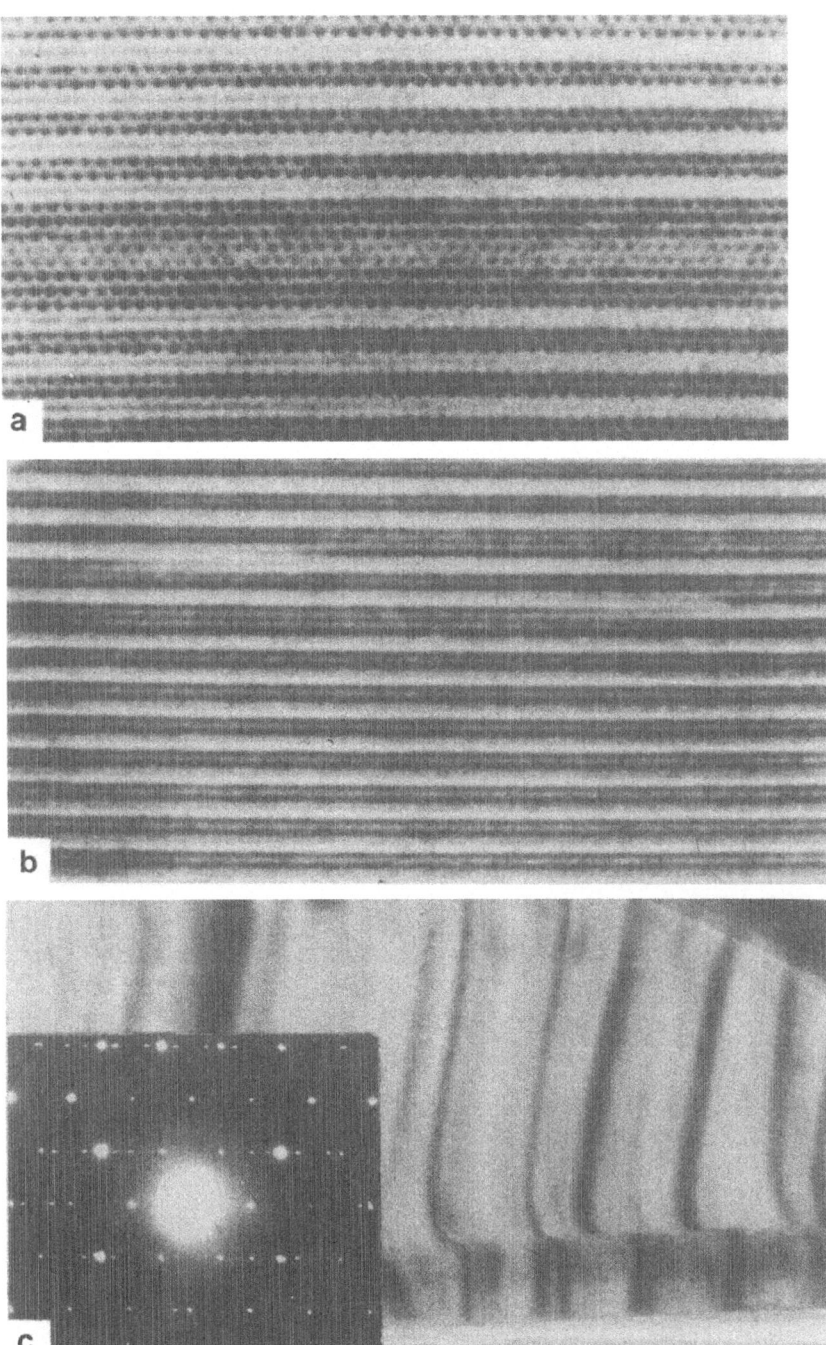

Fig. 119a – c. HREM image of **a** triple-chain silicate containing a nine-chain slab; **b** triple-chain silicate showing transformation of a quadruple-chain slab into a triple-chain one; **c** EM diffraction contrast image of intergrown smectite and triple-chain silicate and SAED pattern obtained from the area containing the phase boundary

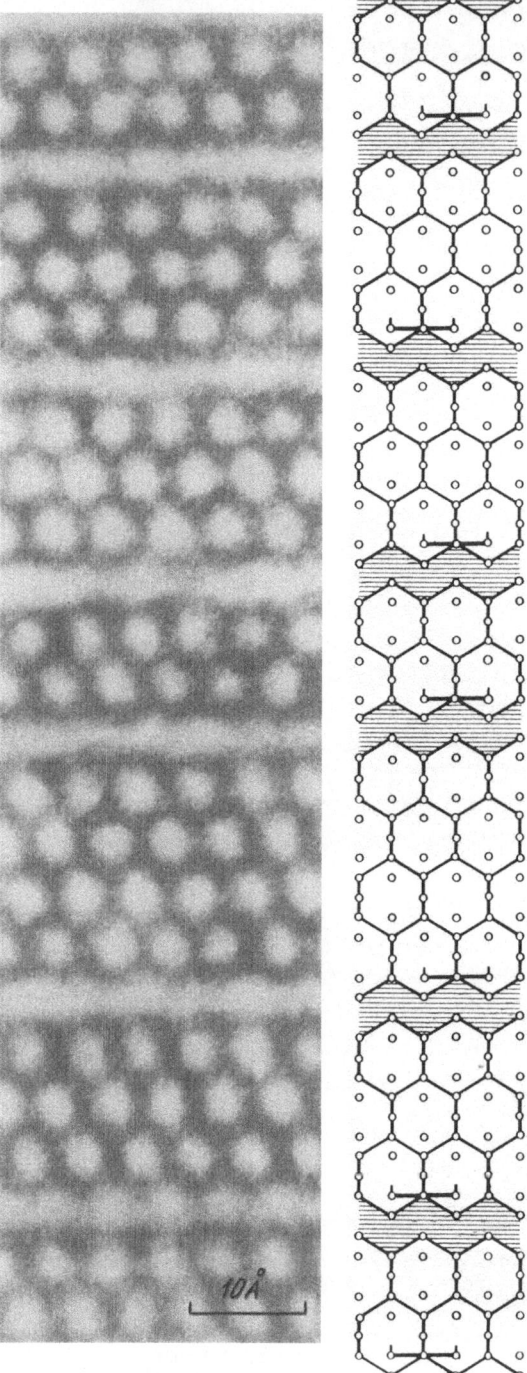

Fig. 120. Relationship between the image and the structure of a chain silicate containing *I*-beams of different widths

pair of adjacent defect slabs are distributed over a centered pattern, which is substantially different from the distribution of bright spots in the image of the basic structure. The above peculiarities are distinctly seen in the images from triple-chain silicate single crystals. Figure 120 shows an example of the direct relationship between the contrast observed and the structure projection down *a* containing, along with triple chain *I*-beams, defect slabs consisting of quadruple-and quintuple-chain *I*-beams.

Assume that the *I*-beams in the basic structure consist of an even number of silicate chains. In the structure projection down *a*, the regions having low charge density will form a hexagonal pattern, so that the distance between the centers is 5.2 Å. This distribution does not distinguish between the regions inside *I*-beam slabs and those at the boundaries. This is clearly seen in the amphibole structure projection (Fig. 114a). Thus the choice of linear fragments in the image that correspond to the boundaries between the neighboring *I*-beams becomes ambiguous. For example, Hutchison et al. (1975), who analyzed the contrast in the image from an amphibole crystallite assumed that the boundaries between *I*-beams were represented by linear rows of grey spots (Fig. 117c, the contrast is inverse). This implied the presence of triple-chain *I*-beam slabs in the amphibole structure. However, if it is assumed, according to the above data, that the boundaries pass through linear rows of dark spots (as the contrast may be reversible), an intrusion of two pyroxene *I*-beam slabs separated by a double-chain slabs is also probable. Then it is seen that a fault in periodicity in the distribution of dark-spot rows not only brings them closer together (down to 4.5 Å) but also shifts one part of the image with respect to the other by $c/2$ along the light boundary. The authors noted both possibilities for the contrast interpretation.

In the case of structures containing even-number chains it is appropriate to obtain images with the incident beam parallel to [101] (Drits et al. 1979). The difference in the image contrast representing the *I*-beams and the boundaries between them are especially expressive.

Analysis of orthogonal chain silicate structures having Wadsley defects shows that the contrast distribution in the projection on (100) should be similar to that for monoclinic varieties. However, the Wadsley defects should not affect the relative orientations for octahedra in the *I*-beams, so that their stacking should be described by either $+ + - - + +$ or $+ - + -$ irrespective of the width in individual *I*-beams.

In some cases, however, the contrast distribution does not obey the above rules. Figure 121a shows an image consisting of alternating bright and grey fringes parallel to *c*. Every two bright fringes are separated by a pair of grey ones with three black zig-zag lines between them. This contrast distribution may seem similar to the one observed for monoclinic triple-chain silicates. The contrast variations can be reasonably well described by a two-dimensional orthogonal lattice with cell dimensions 5.2 Å and 13.5 Å.

Figure 121a shows that linear fringes consisting of the brightest spots represent the boundaries between the projected *I*-beams. Then pairs of adjacent bright fringes separated by 4.5 Å must correspond to two boundary zones between which there is a pyroxene *I*-beam slab. Intrusion of a pyroxene *I*-beam slab will not lead to any relative shifts for the structure fragments on different sides of the

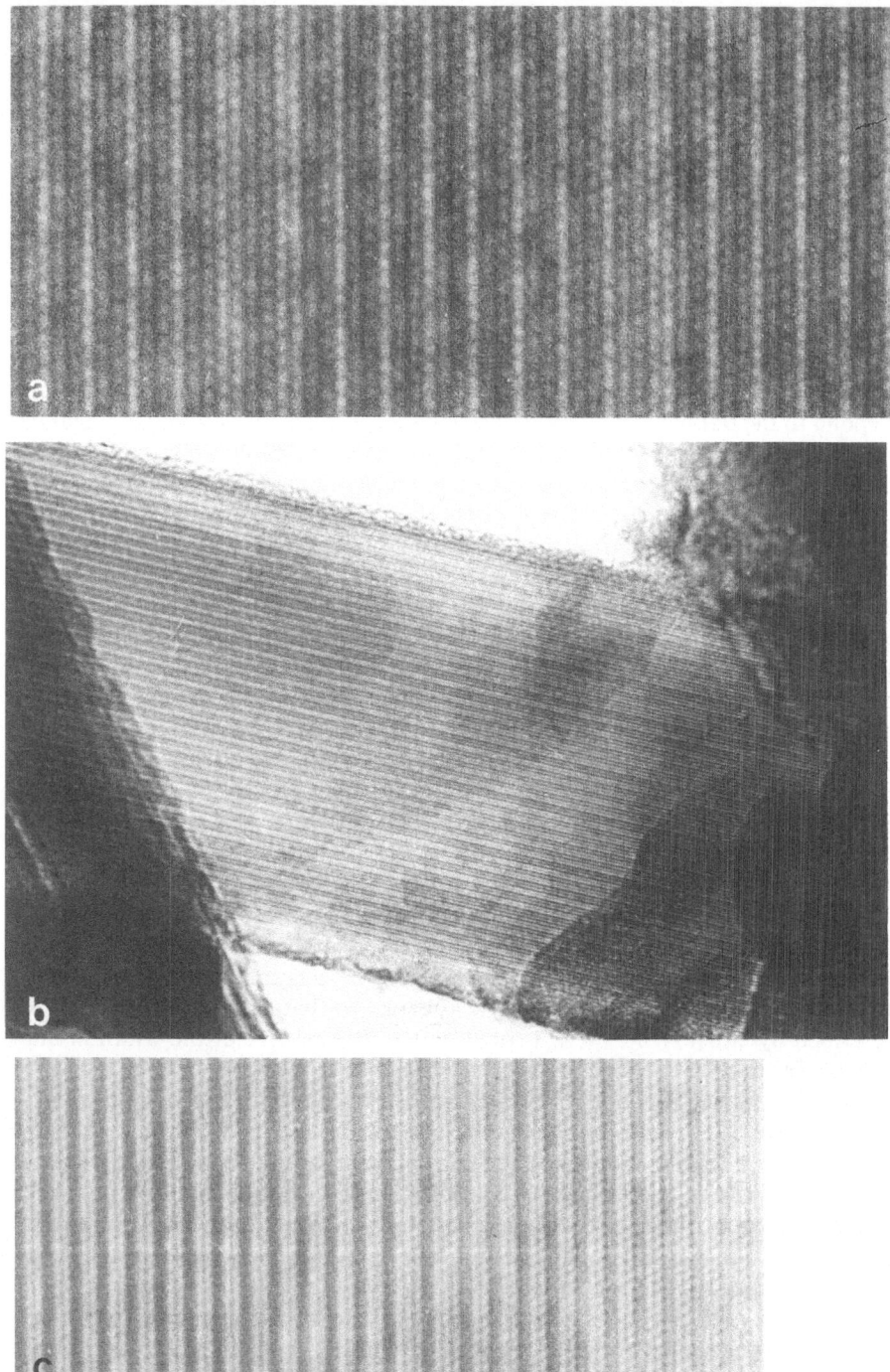

Fig. 121a – c. HREM images of **a** triple-chain silicate containing single-chain slabs; **b** mixed-chain silicate; **c** triple-chain silicate with inverse contrast distribution

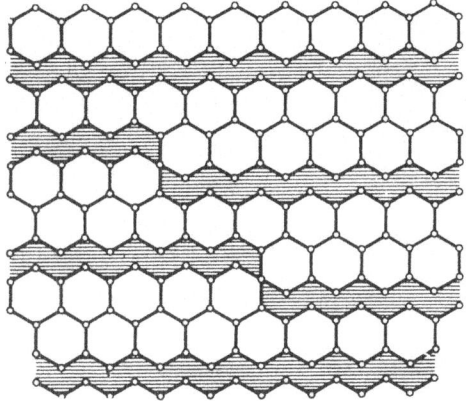

Fig. 122. Mechanism for the transformation of double-chain slab into a triple-chain one, and vice versa

defect region if the relative *I*-beam orientations are not affected. Figure 121a shows that one part of the image is shifted by $c/2$ with respect to the other. This may be associated with the intrusion of a pyroxene *I*-beam having the opposite octahedral orientation with respect to that in the basic triple-chain structure.

Another interesting type of defects in chain-silicate structures that can be observed in *a*-axis HREM images is associated with chain terminations. Figure 119b shows that the fringe contrast representing the intrusion of a pair of quadruple-chain slabs into a triple-chain matrix (in the upper part of the photograph) varies from right to left so that a quadruple chain slab is transformed into a triple-chain one, while the neighboring triple-chain slab is transformed into a quadruple-chain one. The structural interpretations for this effect is given in Fig. 122. Veblen et al. (1977) were the first to describe chain terminations in chain-silicates. Similar structure defects were later observed for actinolite (Jefferson et al. 1978) and synthetic triple-chain silicate (Drits et al. 1979; Zakharov et al. 1979).

11.10 Structural Features of Chain Silicates Revealed in *c*-Axis HREM Images

A fundamentally new level in the study of the nature of defects in chain-silicate structures has been achieved with the aid of HREM images representing structure projections down *c* (Veblen 1981; Veblen and Buseck 1980). Veblen and Buseck (1980) showed that the dark areas in the *c*-axis HREM images represent the *I*-beams, while the bright spots correspond to empty channels separating the adjacent *I*-beams belonging to each given (010) slab (Fig. 123). The authors studied chain-silicates formed by the anthophyllite-talc hydration reaction. Analysis of numerous *c*-axis HREM images has led to a classification of the main defect types in chain-silicates.

One of the common defects is the termination of (010) slabs having a given chain-width. The termination lines are parallel to *c*. The authors note that in the *c*-axis HREM images, the terminated wide-chain lamellae resemble zippers (Fig. 123). Therefore, they refer to these defects as "zipper terminations". Both isolated and cooperative zipper terminations were found in anthophyllite. In the

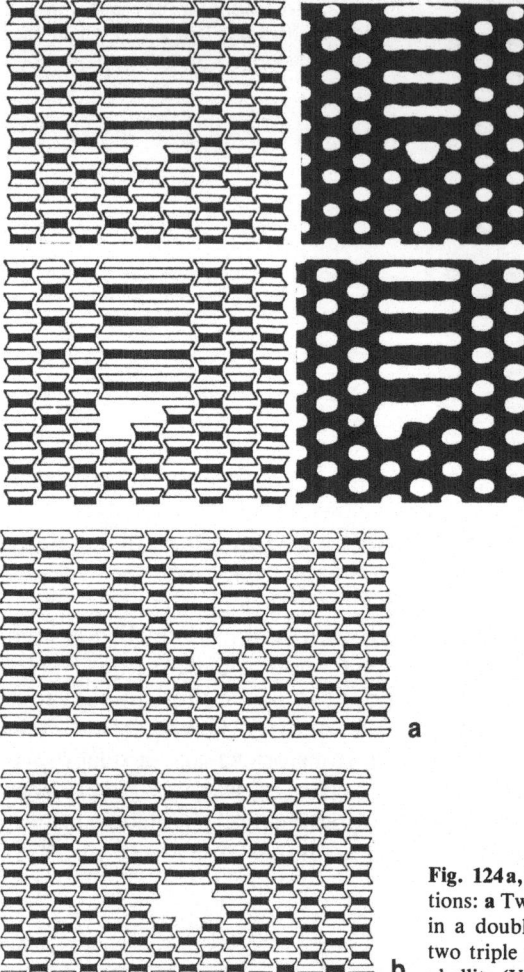

Fig. 123a, b. Simple coherent zipper terminations. **a, b** Two different termination types for sextuple-chain zippers in anthophyllite (*on the left*: models shown in the *I*-beam representation; *on the right*: image calculations based on the models) (Veblen and Buseck 1980)

Fig. 124a, b. Cooperative coherent zipper terminations: **a** Two quadruple zippers terminating together in a double-chain material; **b** One quadruple and two triple zippers terminating coherently in anthophyllite (Veblen and Buseck 1980)

latter case, several adjacent slabs of equal or different widths terminate together at the same place.

Isolated and cooperative zipper terminations can be either coherent or incoherent. In the first case, terminations do not lead to planar defects and are not accompanied by structure distortions except for linear discontinuities parallel to *c* around the termination region. Figure 123 shows an isolated sextuple-chain (010) zipper. The terminated slab is replaced by three amphibole (010) slabs. This is an example of a coherent zipper termination. In the case of the coherent zipper termination shown in Fig. 124a two adjacent quadruple-chain (010) slabs are replaced by four (010) amphibole lamellae (Veblen and Buseck 1980).

In the case of incoherent zipper terminations, substantial structure distortions are observed. For example, Fig. 125 shows a triple-chain (010) slab replaced by an amphibole one. This results in distortions similar to those in the case of edge

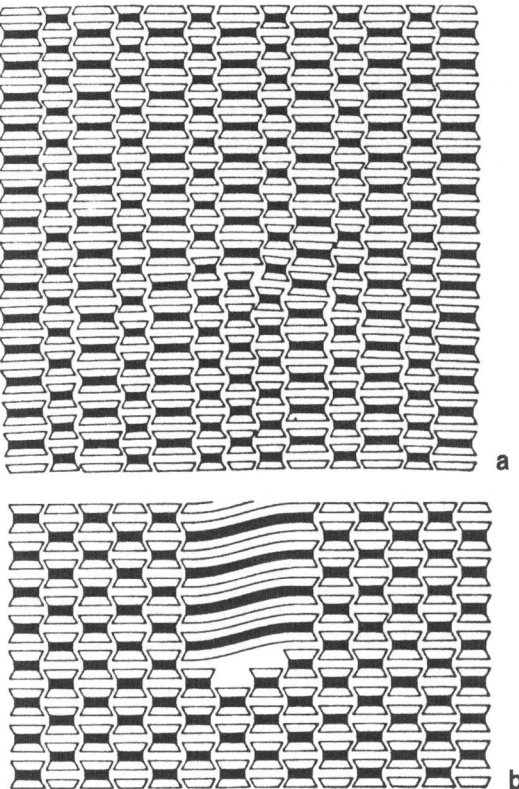

Fig. 125 a b. Violations of chain termination rules: **a** A double-chain slab turning into triple-chain material by partial dislocation; **b** An octuple-chain zipper terminating in anthophyllite (Veblen and Buseck 1980)

dislocations. An incoherent replacement of four amphibole (010) slabs by octuple-chain lamellae leads to the bending of the wide *I*-beams (Fig. 125) (Veblen and Buseck 1980).

Veblen and Buseck (1980) formulated two rules governing coherent zipper terminations. According to Rule 1, the terminating zipper or zippers must be coherently replaced by a structure having the same number of subchains, where the number of subchains is equal to the number of single silicate chains in a tetrahedral ribbon (i.e., two for amphiboles, three for jimthompsonite, etc.). Different number of subchains on different sides of the termination plane leads to Wadsley defects or planar defects accompanied by a projected displacement by $4.5 n$ Å (n is integer) along *b* in the *c*-axis images. Such planar defects appear in the case of a triple-chain zipper terminating inside the amphibole matrix (Fig. 126).

According to Rule 2, for structural coherence at a termination, the number of terminating zippers and that of the zippers in the crystal that replaces it must be either both even or both odd. If an odd number of terminating zippers replaces an even number or vice versa, a planar defect involving a displacement of approximately 4.6 Å along a^* results (in antrophyllite the displacement is a [100]/4. A zipper termination shown in Fig. 124 a is coherent since two quadruple-chain

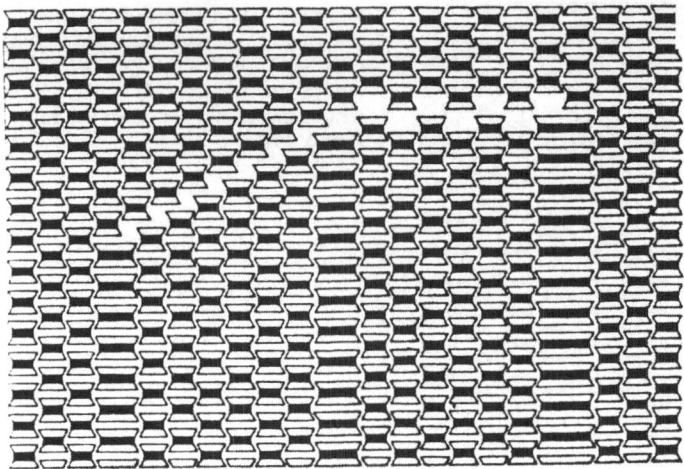

Fig. 126. Both types of displacive fault occurring together at the terminations of a quadruple and two triple zippers (Veblen and Buseck 1980)

(010) slabs are replaced by four amphibole (010) slabs. If a quadruple-chain (010) slab is replaced by two amphibole (010) slabs, Rule 2 is disobeyed and a planar defect parallel to (100) appears that is shown in the right part of Fig. 126. Violation of the above rules leads to incoherent cooperative zipper terminations. In this case, there are several zippers of equal or different width that can be distant from one another, while their terminations are related to the basic crystal matrix by planar defects. For example, Fig. 126 shows an intrusion of two triple-chain terminated zippers and a quadruple-chain one into the anthophyllite matrix. The intrusion of two triple-chain zippers results in an incoherent planar defect displacing the upper part of the anthophyllite structure by $b[010]/4$ with respect to the lower one along the defect plane parallel to (210) (see the left part of Fig. 126). The intrusion of a quadruple-chain zipper leads to a planar defect parallel to (100) that shifts the parts of the crystal on different sides of this plane by $a[100]/4$ with respect to one another.

References

Alario Franco MA, Hutchison JL, Thomas JM (1977) Structural imperfection and morphology of crocidolite. Nature (London) 266:520 – 521

Alcover JF, Gatineau L, Mering J (1973) Exchangeable cation distribution in nickel- and magnesium-vermiculite. Clays Clay Miner 21:131 – 136

Allman R, Lohse HH (1966) Die Kristallstruktur des Sjogrenite und eines Umwandlungsproduktes des Koenenits. Neues Jahrb Mineral H9, 161 – 181

Allman R, Lohse HH, Hellner EB (1968) Die Kristallstruktur des Koenenits. Z. Kristallogr 126:1 – 3, 7 – 22

Allpress JG, Sanders JV (1972) The direct observation of the structure of real crystals by lattice imaging. J Appl Crystallogr 6:165 – 190

Allpress JG, Hewat EA, Moodie AR, Sanders JV (1972). n-beam lattice images, I. Experimental and computed images from $W_4Nb_{26}O_{77}$. Acta Crystallogr A 28:528 – 536

Amelinckx S (1964) The direct observation of dislocations. Academic Press, London New York, 435 p

Amouric M, Baronnet A, Finck C (1978) Polytypisme et désordre dans les micas dioctaèdriques synthétiques. Etude par imagerie de réseau. Mater Res Bull 13:627 – 634

Amouric M, Mercurio G, Baronnet A (1981) On computed and observed HRTEM images of perfect mica polytypes. Bull Mineral 104:298 – 313

Antis GR (1977a) The calculation of electron diffraction intensities by multislice method. Acta Crystallogr A 33:844 – 846

Antis GR (1977b) The calculation of electron diffraction intensities by multislice method. Acta Crystallogr A 33:5, 884 – 886

Avilov AS, Imamov RM, Pinsker ZG, Fedotov AF (1973) The influence of accelerating voltage on the intensity of the forbidden reflection 222Ge in electron diffraction. Kristallografia 18:840 – 844 (in Russian)

Avilov AS, Imamov RM, Semiletov SA (1976) Precise measurement of reflection intensity in electron diffraction patterns from textures and mosaic single crystals. Prib Tekh Exsp 3:214 – 215 (in Russian)

Bailey SW, Brown BE (1962) Chlorite polytypism. I. Regular and semi-random one-layer structures. Am Mineral 47:819 – 850

Belov NV (1971) Essays on structural minerology. Mineral Sb (Lvov) 25:200 – 210 (in Russian)

Berner RA (1962) Tetragonal iron sulfide. Science 137:669 – 671

Besson G (1980) Structure des smectites dioctaèdrique, paramètres conditionnant les fautes d'empliment des feuillets. Thèse Doct, Etat Sci Phys, Orléans

Besson G, Tchoubar C (1972) Détermination du groupe de symétrie du feuillet élémentaire de la beidellite. CR Acad Sci 275:633 – 636

Besson G, Mifsud A, Tchoubar K, Méring J (1974a) Order and disorder relations in the distribution of the substitution in smectites, illites and vermiculites. Clays Clay Miner 22:379 – 384

Besson G, Tchoubar K, Méring J (1974b) Phenomènes de diffraction produits par les systèmes stratifiés à distribution d'atomes partiellement différente de couche à couche. J Appl Crystallogr 87:3, 345 – 350

Besson G, Calle C de la, Rautureau M, Tchoubar C, Tsipursky SI, Drits VA (1982) X-ray and electron diffraction study of the structure of the Garfield nontronite. Proc 7th Int Clay Conf, Bologna Pavia, pp 29 – 40

Bethe H (1928) Theorie der Beugung von Electron an Kristallen. Ann Phys (NY) B7:55 – 85

Blackman M (1939) On the intensities of electron diffraction rings. Proc R Soc London Ser A 173:68 – 82

Bookin AS, Drits VA, Rozhdestvenskaya IV, Semenova TF, Tsipursky SI (1982) Comparison of orientations of OH-bonds in layer silicates by diffraction methods and electrostatic calculations. Clays Clay Miner 30:409 – 414

Brindley GW, Méring J (1951) Diffraction des rayons X par les structures en couches désordonnées. Acta Crystallogr 4:441 – 447

Brown JL, Jackson ML (1973) Chlorite examination by high resolution electron microscopy. Clays Clay Miner 21:1 – 7

Brown JL, Rich CI (1968) High-resolution microscopy of muscovite. Science 161:1135 – 1137

Buerger M (1959) Vector space. Wiley, New York

Bursill LA (1979) Resolution enhancement at 100 kV: Its limitation and some applications. Chem Scr 14:83 – 97

Bursill LA, Wilson AR (1977) Electron-optical imaging of the hollandite structure at 3 Å resolution. Acta Crystallogr A33:672 – 676

Buseck PR (1983) Electron microscopy of minerals. Am Sci 71:175 – 185

Buseck PR (1984) Imaging of mineral with the TEM. EMSA Bull, No 1, 14:47 – 53

Buseck PR, Cowley JM (1983) Modulated and intergrowth structures in minerals and electron microscope methods for their study. Am Mineral 68:18 – 40

Buseck PR, Iijima S (1974) High resolution electron microscopy of silicates. Am Mineral 59:1 – 2, 1 – 21

Buseck PR, Iijima S (1975) High resolution electron microscopy of enstatite. Am Mineral 60:758 – 780

Buseck PR, Iijima S (1976) High resolution electron microscopy of enstatite In: Wenk HR (ed) Electron microscopy in minerology. Springer, Berlin Heidelberg New York

Cameron KL (1975) An experimental study of actinolite-cumingtonite phase relations. Am Mineral 60:373 – 390

Chisholm IE (1973) Planar defects in fibrous amphiboles. J Mater Sci 8:475 – 483

Chukhrov FV, Zvyagin BB (1966) Halloysite a crystallochemically and mineralogically distant species. In: Proc Int Clay Conf, Jerusalem, I, pp 11 – 25

Chukhrov FV, Gorshkov AI, Vitovskaya ES, Drits VA (1980a) Crystallochemical nature of Co-Ni asbolane. Izv Acad Nauk SSSR Ser Geol 6:73 – 81 (in Russian)

Chukhrov FV, Gorshkov AI, Vitovskaya ES, Drits VA (1980b) About the crystallochemical nature of Ni-asbolane. Izv Acad Nauk SSSR Ser Geol 9:108 – 120 (in Russian)

Chukhrov FV, Gorshkov AI, Drits AV, Sivzev AV (1982) New structural variation of asbolane. Izv Acad Nauk Ser Geol 6:69 – 77

Chukhrov FV, Gorshkov AI, Drits AV, Finko VI, Sivzev AI (1983a) Structurally disordered asbolane with the tetrahedral coordination of manganese. Izv Acad Nauk SSSR Ser Geol 12:85 – 95 (in Russian)

Chukhrov FV, Gorshkov AI, Drits AV, Shterenberg LY, Sakharov BA (1983b) Mixed-layer asbolane-buserite and asbolane minerals in the oceanic iron-manganese concretions. Izv Acad Nauk SSSR Ser Geol 5:91 – 100 (in Russian)

Chukhrov FV, Gorshkov AI, Drits VA (1984) Structural models and methods of study of buserite. Izv Acad Nauk SSSR Ser Geol 12:6 – 30 (in Russian)

Cockayane DJH, Porsous JR, Hoelke CW (1971) A study of relationship between lattice fringes and lattice planes of electron microscope images of crystals containing defects. Philos Mag 24:139 – 153

Cowley JM (1967) Crystal structure determination by electron diffraction. Prog Mater Sci 13:267 – 321

Cowley JM (1975a) Implication of non-kinematic and inelastic scattering of electrons for structure analysis. In: Ramaseshan S, Abrahams SC (eds) Anomalous scattering. Proc Int Congr, Copenhagen, pp 113 – 125

Cowley JM (1975b) Diffraction physics. NHPC, Amsterdam

Cowley JM (1978) The imagaing of crystal structures and crystal defects. In: Proc Int Congr Electron Microsc, Toronto, III, pp 207 – 218

Cowley JM, Iijima S (1972) Electron microscopy image contrast for thin crystals. Z Naturforsch 27:445 – 451

Cowley JM, Iijima S (1976) The direct imaging of crystal structures. In: Wenk HR (ed) Electron microscopy of minerology, Springer, Berlin Heidelberg New York

Cowley JM, Moodie AF (1957) The scattering of electrons by atoms and crystals. I. A new theoretical approach. Acta Crystallogr 10:609–619

Cowley JM, Moodie AF (1960) Fourier image $\overline{IY}$. Phase gratings. Proc R Soc London Ser A 76:3378–3384

Cowley JM, Pogany AP (1968) Diffuse scattering in electron diffraction patterns. I. General theory and computational methods. Acta Crystallogr A 24:109–116

Cowley JM, Cohen JB, Salamon MB, Wuensch BJ (1979) Modulated structures. Am Inst Phys, New York

Crawford ES, Jefferson DA, Thomson JM (1977) Electron microscopy and diffraction studies of polytypism in stilpnomelane. Acta Crystallogr A 35:548–553

Cressey BA, Whittaker EJW, Hutchison JL (1982) Morphology and alteration of asbetiform grünerite and anthophyllite, Mineral Mag 46:77–87

Dorling M, Zussman J (1980) Comparative studies of asbetiform and non-asbestiform calcium-rich amphiboles. 4th Int Conf Asbestos, Torino, pp 317–333

Dorset L, Hauptman HA (1976) Direct phase determination for quasi-kinematical electron diffraction intensity data from organic microcrystals, Ultra Microsc 1:195–201

Drits VA (1975) Structural and crystal-chemical features of layer silicates. In: Crystal chemistry of minerals and geological problems. Nauka, Moscow, pp 35–50 (in Russian)

Drits VA (1982) Electron diffraction, high resolution electron microscopy and structural analysis. In: Modern electron microscopy in investigation of materials. Nauka, Moscow, pp 23–40 (in Russian)

Drits VA, Aleksandrova VA (1975) The crystal structure refinement of talc. In: Crystal chemistry of minerals and geological problems. Nauka, Moscow, pp 99–109 (in Russian)

Drits VA, Goncharov YA (1974) A new type of chain silicate. In: 10th Int Mineral Assoc B, 9th General Meeting, West Berlin and Regensburg, Deutsche Mineral Gesell

Drits VA, Kossovskaya AG (1980) Geological crystal chemistry of rock-forming dioctahedral smectites. Litol Polezn Iskop 1:84–112 (in Russian)

Drits VA, Sakharov BA (1976) X-ray structure analysis of mixed-layer minerals. Nauka, Moscow, 255 pp, (in Russian)

Drits VA, Sokolova GA (1971) Structure of palygorscite. Kristallografia 16:228–231 (in Russian)

Drits VA, Dmitrik AL, Goncharov YM, Hadji IP, (1973) Selected-area electron diffraction studies of chain silicates $NaMg_4Si_6O_{16}(OH)_2$ and $Na_{1.3}Mg_{4.7}Si_6O_{16}F_2$. In: Proc 9th USSR Conf Electron Microsc. Nauka, Moscow, pp 91–92 (in Russian)

Drits VA, Goncharov YI, Aleksandrova VA, Dmitrik AL (1974) A new type of ribbon silicates. Kristallografia 19:1186–1193 (in Russian)

Drits VA, Dmitrik AL, Hadji IP (1976a) Selected area electron diffraction study of fibrous chain silicates. In: Proc 10th USSR Conf Electron Microsc. Nauka, Moscow, pp 322–323 (in Russian)

Drits VA, Goncharov YI, Hadji IP (1976b) Formation conditions and physico-chemical properties of triple-chain silicate with the unit Si_6O_{16}. Izv Acad Nauk SSSR Ser Geol 7:32–41 (in Russian)

Drits VA, Korytkova EN, Dmitrik AL, Aleksandrova VA (1978) Synthesis of the mica $Na(Si_{3.5}Mg_{0.5})Mg_{3.0}O_{10}(OH)_2$ having new type of isomorphism, a superperiodicity in silicate layers and talc-like stacking sequence. In: Methods for the study of isomorphous substitutions in silicates. Nauka, Moscow, SO Acad Nauk SSSR, pp 128–143 (in Russian)

Drits VA, Zakharov ND, Hadji IP (1979) Direct observation of the structural motif in chain silicates by high-resolution electron microscopy. Izv Acad Nauk SSSR Ser Geol 11:82–89 (in Russian)

Drits VA, Tsipursky SI, Plançon A (1984) Application of the method for the calculation of intensity distribution to electron diffraction structure analysis. Izv Akad Nauk SSR Ser Phys 2:1708–1713 (in Russian)

Drits VA, Petrova VA, Gorshkov AI, Svalnov VA, Sokolova AL (1985) Manganese minerals Fe–Mn from the microconcretions of the sediments of the central part of Pacific Ocean and their postsedimentational transformations. Litol Polezn Iskop 5:82–88 (in Russian)

Drits VA, Hadji IP, Cherkashin VI, Bookin AS (1986) Application of HREM, electron and X-ray diffractions to the study of defects in chain silicates. Izv Akad Nauk SSSR Ser Phys 3:522–526 (in Russian)

Eggleton RA (1972) The crystal structure of stilpnomelane. Mineral Mag 38:693–711

Eggleton RA, Guggenheim S (1986) A re-examination of the structure of ganophyllite. Mineral Mag 50:307 – 315

Erickson H, Klug A (1971) Measurement and compensation of defocusing and aberrations by Fourier processing of electron micrographs. Philos Trans R Soc London Ser B 261:105 – 118

Evans HT, Allman R (1968) The crystal structure and crystal chemisty of valleriite. Z Kristallogr 127:73 – 93

Fan HF, Zhong ZY, Zheng CD, Li FH (1985) Image processing in high-resolution electron microscopy using the direct method. I. Phase extension. Acta Crystallogr A41:163 – 165

Fejes PL (1977) Approximation for the calculation of high-resolution electron-microscope images of thin films. Acta Crystallogr A33:109 – 113

Fifty Years of Electron Diffraction (1981) Ed. Goodman P, D. Riedel Publishing Company, Dordrecht, Holland, 440 p

Filippenko OS, Pobedimskaya EA, Ponomarev VI, Belov NV (1971) Crystal structure of synthetic BaS_6O_{16}. Dokl Acad Nauk SSSR 196:1337 – 1340 (in Russian)

Franz J, Althaus I (1974) Synthesis and thermal stability of $2\frac{1}{2}$ octahedral sodium mica. Contrib Mineral Petrol 46:227 – 232

Fujimoto F (1959) Dynamical theory of electron diffraction in Laue-case. I. General theory. J Phys Soc Jpn 14(11):1558 – 1568

Fujiwara K (1961) Relativistic dynamic theory of electron diffraction. J Phys Soc Jpn 16(11): 2226 – 2238

Gard JA (1976) Interpretation of electron diffraction patterns. In: Wenk HR (ed) Electron microscopy in minerology. Springer, Berlin Heidelberg New York

Gatineau L, Mering JR (1966) Relations ordre-désordre dans les substitutions isomorphes des mica. Bull Gropue Fr Argiles 18:67 – 74

Goncharov YI, Drits VA, Aleksandrova VA (1973a) A new chain silicate structural type obtained from the system $NaF-MgF_2-MgO-SiO_2$. In: Proc 9th Conf Exp Mineral Petrogr, Ikutsk, pp 217 – 218 (in Russian)

Goncharov YI, Drits VA, Hadji IP, Aleksandrova VA (1973b) Thermal study of artificial hydroxyl-asbestos. In: Proc 5th Conf Therm Anal, Novosibirsk, pp 181 – 182 (in Russian)

Goodman P, Lehmpfuhl G (1967) Electron diffraction study of MgO hoo-systematic interactions. Acta Crystallogr 22:14 – 24

Goodman P, Moodie AF (1974) Numerical evaluation of N-beam wave function in electron scattering by multi-slice method. Acta Crystallogr A30:280 – 290

Gorshkov AI (1970) Use of selected-area electron diffraction for obtaining basal reflections from platy layer silicates. Izv Acad Nauk SSSR Ser Geol 3:115 – 120 (in Russian)

Gorshkov AI, Drits VA (1984) Selected-area electron diffraction of defect mixed-layer structures. Izv Acad Nauk SSSR Ser Phys 9:1678 – 1682 (in Russian)

Gorshkov AI, Drits VA, Sokolova GV (1975) Superperiodicity in mixed-layer structures studied by selected-area electron diffraction. Izv Acad Nauk SSSR Ser Geol 3:76 – 83 (in Russian)

Grigorieva LF, Makarova TA, Korytkova EN, Chigareva OG (1975) Synthetic asbestiform amphiboles. Nauka, Leningrad, 249 pp (in Russian)

Guggenheim S, Balley SW (1982) The superlattice of minnesotaite. Can Mineral 20:579 – 584

Guggenheim S, Eggleton RA (1986a) Structural modulations in iron-rich and magnesium-rich minnesotaite. Can Mineral 24:479 – 497

Guggenheim S, Eggleton RA (1986b) Cation exchange in ganophyllite. Mineral Mag (in press)

Guggenheim S, Bailey SW, Eggleton RA, Wilkes P (1982) Structural aspects of greenalite and related minerals. Can Mineral 20:1 – 18

Güven N (1971) The crystal structure of $2M_1$-phengite and $2M_1$-muscovite. Z Kristallogr 134:487 – 490

Güven N (1974) Factors affecting selected-area electron diffraction patterns of micas. Clays Clay Mineral 22:97 – 106

Güven N, Burnham CW (1967) The crystal structure of 3 T muscovite. Z Kristallogr 125:1 – 6

Hadji IP, Drits VA, Dmitrik AL (1978) A new polymorph of fibrous fluoroamphibole. In: New data in minerals of the USSR. Nauka, Moscow, pp 153 – 162

Harris DC, Vaughan DJ (1972) Two fibrous iron sulfides and vallerite from Cyprus with new data on vallerite. Am Mineral 57:1037 – 1053

Hashimoto H, Manuami M, Naki I (1961) Dynamical theory of electron diffraction for the electron-microscopic image of crystal lattices. Philos Trans R Soc London 253:459 – 489

Hashimoto H, Endoh H, Ono A, Watanabe E (1977) Direct observation of fine structure within images of atoms in crystals by transmission electron microscopy. J Phys Soc Jpn 42(3):1073–1074

Hashimoto H, Endoh H, Takai Y, Tomioka H, Yokota Y (1979) Identification of atoms in crystals by the fine structure of their images. Chem Scr 14:23–31

Hauptman HA (1972) Crystal structure determination. The role of the Cosine Seminvariants. Plenum, New York

Heidenreich RD (1964) Fundumentals of transmission electron microscopy. Wiley Interscience, New York

Hirsh PB, Howie A, Nicklson R, Pachley D, Whelan M (1977) Electron microscopy of thin crystals. Krieger Huntington, New York

Horiuchi S (1979) Visualizing Atoms in inorganic compounds by IMV HRTEM. Chem Scr 15: 75–81

Horiuchi S, Kikuchi I, Goto M (1977) Structure determination of mixed-layer bismuth-titanate by super high resolution electron microscopy. Acta Crystallogr A33:701–703

Howie A, Whelan MJ (1961) Diffraction contrast of electron microscope images of crystal lattice defects. II. The development of a dyanmic theory. Proc R Soc London Ser A263:217–237

Hutchison IL, Iresteta MG, Wittaker JW (1975) High-resolution electron microscopy and diffraction studies of fibrous amphiboles. Acta Crystallogr A31:794–799

Iijima S (1971) High-resolution electron microscopy of crystal lattice of titanium-niobium oxide. J Appl Phys 42(13):5891–5893

Iijima S (1973) Direct observation of lattice defects in $H-Nb_2O_5$ by high resolution electron microscopy. Acta Crystallogr A29:18–24

Iijima S (1975a) High resolution electron microscopy of crystallographic shear structures in tungstenoxides. J Solid State Chem 14:53–65

Iijima S (1975b) Ordering of the point defects in nonstoichiometric crystals of $Nb_{12}O_{29}$. Acta Crystallogr A31:784–790

Iijima S (1978) Many-beam lattice images from thicker crystals. In: Proc Int Congr Electron Microsc, Toronto, III, pp 207–218

Iijima S, Allpress JG (1974) Structural studies by high resolution electron microscopy: tetragonal tungsten bronze-type structures in the system Nb_2O_5-WO_3. Acta Crystallogr 30:20–36

Iijima S, Buseck PR (1978) Experimental study of disordered mica structures by high-resolution electron microscopy. Acta Crystallogr A34:709–719

Iijima S, Zhu J (1982) Electron microscopy of a muscovite-biotite interface. Am Mineral 67:1195–1205

Iijima S, Kimura S, Goto M (1973) Direct observation of point defects in $Nb_{12}O_{29}$ by high resolution electron microscopy. Acta Crystallogr A29:632–636

Iijima S, Kimura S, Goto W (1974) High-resolution microscopy of nonstoichiometric $Nb_{22}O_{54}$ crystals: point defects and structural defects. Acta Crystallogr A30:251–257

Imamov RM (1977) Electron diffraction and its application to determination of semiconductor crystal structures. Thesis, Inst Crystallogr, Moscow, 210 pp (in Russian)

Imamov RM, Pinsher ZG (1965) Electron diffraction analysis of AgTlSe. Kristallografia 10:199–204 (in Russian)

Imamov RM, Avilov AS, Semilietov SA (1982) Electron diffraction structure analysis: advantages and prospects. Modern electron microscopy in investigation of materials. Nauka, Moscow, pp 73–79 (in Russian)

International Tables for X-ray Crystallography (1962) Kinoch, Birmingham, pp 1–3

Ishizuka K, Uyeda N (1977) A new theoretical and practical approach to the multislice method. Acta Crystallogr A33:740–749

Ishizuka K, Miyazaki M, Uyeda N (1982) Improvement of electron microscope images by the direct phasing method. Acta Crystallogr A38:408–413

Jambor IL (1969) Coalingite from the Muskok Intrusion. Am Mineral 54:437–448

Jefferson DA, Millinson LG, Hutchison JL, Thomas JM (1978) Multiple-chain and other unusual faults in amphiboles. Contrib Mineral Petrol 66:1–4

Kakinoki J, Komura Y (1952) Intensity of X-ray by an one-dimensionally disordered crystal. J Phys Soc Jap 7:30–35

Klug A (1979) Image analysis and reconstruction in the electron microscopy of biological macromolecules. Chem Scr 14:245–256

Kodama H (1977) An electron diffraction study of a microcrystalline muscovite and its vermiculitized products. Mineral Mag 41:461–468

Korytkova EN, Drits VA (1977) Micas with a new type of isomorphism. In: Problems of isomorphism and genesis of mineral species. Elista Univ Press, Elista, pp 219–224 (in Russian)

Kumao A, Hashimoto H, Hissen HI, Endon H (1981) Ca and Na positions in labrodite feldspar as derived from high-resolution electron microscopy and optical diffraction. Acta Crystallogr A37:229–238

Kunze G (1956) Die gewellte Struktur des Antigorits. Z Kristallogr 108:82–107

Landuyt J van, Amelinckx S (1975) Multiple beam direct lattice imgaing of new mixed-layer compounds of the bastnaesite-synchisite series. Am Mineral 60(5–6):351–358

Lee SY, Jackson ML, Brown JL (1975a) Micaceous occlusions in kaolinite observed by high-resolution electron microscopy. Clays Clay Mineral 23:125–129

Lee SY, Jackson ML, Brown JL (1975b) Micaceous vermiculite, glauconite, and mixed-layer kaolinite-montmorillonite examination by high resolution electron microscopy. Soil Sci Soc Am Proc 39(4):793–800

Le Poole JB (1947) Ein neues Electronen-Mikroskop mit stetig regelbarer Vergrößerung. Philips Techn Rundsch 9:33–43

Liebau F, Hesse KF (1975) The structure of $Ba_6Si_{10}O_{26}$ a silicate with quintuple tetrahedral chains. Acta Crystallogr A31:74–75

Li FH, Hashimoto H (1984) Use of the dynamical scattering in the structure determination of a minute fluorocarbonate mineral cebaite $Be_3Ce_2(CO_3)_5F_2$ by high resolution electron microscopy. Acta Crystallogr B40:454–461

Li FH, Tang D (1985) Pseudo-weak-phase object approximation in high resolution electron microscopy. I. Theory. Acta Crystallogr A41:376–382

Lynch DE, O'Keefe MA (1972) n-beam lattice images. II. Methods of calculation. Acta Crystallogr 28:536–548

Lynch DF, Moodie AF, O'Keefe MA (1975) n-beam lattice images. V. The use of the charge-density approximation in the interpretation of lattice images. Acta Crystallogr A31:300–307

Mamy J, Gaultier JP (1976) les phenomènes de diffraction de rayonnements X et électronique par les réseaux atomiques: application à l' étude de l'ordre cristalline dans les miréraux argileux. Ann Agron 27:1–16

Marinder BO, Sundberg M (1984) The structure of $NaNb_7O_{18}$ as deduced from HREM images and X-ray powder diffraction data. Acta Crystallogr B40:82–86

Menter JW (1956) The direct study by electron microscopy of crystal lattices and their imperfections. Proc R Soc London A236:119–135

Méring J, Oberlin A (1971) Smectite. In: The electronoptical investigations of clays. Mineral Soc, London

Millinson LC (1980) Termination of planar defects in the amphibole mineral nephrite observed by high-resolution electron microscopy. Acta Crystallogr A36:378–381

Millinson LC, Hutchison JL, Jefferon DA, Thomas JM (1977) Discovery of new types of chain silicates by high-resolution electron microscopy. J Chem Soc London Chem Commun 22:910–911

Millinson LC, Thomas JM, Jefferson JM, Hutchison JL (1980) The internal structure of nephrite experimental and computational evidence for the coexistance of multiple-chain silicates within an amphibole host. Philos Trans R Soc London Ser A295:537–552

Miscell DL (1978) The phase problem in electron microscopy. In: Adv Opt Electron Microsc 7:185–279

Moret R, Huber M, Comes R (1976) Diffuse scattering and titanium short-range order in $Ti_{1+x}S_2$. Phys Status Solidi 38:695–700

Nakajima Y, Ribbe PH (1980) Alteration of pyroxenes from Hokkaido, Japan to amphibole, clays and other biopyriboles. Neues Jahrb Mineral Monatsh 6:258–268

Nakazawa H, Morimoto N, Watanabe E (1974) Direct observation of the non-stoichiometric pyrrhotite. In: Proc 8th Int Congr Electron Microsc, Canberra 1974, pp 498–499

Nissen HU, Wessicken R, Woensdrept CF, Pfeifer HR (1979) Disordered intermediates between jimthompsonite and anthophyllite from the Swiss Alps. In: Mulvey T (ed) Electron microscopy and analysis. Conf Ser 52, Inst Phys, Bristol, pp 99–100

Novikov VM, Berkhin SI, Gorshkov AI, Drits AV, Organova NI, Rudnitskaya ES (1973) Mixed-layer chlorite-swelling chlorite. Izv Acad Nauk SSSR Ser Geol 8:38–47 (in Russian)

O'Keefe MA (1973) n-beam lattice images. IY. Computed two-dimensional images. Acta Crystallogr A29:389–401

O'Keefe MA, Buseck PR (1979) Computation of high-resolution TEM images of minerals. Trans Am Crystallogr Assoc 15:27–44

O'Keefe MA, Sanders JV (1975) n-beam lattice images. IV. Degradation of images resolution by a combination of incident-beam divergence and spherical aberration. Acta Crystallogr A31:307–310

O'Keefe MA, Buseck PA, Iijima S (1978) Computed crystal structure images for high-resolution electron microscopy. Nature (London) 274:322–324

Olives J, Amouric M (1984) Biotite chloritization by interlayer brucitization as seen by HREM. Am Mineral 69:869–871

Olives J, Amouric M, Fouquet C, Barronnet A (1983) Interlayering and interlayer slip in biotite as seen by HRTEM. Am Mineral 68:754–758

Organova NI (1972) Diffraction study of hybrid-structure ore minerals. Thesis, IGEM Acad Nauk SSSR, Moscow 24 pp (in Russian)

Organova NI, Drits VA, Genkin AO, Distler VV, Yevstigneyeva TL, Dmitrik Al (1971a) Crystal chemistry of hybrid-structure ore minerals. In: Proc 10th Conf Appl X-rays Study mater, Nauka, Moscow, 50 pp (in Russian)

Organova NI, Genkin AD, Drits VA, Dmitrik AL, Kuzmina OV (1971b) Tochilinite, a new iron-magnesium sulphide-hydroxide. Zap Vses Mineral Ova 100:477–487 (in Russian)

Organova NI, Drits VA, Dmitrik AL (1972) Crystal structure of tochilinite. I. Isometrical variety. Kristallografia 17:761–767 (in Russian)

Organova NI, Drits VA, Dmitrik AL (1973a) On one-layer valleriite. Dokl Acad Nauk SSSR 212:192–195 (in Russian)

Organova NI, Drits VA, Dmitrik AL (1983b) Crystal structure of tochilinite. 2. Acicular variety. Unusual diffraction patterns. Kristallografia 18:960–965 (in Russian)

Organova NI, Drits VA, Dmitrik AL (1974) Selected-area electron diffraction study of a type II "valleriite-like" mineral. Am Mineral 59(1–2):190–200

Papike JJ, Cameron M (1976) Crystal chemistry of silicate mineral of geophysical interest. Rev Geophys Space Phys 1:37–80

Papike JJ, Ross M (1973) Gedrite: crystal structures and intracrystalline cation distribution. Am Mineral 55:N11–12 1945–1972

Pierce L, Buseck PR (1976) A comparison of bright field and dark field imaging of pyrrhotine structures. In: Wenk HR (ed) Electron microscopy in minerology. Springer, Berlin Heidelberg New York

Pinsker ZG (1949) Electron diffraction. Akad Nauk SSSR, Moscow-Leningrad, 350 pp (in Russian)

Plançon A (1980) The calculation of intensities diffracted by a partially oriented powder with a layer structure. J Appl Crystallogr 13:524–528

Plançon A (1981) Diffraction by layer structures containing different kinds of layers and stacking faults. J Appl Crystallogr 14:300–304

Plançon A, Tsipurski SI, Drits VA (1985) Calculation of intensity distribution in these case of oblique texture electron diffraction. J Appl Crystallogr 18:191–196

Proc Int Congr Electron Microsc, Kyoto 1986, vols I–III

Ridder R, Tandelod G van, Dyck D van, Amelinckx S (1976) A cluster model for the transition state and its study be means of electron diffraction. Phys Status Solidi 38:663–673

Ridder R, Dyck D van, Tandelod G van, Amelinckx S (1977a) A cluster model for transition state and its study be means of electron diffraction. Application for some particular systems. Phys Status Solidi 40:669–683

Ridder R, Dyck D van, Tandelod G van, Amelinckx S (1977b) Microstructural study of copper-intercalated niobium disulphide and tantalum disulphide. Phys Status Solidi 39:383–399

Rothbauer R (1971) Untersuchung eines $2 M_1$ Muskovits mit Neutronenstrahlen. Neues Jahrb Mineral Monatsh 4:143–154

Sakharov BA, Naumov AS, Drits VA (1982a) X-ray diffraction by mixed-layer structures with random distribution of stacking faults. Dokl Akad Nauk SSSR 265:339–343 (in Russian)

Sakharov BA, Naumov AS, Drits VA (1982b) X-ray intensities scattered by layer structures with short range ordering parameters SI and GI. Dokl Acad Nauk SSSR 265:871–874 (in Russian)

Scherzer O (1949) Theoretical resolution limit of the electron microscope. J Appl Phys 20:20–30

Self PG, O'Keefe MA, Buseck PR, Spargo AEC (1983) Practical computation of amplitudes and phases in electron diffraction, Ultramicroscopy 11:35–52

Sinclair R (1978) Application of lattice fringe imaging. In: Proc Int Congr Electron Microsc, Trononto, III, pp 140 – 146

Skarnulis AJ, Cowley JM (1976) Refinement of the defect structure of $GeNb_9O_{25}$ by high-resolution electron microscopy. Acta Crystallogr A 32:799 – 803

Smith PPR (1977) An electron microscope study of amphibole lamellae in augite. Contrib Mineral Petrol 59:317 – 322

Smith PPR, Parise JB (1985) Structure determination of $SnSb_2S_4$ and $SnSb_2Se_4$ by HREM. Acta Crystallogr B 41:84 – 87

Spinner GE, Self PG, Iijima S, Buseck PR (1984) Stacking disorder in clinochlore chlorite. Am Mineral 69:252 – 263

Stewart JM, Kruger GJ, Ammon HL, Dickinson CW, Hall SR (1972) The X-RAY72 system – version of June 1972. Tech Rep TR-192. Computer Science Center, Univ Maryland, College Park, Maryland

Sturkey L (1957) The use of the electron diffraction intensities in structure determination. Acta Crystallogr 10:858 – 859

Sturkey L (1962) The calculation of electron diffraction intensities. Proc R Soc London Ser 80:321 – 354

Suito E, Arakawa M, Yoshida T (1969) Electron microscopic observation of the layer of organo-montmorillonite. In: Proc Clay Conf, Tokyo, pp 757 – 765

Suquet H (1978) Propriétés de conélement à structure de la saponite: comparison avec la vermiculite. These Doct, Etat Univ Pierre et Marie Curie

Tanaka M, Jouffrey B (1980) Many-beam lattice images calculated at 100 kV and 1000 kV. Acta Crystallogr A 36:1033 – 1041

Tanaka M, Jouffrey B (1984) Lattice-image interpretation of a relatively-small-unit-cell crystal. Acta Crystallogr A 40:143 – 151

Tomeoka K, Buseck PR (1985) Indicators of aqueous aleration in CM carbonaceous chondrites. Geochim et Cosmochim Acta 49:2149 – 2164

Tsipursky SI, Drits VA (1977a) Determination of (Si, Al)-distribution in the structure of a 1 M diocta-hedral mineral. In: Problems of isomorphism and genesis of mineral species and complexes. Proc 8th USSR Symp Isomorph 2:81 – 89 (in Russian)

Tsipursky SI, Drits VA (1977b) Effectivity of the electronic method of intensity measurement in structural investigation by electron diffraction. Izv Acad Nauk SSSR Ser Phys 1:2263 – 2271 (in Russian)

Tsipursky SI, Drits VA (1984) The distribution of octahedral cations in the 2:1 layers at dioctahedral smectites by oblique texture electron diffraction. Clay Mineral 19:177 – 192

Tsipursky SI, Drits VA, Chekin SS (1978) Revealing of the structural ordering of nontronites by oblique-texture electron diffraction. Izv Acad Nauk SSSR Ser Geol 10:105 – 113

Uyeda N, Kobayashi T, Ishizuka K, Fujiyoshi Y (1979) High voltage electron microscopy for image discrimination of constituent atoms in crystals and molecules. Chem Scr 14:47 – 61

Utevsky LM (1973) Diffractional electron microscopy in metal sciences. Metallurgiya, Moscow, 580 pp (in Russian)

Vainshtein BK (1956) Structural electronography. Akad Nauk SSSR, Moscow, 314 pp (English translation)

Vainshtein BK (1961) An empirical law of scattering in electron diffraction analysis. Kristallografia 6:965 – 967 (in Russian)

Vainshtein BK, Lobachev AN (1961) Dynamic scattering and its application to electron diffraction structure analysis. Kristallografia 6:763 – 765 (in Russian)

Veblen DR (1980) Anthophyllite asbestos: microstructures, intergrown sheet silicates, and mechanism of fiberformation. Am Mineral 65:1075 – 1086

Veblen DR (1981) Non-classical pyriboles and polysomatic reactions in biopyriboles. In: Veblen DR (ed) Amphiboles and other hydrous pyriboles. Rev Mineral 9A Mineral Soc Am Washington, pp 189 – 236

Veblen DR (1983a) Exsolution and crystal chemistry of the sodium mica wonesite. Am Mineral 68:554 – 565

Veblen DR (1983b) Microstructures and mixed-layering in intergrown wonesite, chlorite, talc, kaolinite, and biotite. Am Mineral 68:566 – 580

Veblen DR, Buseck PR (1979a) Chain-width order and disorder in biopyriboles. Am Mineral 64:687 – 700

Veblen DR, Buseck PR (1979b) New ordering schemes in mixed-chain silicates. In: Cowley JM, Salamon MB, Wuensch BJ (eds) Modulated structure N4. Am Inst Phys, pp 321 – 323

Veblen DR, Buseck PR (1980) Microstructures and reaction mechanisms in biopyriboles. Am Mineral 65:599 – 623

Veblen DR, Buseck PR (1981) Hydrous pyriboles and sheet silicates in pyroxenes and uralites: intergrowth microstructures and reaction mechanism. Am Mineral 66:1107 – 1134

Veblen DR, Ferry JM (1983) A TEM study of biotite-chlorite reaction and comparison with petrologic observations. Am Mineral 68:1160 – 1168

Veblen DR, Buseck PR, Burnham CW (1977) Asbestiform chain silicates: new minerals and structural groups. Science 198:4315

Vrublevskaya ZV, Zvyagin BB, Fedotov AF (1974) Comparison of triclinic unit cells for the given coordinate lattice plane. Kristolografia 19:730 – 737 (in Russian)

Wadsley AD (1952) The structure of lithiophorite. Acta Crystallogr 5:676 – 680

Warren B (1930) The structure of tremolite. Z Kristallogr 72:42 – 57

Warren B, Bragg L (1928) The structure of diopside. Z Kristallogr 69:1068 – 1193

Wenk HR (ed) (1976) Electron microscopy in minerology. Springer Berlin Heidelberg New York

Wenk HR (1978) The Electron microscopy in earth sciences. In: Proc 9th Int Congr Electron Microsc, Toronto, III, pp 409 – 419

Whittaker EJ, Cressey BA, Hutchison JL (1981) Transformations of multiple-chain lamellae in grünerite asbestos. Mineral Mag 44:27 – 35

Wilson D (1949) X-ray optics. London, 144 pp

Woolfson MM (1961) Direct methods in crystallography, Oxford London

Yoshida T (1973) Elementary layers in the interstratified clays minerals as revealed by electron microscopy. Clays Clay Mineral 22:413 – 420

Yoshida T (1976) Study of microstructure of mica and montmorillonite by high-resolution electron microscopy. Clay Sci 5, pp 1 – 7

Zakharov ND, Hadji IP, Rozhanskiy VN (1979) Structural and compositional heterogeneity in chain silicates observed by many-beam high-resolution electron microscopy. Dokl Acad Nauk SSSR 249:359 – 362 (in Russian)

Zolensky ME (1984) Hydrothermal alteration of CM carbonaceous chondrites. Meteoritics 19:346 – 347

Zvyagin BB (1967) Electron diffraction analysis of clay mineral structures. Plenum, New York, 320 pp

Zvyagin BB (1968) Indexing of electron diffraction patterns and construction of structure projections for an arbitrary crystal orientation. Kristallografia 3:603 – 605 (in Russian)

Zvyagin BB (1986) Structural diversity and complete symbolic description of chain-ribbon silicates. Kristallografia 31:1124 – 1129 (in Russian)

Zvyagin BB, Fedotov AF (1974) Orientations of single crystals to obtain definite selected-area electron-diffraction patterns in the general case of triclinci lattices. Izv Acad Nauk SSSR Ser Phys 38:2296 – 2299 (in Russian)

Zvyagin BB, Fedotov AF (1975) Advantages of a high-voltage electron diffractometer for indexing of diffraction patterns. In: Crystal chemistry of minerals and geological problems. Nauka, Moscow, pp 126 – 128

Zvyagin BB, Gorshkov AI (1969) Electron microscopy and electron diffraction. In: Methods for electron microscopy for minerals. Nauka, Moscow, pp 207 – 310

Zvyagin BB, Pinsker ZG (1949) Electron diffraction study of the montmorillonite structure. Dokl Acad Nauk SSSR 68:30 – 35

Zvyagin BB, Vrublevskaya ZV (1972) Crystal lattice planes normal to the incident beams as represented in electron diffraction patterns. Kristallografia 17:1056 – 1057 (in Russian)

Zvyagin BB, Vrublevskaya ZV (1974) Interpretation and simulation of diffraction patterns from triclinic lattices. Kristallografia 19:865 – 866 (in Russian)

Zvyagin BB, Vrublevskaya ZV, Soboleva SV, Sidorenko OV, Zhukhlistov AP (1979) High-voltage electron diffraction analysis of layer silicates. Nauka, Moscow, 223 pp

Zworikin VR, Morton GA, Ramberg EG, Hiliery J (1945) Electron optics and electron microscopy. New York

Subject Index